With gratitude and
appreciation to my
wonderful friends Moses
and Dinah, and their
talented and creative children
Zara, Johan, and Glory. Thank
you for being part of my life.

Kenny
August 11, 2019

GERMS GONE WILD

HOW THE UNCHECKED DEVELOPMENT OF DOMESTIC BIODEFENSE THREATENS AMERICA

KENNETH KING

PEGASUS BOOKS
NEW YORK

This book is dedicated to the memories of:
My wife of 29 years, Betty Segars King
My father, Elwood S. King
My mentor, James Baker Hall

————

The publisher fondly dedicates this volume to the talented Maria Fernandez and
Phil Gaskill, without whom this book would not have been possible.

GERMS GONE WILD

Pegasus Books LLC
80 Broad Street, 5th Floor
New York, NY 10004

Copyright © 2010 by Kenneth King

First Pegasus Books cloth edition 2010

Interior design by Maria Fernandez

ISBN: 978-1-60598-100-0

10 9 8 7 8 6 5 4 3 2 1

Printed in the United States of America
Distributed by W. W. Norton & Company, Inc.

Contents

CHAPTER ONE

A Biodefense Juggernaut

An Invasion, and the Resistance

On April 23, 2007, America's "war on terror" sent emissaries to a rural pasture in south central Kentucky. They arrived by military helicopter and a van and car convoy escorted by sheriff's deputies and state police. The convoy passed a cluster of protesters holding signs such as "Hal No! No Bio-Lab!" and "The Chamber of Commerce Is Not the Community."

Nearly six thousand people from this heavily Republican, largely rural district had signed petitions asking the Department of Homeland Security (DHS) *not* to bring the country's second biggest biodefense facility to this particular plot of land, despite the fact that their congressman, county judge, mayor, city council, fiscal court, Chamber of Commerce, and local newspaper had spoken on their behalf and told DHS what a perfect spot this would be for a $550 million biobonanza.

Like people in other rural communities alarmed by NBAF (the National Bio and Agro-Defense Facility), they had been told by the local "influentsia" that they were backward and ignorant for worrying about this Homeland Security-controlled facility and its Pandoric tinkerings with a yet-to-be-specified list of incurable pathogens. It would be "as safe as going to Wal-Mart," their congressman said. It would be "state of the art." It would be a "quantum leap."

Despite those exhortations, 2800 people had signed petitions opposing the facility during two weeks in March 2006. Another 2000 people had signed during the next two months. The rest had trickled in afterward, without any major efforts to sustain the initial drive. In the meantime, opponents had held two major rallies, established a Web site with Fact Sheets and a video, appeared on radio and television, and written a slew of letters and op-eds to local and state newspapers.

Now, over a hundred of them had gathered, on short notice, to confront the Homeland Security inspectors as they tried to slip quietly into the county and out. Channel 36, a Lexington television station, filmed the protesters, the stone-faced passage of the official convoy, and the setdown of the military chopper. After the convoy had passed, the protesters followed them part of the way back, stopping at the boundary of the proposed lab site.

For thirty minutes or so, as the inspectors assessed the biotech potential of Kentucky pasture, the protesters chanted slogans: "No Bio Lab!" "No Fort Detrick!" "No Plum Island!" Some people shouted more spontaneous comments: "Where's Hal Rogers?" (Rogers was the U.S. Representative chiefly responsible for DHS's interest in Kentucky). "Look at this beautiful country! You ought to be ashamed of yourselves!" "This is our home! You go back to yours!" A large state trooper loomed near the boundary line, screening the protesters from the entourage of inspectors and inspector-handlers. That evening, Channel 36 spliced together a two-minute newscast. Prominent in the newscast was the military helicopter, dropping out of the sky like a predatory bird, like the shock troops for an armed invasion.

Biodefense Shucking and Jiving

I had worked feverishly the week before, trying to help plan the protest, writing press releases and a speech, and sending E-mails to people who had signed online petitions. Four months earlier, my wife of 29 years had died suddenly, less than a year after our divorce. My mother had torn her knee ligaments at almost the same time, and in the summer following the protest, we learned that my father had a rare form of lymphoma. I was not the only local NBAF opponent affected by personal tragedies during this period. Among the small steering group for Citizens Against a Kentucky Biolab, Floyd Lovins lost his longtime friend and employer, Phil Cash of Melody Music, the same week the congressman announced his NBAF efforts. David Taylor would lose his father to cancer a few months later. The local NBAF propaganda effort rolled right along regardless.

By the time that military helicopter descended onto the Pulaski County pasture, the war on terror already had beachheads in our county, thanks to the wheeling and dealing of our congressman, U.S. Representative Harold "Hal" Rogers. Rogers had chaired the House Appropriations homeland security subcommittee until the November 2006 elections conveyed the position to a Democrat. His achievements in that position illustrate the way homeland security and biodefense have become trendy new pork barrels.

In 2004, Rogers established—in Pulaski County—the National Institute for Hometown Security and the Kentucky Homeland Security University Consortium. Both were funded by DHS, the agency funded by Rogers's subcommittee. In a synergistic fashion typical of Rogers, both organizations were housed in an earlier pork barrel project, the Southern Kentucky Rural Development Center. (Rogers had helped build SKRDC with $15.5 million of public money in the 1990s, and the Center now operates with an $18 million annual budget, mostly derived from public funds.) A December 2005 *Washington Post* article noted that the Institute and Consortium had received, in the first year of their existence, $34 million in combined grants from DHS's science and technology directorate, far outstripping the biodefense proceeds of better-known consortia at Texas A&M, Johns Hopkins, Minnesota, Southern Cal, and Maryland.[1]

Few would have considered Pulaski County a central battleground in the war on terror, but former Homeland Security Director Ridge said DHS funded the National Institute for Home*town* Security under the "unique notion that the homeland is not secure until the hometown is secure."[2] Science Applications International Corporation, a leading biodefense contractor, explained its opening of operations in the county not as a subtle form of influence-peddling, but as a spiritual quest to soak up the Hal Rogers zeitgeist: "Being close to leadership helps us understand trends in government."[3] Apparently standard Washington lobbyist come-ons weren't intimate enough for SAIC.

Once the congressman had established his Homeland Security beachheads, however, he wanted more. And so he went after NBAF, a proposed human, animal, and zoonotic disease supercenter surpassed in size only by the new "biodefense campus" at Fort Detrick in Maryland. The new NBAF will be controlled by DHS, and like Detrick, will include Biosafety Level Four (BSL-4) labs—studying diseases for which no vaccine or cure exists—as a major component.

NBAF was announced in 2005 as a replacement for an existing offshore facility at Plum Island, New York. Plum Island might not have a BSL-4 rating but it does have a troubled history involving OSHA (Occupational Safety and Health Administration) and EPA (Environmental Protection Agency) safety citations, a critical investigation by the Government Accountability Office (GAO), several foot-and-mouth outbreaks, and a three-hour total power loss which zapped the negative air pressure, the most important fail-safe mechanism in "high containment" labs. Michael Carroll's 2004 bestseller, *Lab 257*, recounts Plum Island's troubled past and also suggests that it may accidentally have introduced Lyme disease and West Nile virus into the U.S. New York congressional delegations—Senator Clinton among them—have consistently resisted federal efforts to make Plum Island a BSL-4 facility handling incurable human diseases.

Rogers, on the other hand, formed a new consortium composed of the Universities of Kentucky, Tennessee, and Louisville, and the Oak Ridge National Laboratory to solicit NBAF for Pulaski County. The *Somerset Commonwealth Journal*, the sole newspaper in Pulaski County, promoted the facility with twenty-one articles in six weeks, carrying headlines like IT WILL TRANSFORM US; LAB'S IMPACT: MIND-BOGGLING; and HAZMAT PERSONNEL ENDORSE NEW LAB.

Early on, several Pulaski County residents—one of them a public health doctor, and another a librarian with a biologist daughter—discovered *Lab 257*, the book on Plum Island. And the *Lexington*

Herald-Leader's article on the congressman's NBAF dreams referenced safety lapses documented at various facilities by the Council for Responsible Genetics and the Sunshine Project.[4] Such alternative sources of information were crucial in light of the *Commonwealth Journal's* propaganda campaign. The local newspaper's chief safety criterion, apparently, was that if Hal says it's safe, it's safe: *Congressman Hal Rogers has worked for a quarter-century to improve things for the 5th District. He would not bring something that is dangerous or negative to Pulaski County. He has earned our trust and we have faith in his ventures.*[5]

When the congressman launched his NBAF efforts in 2006, he apparently assumed we'd all follow the paper's lead, bowing our heads and confessing his goodness and omniscience. Otherwise he might have followed the strategy of NBAF suitors like Kansas and North Carolina and kept a low profile. Since DHS had said "community acceptance" would be a consideration in the site selection process, though, launching a big public-relations effort to get the community behind him seemed sensible. The *Commonwealth Journal* obligingly went on its 21-stories-in-six-weeks binge. Readers were told again and again of the biotech bonanza that might be theirs, and of the absolute safety of a facility featuring world-class scientists.

EXPERT PROCLAIMS BIO-LAB "SAFE," read one of the headlines. The expert in question was the Veterinary Dean at the University of Tennessee, one of the members of the congressman's consortium. The subtitle for his article admonished readers to *"Listen to statistics, not fear-mongers."* Unfortunately, the dean got so caught up in superlatives he left his statistics back in Tennessee: "The track record for these facilities around the nation is astounding. . . . In terms of safety, you couldn't get any better."[6]

The *Commonwealth Journal* did get hold of something it called facts: "In over 50 years of BSL-4 labs," it assured readers, "there has never been an accident of any type."[7] An accompanying sidebar, "You Need to Know," said "Bio-Safety Level-4 labs have been scattered throughout the United States for more than 50 years and there has never been a single reported incident of agent release or contamination."

When I became involved with the opposition three months later, I discovered that these comforting "facts" were simply lies. BSL-4 labs hadn't been "scattered throughout the United States for more than 50 years." Until the 1990s, the only U.S. BSL-4 labs had been at the Centers for Disease Control and the USAMRIID facility at Fort Detrick—each constructed in the 1970s.

It was also a lie to say there had never been an accident, or an "agent release or contamination." Given the massive secrecy surrounding such complexes, news of accidents often comes only from confidential informants—a form of patriotism made dangerous by current PATRIOT Act provisions. But it was easy to find public documentation of incidents involving Fort Detrick—the country's chief bioweapons facility prior to 1969, and its chief biodefense facility since. A series of recent headlines, from the Associated Press and elsewhere, suggested that our local journalists were the truly ignorant ones:

> *Infected Researcher Broke Safety Rule at Army Lab* (May 2000)
> *Fort Detrick Waste Cleanup Cost Grows* (May 2001)
> *Army Lost Track of Anthrax Bacteria; Specimens at Md.'s Fort Detrick May Have Been Misplaced or Stolen* (Jan. 2002)
> *Fort Detrick to Remove Radioactive Sludge Stored Near School* (March 2002)
> *2nd Leak of Anthrax Found at Army Lab* (April 2002)
> *Infectious Germs Halt Cleanup of Fort Detrick Dump* (April 2002)
> *FBI Probes Possibility Anthrax Was Smuggled Out of Fort Detrick Maryland* (June 2002)
> *Army Aims to Correct 'Sloppy Methods' after Accidental Release of Anthrax Spores* (July 2002)
> *Fort Detrick Unearths Hazardous Surprises* (May 2003)
> *Chemical Dump Site at Fort Detrick Cleaned Up* (June 2004)
> *Fort Detrick Had Multiple Anthrax Leaks in 2001-2002, Report Finds* (April 2006)

In August 2006, I would be booed at the local Chamber of Commerce for including these headlines in my questioning of Ewell Balltrip, director of the National Institute for Hometown Security, following his latest hooray-for-the-biolab presentation. He had asked for questions, and I had risen to offer some. I didn't get to finish them. The Chamber was interested in money, not questions.

It annoyed the hell out of me when then-Governor Ernie Fletcher came to town after the protest and said, "Those who don't support, education could help them understand."[8] I had a Ph.D. and a Vanderbilt law degree, and by then had read hundreds of books and articles on bioweapons and biodefense. I doubted the governor had read anything but press releases, so I felt maybe I could waive his proposed remediation. Apparently, though, Fletcher thought my farm had more capacity for sustained thought than I did, since "my farm" at least recognized a good thing when it saw it: "I think the farms around will realize that their land values will go up very, very much."[9]

Around the country, it became a favorite ploy of NBAF boosters to claim that everyone supported the project but a few ignorant bumpkins. In September 2007, the mayor of Flora, Mississippi—whose town was then a finalist for the DHS facility, and who didn't describe his own biolab course of study—said, "Education is the whole key to it . . . You have to find the people who are concerned and educate them. In the end, you're still going to have a few idiots."[10] I was inclined to attribute idiocy instead to people who place blind faith in politicians and biodefense researchers, like the Pulaski County dentist who wrote that "Folks circulating petitions for people to sign are not qualified to speak in science and research arenas." We should understand, he said, that "Scientists certainly do not work in conditions that are not safe. The kind of folks who work in labs, like the one proposed for Somerset, are dead [sic] serious-minded people. Scientists of this caliber function on a genius level."[11]

One would have thought opponents of a new BSL-4 lab in Boston could have escaped this sort of condescension, since 150 Massachusetts university professors, including two Harvard Nobel laureates, signed a letter opposing that facility. But no, Senator Edward Kennedy, who joined Boston mayor Menino and then-governor Mitt Romney as prominent supporters of the controversial NIH project, told newspapers in 2003 (admittedly, a year before the academics' letter) that the concerns of neighborhood groups could be "addressed" with a proper outreach campaign.[12] Kennedy's language is more respectful than the Mississippi mayor's, but the assumptions are the same: any concerns about safety are unfounded and will diminish once opponents are properly educated.

A Mindless Proliferation of Deadly Pathogens

DHS's NBAF project is part of a huge explosion in biodefense funding and construction following the anthrax letter attacks of 2001. That event has led to a seven-fold increase in biodefense funding,[13] a twelvefold increase in BSL-4 lab space,[14] and a mushrooming of BSL-3 labs so vast no one in the country knows how many are actually out there. The only tracking of numbers is through the CDC and USDA "Select Agent" programs—labs working with pathogens on the CDC's select agent list are required to register—and there were 1356 of those BSL-3 labs in October 2007.[15]

BSL-4 labs study the world's deadliest diseases, things like Ebola and the Marburg virus for which no cure or effective treatment currently exists. Researchers work out of pressurized space suits with their own oxygen supply, to avoid breathing or otherwise coming into contact with death-penalty pathogens. They hope to avoid needle pricks or tears in their suits: otherwise they might become casualties of their own work, or, even worse, carry disease out of the fortress-like labs into the community at large—something that happened with high-containment SARS labs in Asia in 2003. Until the 1990s the only BSL-4 facilities in the US were USAMRIID at Fort Detrick and the CDC in Atlanta. By 2000 there were four operational BSL-4 labs.[16] Counting those already constructed and others on the books, there will soon be fifteen.[17]

BSL-3 labs study diseases that are deadly enough, but for which there is some possibility of vaccination or treatment. They feature the same safety protocols as BSL-4 labs except for the space suits. Again, no one knows how many BSL-3 labs are out there, except that the number exceeds 1356. In February 2008, there were only eight times that many Starbucks in the country,[18] and perhaps the Hazelnut Latte isn't the appropriate business model for biowarfare agents. (Starbucks closed 600 stores a few months later; America's biodefense program has gone the other direction.) Work with anthrax requires only a BSL-2 rating, or BSL-3 if there is a chance of the anthrax becoming aerosolized: there is some sort of vaccine, and the disease can be successfully treated if the right antibiotics are started soon after exposure. This assumes early symptoms aren't confused with the flu and other respiratory ailments. Given the proliferation of biodefense research, probably more U.S. labs have access to anthrax now than at any time in our history (more than 350, according to a 2004 *San Francisco Chronicle* article).[19] The same is true of other biowarfare agents, and that ought to trouble us, given that DNA testing long ago narrowed the source of the 2001 anthrax to four or five existing biodefense facilities.

The new fixation on biodefense has distorted the focus and research of such traditional public health agencies as the Centers for Disease Control and the National Institutes of Health, turning them into biodefense apologists and infusing them with cultural norms more traditionally associated with defense and intelligence agencies. Though there are certainly scientists eager to capitalize on the new funding cornucopia, others see the new focus as harmful and self-defeating. In 2005, 758 of the 1143 scientists receiving NIH funding for microbiology research signed an open letter stating that "The diversion of research funds from projects of public-health importance to projects of high biodefense relevance but low public-health importance represents a misdirection of NIH priorities and a crisis for NIH-supported microbiological research."[20]

In October 2007, the Oversight and Investigations subcommittee of the House Energy and Commerce committee conducted a hearing titled "Germs, Viruses, and Secrets: The Silent Proliferation of Bio-Laboratories in the United States." The hearing's star witness was Dr. Keith Rhodes, Chief Technologist of the Government Accountability Office, who had been investigating the proliferation for two different congressional subcommittees. Also invited were representatives from the CDC, NIH, watchdog groups, and biodefense think tanks, and the president of Texas A&M University, where all "select agent" research had just been suspended by the CDC. Chairman Bart Stupak pointedly noted that the Department of Homeland Security had declined an invitation to appear.[21] The fact that DHS acts like a law unto itself is one of the things that concerns potential neighbors of its biodefense projects.

Edward Hammond of the Sunshine Project testified that over

20,000 people are currently involved in biodefense research, with a twelvefold increase in BSL-4 lab space just since May 2004. Even more BSL-3 lab space has been added; NIH alone (through the National Institute of Allergies and Infectious Diseases, or NIAID) is building fourteen new Regional Biocontainment Laboratories (one of them at the University of Louisville, in my home state). The new construction represented just by major projects, he said, constitutes the equivalent of 36 super Wal-Marts.[22]

Alan Pearson of the Center for Arms Control and Non-Proliferation said annual U.S. bioweapons-related spending rose from $1.327 billion in 2001 to a high of $9.509 billion in 2005. Over $40 billion of new biodefense money had been spent since the 2001 anthrax attacks.[23] (That number has since risen to more than $57 billion.)

Contrary to the bland assurances of biodefense boosters, Dr. Rhodes told the subcommittee there is a "baseline risk" associated with any high-containment germ facility, attributable to human error. The risk is increasing with the expansion, and is greatest at new facilities lacking experience with standard safety protocols.[24] Several witnesses indicated concern that the speed and size of the current expansion are completely overwhelming the supply of trained personnel and the capacity for training others. A lot of new people are researching a lot of new germs: only 15 of the 435 researchers who received NIAID funding from 2001 to 2005 for work on bioweapons agents anthrax, brucellosis, glanders, plague, meliodosis, and tularemia had received funding for the agents before 2001.[25] This resembles suddenly placing crop-duster pilots into the cockpits of Boeing 747s, just because the crop dusters got together and decided to write a grant. And some of the new research—attempts to genetically engineer new pathogens, "aerosol challenges," and "threat assessment research" involving simulated bioweapons attacks—is especially dangerous.

Even more troubling, the expansion is proceeding more or less mindlessly, with little objective oversight. Several different federal agencies are involved in biodefense research; each has been on its own spending spree, with no one conducting a comprehensive needs-assessment or risk-benefit analysis. A huge new "biodefense campus" at Fort Detrick will feature expanded facilities for the Army and new facilities for Homeland Security, NIAID, CDC, and USDA. The rationale for this BSL-4 megaplex was that having the facilities together in one place would eliminate duplication. Yet the CDC is constructing its own vast new suite of BSL-4 labs; NIAID is building new BSL-4 facilities in Boston and Galveston, and the fourteen Regional Biocontainment laboratories; and DHS and USDA are working together on NBAF. Meanwhile various federal agencies fund a proliferation of biodefense research at our universities, posing a serious threat to whatever academic integrity still remains there, and vastly increasing the odds that a rogue scientist, a foreign terrorist-infiltrator, or a Timothy McVeigh/Unabomber type will get knowledge and state-of-the-art bioweapon agents from our own facilities.

Rhodes said the decentralized, unregulated nature of the expansion means no single federal agency has the mission of tracking the number of new labs, or their aggregate risk. No one is ensuring "that sufficient but not superfluous capacity—that brings with it additional, unnecessary risk—is being created."[26]

All these new facilities get minimal regulatory attention once they're up and running. No single agency in the federal government has primary oversight responsibility. If a facility deals with any of the 72 germs listed by the CDC as select agents, it is subject to some regulation by the CDC or USDA. If a facility engages in genetic engineering of pathogens—recombinant DNA research—and receives NIH funding, it is lightly regulated by the National Institutes of Health. It is expected, then—in theory, if not in practice—to have the recombinant research reviewed by an institutional committee, and to keep minutes of the committee meetings. If a facility doesn't fall within one of these two situations, it may not be regulated at all.

The primary agencies involved with regulation—the CDC, NIH, and USDA—all operate research facilities themselves. This sets up a situation in which the chief regulatory agencies are tempted to downplay the seriousness of safety lapses, to forestall public concern about the safety of their own facilities. This conflict of interest exhibited itself at the subcommittee hearing, where CDC representatives fielded questions both about their oversight of Texas A&M and about a June power loss at the CDC's own suite of new BSL-4 labs. Predictably, both the CDC and NIH representatives suggested the current state of affairs is actually quite safe, thank you.

Rhodes disagreed. Pressed by a Texas congressman who apparently expected a different answer, Rhodes revealed that, in his opinion, the biodefense expansion has made the country less safe than it was before the attacks.[27]

Biodefense: The New Military-Industrial-Academic Complex

Even before the FBI formally announced that the 2001 anthrax attacks had been launched from one of our own biodefense facilities, the biodefense boom already seemed yet another knee-jerk war-on-terror overreaction like the Patriot Act and invading Iraq. It is good that a few congressmen examined what biodefense proliferation has wrought, but a strong array of forces remains interested in stoking the bioterror fires and grabbing the funds that are out there: public health, homeland security, and defense agencies enjoying the infusions of new monies; politicians looking for economic development windfalls; research universities already knee-deep in biotech-for-profit conflicts of interest; hometown newspapers who seem to think snagging a biodefense lab is some sort of macho athletic competition. (*Maybe we can mess with Texas for lab site*, the *Athens Banner-Herald* mused just after Athens, Georgia was announced as an NBAF finalist.[28] *This exercise has shown that Kansas can compete with the "big boys,"* said the *Lawrence Journal-World*.[29] Kansas Senator Pat Roberts joined Kentucky Congressman Rogers in likening the NBAF selection process to an NCAA basketball tournament: *I think that it's a lot like a Final Four: I think we'll make the cut, and I think we're very well-suited*.[30])

The economic impact estimates may well be inflated, since they seem to assume pharmaceutical and biotech companies will flock to the occupied regions to exploit the commercial potential of biodefense research. There may be limited demand, though, for vaccines against diseases rarely encountered outside of bioterror scenarios. (Of course, the companies could always take a page out of the anthrax terrorists' playbook and fluff the market with a small bioterror demonstration.) The figures being floated by promoters do demonstrate why this new military-industrial-academic complex will spend any monies, make any outrageous statements, to promote the projects. A University of Georgia study (prepared on behalf of the university's NBAF solicitation) estimated that the overall twenty-year impact of NBAF would be $3.5 billion to $6 billion.[31]

NBAF contenders were willing to plunk down big sums to get a piece of all that. An August 2007 article in the *San Antonio Current*, an alternative weekly, reported that the Texas NBAF consortium had already spent $500,000 on lawyers and public relations specialists.[32] A December 6, 2006 newspaper article revealed that North Carolina consortium representatives were making twice-weekly trips to Washington to lobby for the NC bid, that they had hired a public relations agency, and that they planned to file as a 501(c)3 nonprofit to seek donations to help cover recruiting costs.[33] Other NBAF finalists probably made similar efforts. Kansas, for instance, set aside $250,000 for lobbying early on,[34] plus another million to defend against a lawsuit brought by the Texas consortium.[35]

Culture of Deception

I became involved in the local struggle against NBAF primarily out of disgust at the deceptions of local boosters and their efforts to ridicule the concerns of opponents. Little held me to Pulaski County following my divorce, and I expected to be long gone by the time NBAF opened for business.

Perhaps, had the lies not been so blatant, I would have posted a "For Sale" sign and occupied my mind with more pleasant subjects. I'd spent seven weeks of the previous summer teaching in Paris; I had things I wanted to write about France, about the fiction of Wendell Berry, and the poetry of W. S. Merwin; and a book's worth of poems I needed to submit to magazines. Perhaps, even if my fellow Pulaski Countians knew the real dangers of NBAF, some might have welcomed it nonetheless. But the local paper wasn't offering even a hint of the danger, just a bunch of Pollyanna-ish drivel. I knew how to research, I knew how to write, and it seemed no one else was situated to do this.

At first I thought the local deceptions arose out of our peculiar situation: a powerful congressman, a worshipful business community, and a newspaper that kowtowed to both. I eventually learned, however, that similar forces operated in other parts of the country, that Pulaski County had no monopoly on prevarication. This became abundantly clear when I began researching the NBAF finalists. Lies marked the claims not only of politicians and economic development types, but the communications of university academics and the Department of Homeland Security itself.

None of this will surprise anyone who has explored the history of U.S. bioweapons research. Moral norms that arose during World War II and the Cold War are still with us, and fit perfectly into a war-on-terror mindset that sanctioned the ignoring of habeas corpus, the use of secret renditions and torture, and pervasive spying on American citizens. And recent coverups of safety breaches and researcher infections at Boston University and Texas A&M, occurring even as the institutions solicited major new biodefense projects, show clearly that contemporary research universities mirror their biodefense partners' lack of a moral compass.

NBAF public relations efforts have purveyed falsehood in especially reckless fashion. Apparently, the project's huge price tag tempted its promoters to play fast and loose with the truth, and biodefense research itself lacked—and lacks—any culture of integrity that would restrain them.

In August 2008, Americans were reminded that the FBI believed the anthrax attacks of 2001—the events that prompted the current proliferation of new high-risk germ labs—had been launched from within the country's *own* biodefense complex. Not everyone believed those attacks were simply the project of a single crazed researcher.

PLUM ISLAND PRELUDE

THE MORE THINGS CHANGE, THE MORE THEY STAY THE SAME.
Build it, and they will learn to live with it. In the 1950s, the USDA constructed its Plum Island research facility (85 miles from New York City and only 2 miles from the east end of Long Island), over the objection of local residents, and in conjunction with the Army, which wanted to develop methods of destroying enemy food supplies. The USDA had first been interested in Prudence Island off the coast of Rhode Island, but was forestalled by wealthy Newporters and their Anti-Prudence Island Laboratory Committee.[1] Apparently the Newporters didn't trust the assurances of their state-of-the-art government. This may have made them America's first "biodefense activists."

When the USDA turned its attention to New York, that state's senators demanded a provision in the appropriations bill requiring full hearings, following reasonable public notice to those living within twenty-five miles of the facility. Michael Carroll's *Lab 257* describes how USDA's notice consisted of newspaper ads one week before the hearings, and how, despite the short notice, 1544 people objected through sixteen different petitions, written statements, and telegrams. Recorded opinions ran three to one against the laboratory. Local farmers and oyster growers were even joined in their concerns by a business organization, the Long Island Association.[2] (The USDA had not yet acquired DHS's Pavlovian skill in making chambers of commerce salivate on command at the mention of construction and real estate dollars.)

The hearings were all for show, apparently. Because only one percent of the residents filed objections, USDA assumed that the other 99% supported the facility. In fact, as Carroll points out, the others had probably missed the newspaper ad and weren't even aware what USDA was up to.[3] Over fifty years later, DHS would follow USDA's precedent in assessing "community acceptance" for NBAF. If there wasn't a huge stink of the sort that occurred around Butner, North Carolina—where opponents' notions of "Whatever It Takes" included the possibilities of civil disobedience and equipment sabotage, where not even one supporter dared speak at the 2008 Draft Environmental Impact Statement (DEIS) hearing, and where politicians of all sorts were backpedaling and withdrawing support while they still could—DHS was happy to assume "the community" wanted the thing. Toto, is the NBAF in Kansas yet?

At Plum Island, in 1952, the Army had in fact already awarded a secret contract to construct a germ-warfare facility, a full month before selection hearings began. Even as it began adapting an old mine storage facility as Lab 257, however, it was having second thoughts about research focused on biowarfare against enemy food supplies. So, in 1954, the Army formally left the island and transferred control to USDA.[4]

Carroll argues that Plum Island continued to have close ties with the Army's bioweapons facility at Fort Detrick, that the USDA took over some of the Army's planned germ warfare research and performed other research on contract for the Army. Plum Island officials had consistently denied any involvement with biological warfare research until 1993, when *Newsday* unearthed previously classified documents detailing plans to disrupt the Soviet economy by spreading livestock diseases.[5]

An interesting—and so far apparently unnoticed—footnote in a

2003 GAO report lends further credence to Carroll's allegations: "Out of concern that Iraqi scientists were trying to manipulate camel pox for possible warfare use, USDA conducted work for the Department of Defense to determine if camel pox could be manipulated into an agent similar to smallpox."[6] Department of Defense? Genetic engineering of camel pox to make it resemble smallpox? Sounds like biowarfare research, doesn't it?

The "no biowarfare research" claim would not be the last deception perpetrated by Plum Island officialdom, and those deceptions, and the facility's military origin and connections, have made Plum Island neighbors distrustful and helped establish plausibility for Carroll's suggestions that Plum Island mishaps unleashed Lyme disease and the West Nile virus on the U.S. The transfer of Plum Island to DHS in 2003 would reignite such concerns, about both Plum Island and DHS's budding NBAF project, which proposed to add BSL-4 labs and human and zoonotic disease research to Plum Island's animal disease mission.

Because Carroll's Lyme disease and West Nile virus theories are based on circumstantial evidence, NBAF proponents and DHS spokespeople have rushed to characterize the book as "science fiction," even though the Final Environmental Impact Statement for NBAF sensibly cited Carroll's book as a reliable source of information.[7] The NBAF propagandists desperately want to discredit the book, however, because it amply demonstrates a continuing pattern of safety breakdowns and carelessness. Potential NBAF neighbors naturally assumed that NBAF, a supersized version of Plum Island dealing with more dangerous diseases, might have supersize safety breakdowns. One response by the NBAF public relations elves has been to tell fairy tales about a mythical biodefense heaven where infallible state-of-the-art technology protects us even from the errors of fallible humans. The other response has been to smear Carroll's book, to suggest that the incidents he describes never happened.

Most of those stories are firmly documented in public news sources, however, even though Carroll did obtain additional details from Freedom of Information documents and personal interviews. When the propagandists condescend to acknowledge the Plum Island problems at all, they consign them to the realm of ancient history, insisting that lapses simply can't happen in a state-of-the-art NBAF. Eerily, however, both of the most serious Plum Island incidents—the 1978 foot-and-mouth outbreak, and 1991 and 2002 power failures resulting in the shutdown of the negative air pressure system— repeated themselves elsewhere at "world-class" facilities in the summer of 2007, just in time to feature prominently in the October 2007 GAO report to Congress.

In the 1978 Plum Island foot-and-mouth outbreak—the first appearance of the highly contagious and economically destructive FMD in the United States since 1929—healthy animals being held outside the lab containment areas inexplicably became infected with FMD. The cause was never discovered, but Plum Island workers hastily slaughtered all the livestock on the island and frantically fanned out in teams to mainland farms making sure the disease hadn't spread off the island. One case on the mainland would have meant the immediate shutdown of U.S. meat exports at a cost of billions of dollars. (One of Plum Island's own directors put the estimated tag at $60 billion.) According to the GAO, only the facility's island location kept the Office International des Epizooties from revoking the U.S.'s FMD-free designation.

After the 1978 outbreak, Plum Island officials stopped keeping livestock outside the containment areas. According to the GAO, there have been six other releases of FMD virus within the facility between 1971 and 2004. In these other incidents, FMD spread outside desig- nated FMD research areas and infected livestock elsewhere in the facility. Technically, FMD did not escape outside the lab buildings themselves, as had happened in 1978. The GAO pointed out, how- ever, that many of these incidents were not related to the facility's age, and could just as easily have occurred in a shiny new NBAF.

In 2007, an FMD outbreak also occurred at Britain's renowned Pir- bright research complex, infecting livestock on several nearby farms and forcing the slaughter of over 2,000 animals and an estimated eco- nomic loss of 47 million pounds. Investigators believe a decaying drain system allowed inadequately treated pathogens to escape into the surrounding soil, and that the disease was then spread by vehicle traffic to the vicinities of the affected farms. NBAF will be a shiny new lab, of course, but shiny new labs age just as shiny new cars do, and future budget problems may well result in skimping on maintenance. The GAO warns that typically, high-containment germ labs are built under grants from one or more federal agencies, but that maintenance then becomes the responsibility of some other entity—such as a cash-strapped university. Indeed, according to Carroll, many of Plum Island's problems stem from inadequate funding for maintenance.

The late eighties and nineties saw Plum Island getting a lot of negative attention from federal regulatory authorities. A 1988 OSHA inspection turned up 139 violations, "covering everything from exposed electrical cables to open incinerator pits and untested fire alarms."[8] OSHA also found that workers "required" to wear respirators in the vicinity of harmful viruses had not even been fitted for such respirators. In 1993, OSHA charged the center with 25 "serious violations," including improper disposal of needles. 1995 brought a $111,000 fine for illegally storing hazardous materials.[9]

1992 *New York Times* articles reported that staff cutbacks after a private company assumed management (under the Reagan-era privatization push) had "undermined the traditional safety precautions at the island and that repairs [were] expected to cost $60 million." An internal USDA memo described long-standing environmental violations.[10] And because of rising operating costs, spending for actual research—as opposed to facility maintenance—had declined 20%.[11]

In August 1991, Hurricane Bob shut down both regular and backup electrical systems for 18 hours, disabling the negative air pressure.[12] Carroll, quoting one of PI's employees, describes the event as a "biological meltdown," complete with an overflowing sewage holding tank; thawing, oozing freezers of pathogens; failing door gaskets; and air vents stuck in open position and thus allowing mosquitoes, flies, and moths unfettered access to and from the raw sewage and diseased animals.[13]

In 1994, new Plum Island director Roger Breeze attempted to improve the facility's public image (according to the *New York Times*, Plum Island was "an object of suspicion, derision and outright hostility") by opening it up for public tours.[14]

In 1999 Plum Island again tried to upgrade its image, as it sought $225 million in federal funds for a BSL-4 upgrade allowing it to study diseases deadly to humans. So the Plum Island PR elves offered a tour for reporters in which they "tried to dispel some of the wilder rumors about the place," such as the idea of aliens living "in the Army bunkers that ring the island."[15] Cute, but no sale to local residents, most of whom got their concerns about Plum Island not from outer space but from the *New York Times*, *Hartford Courant*, and other earthbound publications which chronicled the various safety break-downs, environmental violations, and power losses.[16]

U.S. Congressman Michael Forbes successfully lobbied the Clinton administration to remove an initial $24 million appropriation for the Plum Island upgrade from the 2001 budget,[17] but USDA continued to moan about Plum Island's poor press and to lobby for a BSL-4 facility into the first half of 2001. William Smith, executive director of Fish Unlimited, blamed the facility itself for the poor relations. "It's their fault because they have misled the public for years about the true nature of the facility. . . . They have minimized their safety problems and tried to cover up accidental releases."[18]

After the Pentagon was attacked on 9/11, Plum Island scientists imitated their germ lab colleagues at Detrick and the CDC by getting the hell out of there,[19] even though, years later, NBAF propagandists would ridicule local residents worried that NBAF might make a tempting terrorist target.

In August 2002, 76 Plum Island maintenance and operations workers went on strike.[20] Two months later, with the strike still unre-solved, Senators Clinton and Schumer accused the Bush administra-tion of placing union-busting ahead of national security. According to the *New York Times*, union members, workers on the island, and local officials had all expressed concern about the ability of the center to function effectively, feeling that the strike-busting replacement workers "have not been adequately screened and do not have suffi-cient training to handle an emergency."[21]

A later *Times* story revealed that one of the replacement workers, a computer technician who had control of the all-important ventilation systems in the Plum Island containment areas, had been arrested three times for assault. The worker had left the job without notice for three days in late October, and had then been dismissed. But he had taken with him not just his notice of dismissal, but a Plum Island laptop containing important information about the facility's computer system. According to a government official who insisted on anonymity: "It is my understanding that from a computer, at a remote location, he could have used his dial-up password to access the system and change the air pressure in the containment areas so that the contaminated air could be forced out into the environment."[22] The human factor rears its persistent head again.

With the strike still underway, two separate power failures occurred in December 2002. The first lasted for three hours and included the failure of all three backup generators. As door seals collapsed, the workers sealed the doors with duct tape—a technology more often associated with backyard tinkerers than with state-of-the-art biolabs. But the CDC would also go the duct tape route when its state-of-the-art biolab suffered a complete power failure a few years later. At Plum Island, as at the CDC later, the power loss meant that the most important biolab safety mechanism, negative air pressure, was most likely lost. According to the *Times*, the incident "raised fears for the first time that the containment of infectious pathogens could have been seriously compromised at the laboratory." Senator Clinton called—in vain—for the lab to cease all operations until an independent safety review could be conducted.[23]

Plum Island apparently found another use for duct tape: sealing loose lips. The public learned of the power failure only after a replacement worker alerted members of Senator Clinton's staff. Insisting on anonymity, the worker said, "The reason I am coming forward is because what I have seen at the center is really out of hand and something needs to be done about it."[24] Duct tape is notoriously nontransparent, and Plum Island preserved its duct tape ambience intact by denying the *New York Times'* requests to visit the island after the worker's whistle-blowing.[25]

In February 2003, at a public forum in Mattituck, New York, Clinton said even she had been unable to identify and contact officials of LL & B, the private company managing the lab. She said she feared a "total stonewall" after the forthcoming transfer of the facility to Homeland Security's oversight.[26]

Some 500 to 600 people concerned about the strike, the recent safety breakdowns, and the impending transfer to DHS attended the

February forum. The transfer was causing understandable speculation that Plum Island under DHS might be refocused "on the fight
against bioterrorism," involving a BSL-4 upgrade and research on
deadly human diseases. Clinton said the residents of eastern Long
Island had a right to know if the Bush administration planned to alter
the mission of Plum Island: "We cannot allow the challenge of security to create a secret government in the United States."[27]

The takeover was the subject of a long article in the June 1, 2003
New York Times. The *Times* first laid out the reason why people were
concerned:

> As the Department of Homeland Security prepared to formally
> take over the laboratory today, there was every expectation
> that the curtain of secrecy would close tighter than ever.
>
> This time the secrecy will be official, and indeed a matter
> of law, because the Department of Homeland Security has a
> mandate to conduct classified research. Some people say that
> this could push the laboratory in a new, ultra-hush-hush
> direction, and that any of the 200 or so administrators, scien-
> tists or employees who say too much about what goes on at
> the 850-acre island, which is less than two miles off the North
> Fork, will risk committing a criminal act.[28]

But no: Dr. Maureen McCarthy, acting director for research and
development in DHS's Science and Technology Directorate, assured
readers that "openness and closer community ties were a high priority." She even said she would support the creation of a community
advisory board.[29]

A close look at her precise language might have raised questions
about how far the openness and communication would extend. "It is
critical," she said, "to hear and address the concerns of the local community as we develop our plans and move forward."[30] Potential
biodefense neighbors—certainly potential NBAF neighbors—would
get a lot of "hearing and addressing" in the coming months and years.
What it amounted to was the right to state one's concern at a designated time and place—usually, the formal environmental impact
process, and if one raised enough stink, perhaps a supplemental

public relations event—and then to receive in return the appropriate designated public relations response, containing a carefully adjusted conglomeration of such phrases as "state-of-the-art"; "highest level of security"; "world-class scientists"; and the popular "crucial to our nation's struggle against bioterrorism." There, you've been addressed, and isn't it wonderful how open and transparent we all are?

The openness assurances followed a May 27 letter to DHS Secretary Ridge from Senator Clinton and Representative Bishop, declaring that better communications were essential. The letter also urged Ridge to reject any plans to seek a BSL-4 designation for Plum Island. The letter raised concerns that, when echoed later by NBAF opponents, would cause them to be ridiculed as some sort of lunatic fringe. "Any upgrade," Clinton and Bishop said, "would potentially jeopardize the safety of the millions of residents residing in surrounding communities."[31]

McCarthy's response? "I can state definitively that we have no plans, either near-term or long-term, for BSL4 containment labs."[32] And four days later, in a meeting with Clinton and Bishop, Ridge reportedly offered "absolute assurances" that DHS would not seek a BSL-4 designation or use Plum Island to test human diseases.[33] (Two years later, DHS would suddenly announce near-term plans for a new BSL-4 facility to "replace" Plum Island, and probably to be built elsewhere.)

In September 2003, the GAO (then the General Accounting Office) issued a report corroborating some of the concerns of Clinton and Bishop and the striking workers. The report said "fundamental concerns" left the facility "vulnerable to security breaches."[34] Following the September 2001 attacks, USDA had contracted with Sandia National Laboratories to evaluate its security program,[35] yet alarms and door sensors Sandia recommended were not fully operational; there was inadequate lighting to support the outside security cameras; and there was inadequate physical security for "certain assets."[36] Plum Island held the only FMD vaccine bank in North America, for instance, representing years of cooperative research by Canada, Mexico, and the United States, yet the room containing it had "a window opening covered with only plywood."[37]

Plum Island officials had also been negligent in controlling access to the pathogens. Eight foreign scientists had been permitted into the biocontainment area without the required escort, and despite incomplete background checks. Background checks were not even conducted on students who regularly attended classes in the biocontainment area, nor were janitors and other personnel who entered the containment area for nonlaboratory purposes escorted as stipulated by regulation.[38]

In something of a dramatic irony, given the FBI's later allegations about Bruce Ivins, the GAO cited with approval USAMRIID's security practices, in which "background checks are required to be updated regularly to evaluate the continued suitability and reliability of employees working with pathogens."[39]

The facility's incident response capability was also inadequate: the guard force had been operating without formal arrest authority and without a policy on the use of weapons; there were too few guards; arrangements for local law enforcement support were limited; and there was no road map for actions to be taken in the event of a terrorist attack.[40] Furthermore, officials had paid too little attention to how the ongoing strike and the hostility it engendered increased the level of risk.[41]

USDA objected to parts of the report, but DHS said the report was factually accurate and promised to implement its recommendations. In the months ahead, it replaced the maintenance contractor involved in the labor dispute with a new company, and also appointed a new facility director. A December 2003 Associated Press article indicated Clinton and some local officials were cautiously optimistic after the changes, but that they still had major concerns.[42]

In February 2004, Michael Carroll's *Lab 257* was released. Carroll's book, researched for seven years, argued that Plum Island was "a biological time bomb with an appalling safety record, a tempting target known to terrorists and a grave but little-recognized threat to the largest population center in the United States."[43] Along with the aforementioned suggestions that lab experiments might have caused the Lyme disease and West Nile outbreaks—whose epicenters were suspiciously nearby on the U.S. mainland—Carroll told the stories of the more notorious safety breakdowns at the facility and of others not so well known. He was especially critical of the privatization of support services in 1991, which he said had caused plunging workforce morale and a major decline in security.

A few months earlier, DHS had accepted a critical GAO report as factually accurate. Now it—and former Plum Island director Roger Breeze—rushed to discredit Carroll's book. Breeze suggested that Carroll might have been led astray by disgruntled striking workers, to which Carroll retorted that "If I was led astray, I was led astray by government documents yielded by seven years of requests, national archives research and hundreds of hours of interviews, including with Dr. Breeze."[44]

Carroll said the real problem was "that Plum Island is a kingdom unto itself. There is zero public oversight." Representative Bishop, who had frequently raised concerns about the facility, now took issue with Carroll's assertions that things were out of control. "I believe we have a fairly good handle on what's going on there and that the administrators are pretty open about it."[45] Famous last words.

DHS scheduled a quick public-relations pick-me-up for about a dozen journalists, pointing out all the agency had done to improve security since taking over the facility a few months earlier. Visitors were met by security guards now. The guards actually checked IDs. And people who wanted to get into the biocontainment area had to punch in a code first.[46]

Even more impressive, however, the road to biodefense heaven had been rediscovered. "[P]athogens were contained and controlled at the lab, and scientists described the elaborate precautions required for transporting blood and tissue samples from animal pens to laboratories."[47] All the Plum Island sins of the past had been redeemed and forgiven; it was as though they had never happened; and Carroll's book had just missed the mark.

No sale to Debbie O'Kane, director of the North Fork Environmental Council, who wasn't willing to take Homeland Security officials at their word. Despite all the promises of a new openness, there still had been "no direct communication with the local community." "We need," she said, "assurances that the concerns that were cited in the G.A.O. report were addressed systematically."[48] So far, such assurances were in short supply.

A May 2004 *New York Times* article indicated that DHS was "farming out" Plum Island's animal disease work to agro-security "centers of excellence" at Texas A&M and the University of Minnesota, as well as to the University of California at Davis, the University of Southern California, and the University of Maryland. Some saw this as a sign that the facility was being "shunted aside" and might close in a few years.[49] A little over a year later, DHS would announce its intent to replace Plum Island with a new facility, probably to be built on the mainland, where it would lack the extra safeguards provided by an ocean barrier.

Meanwhile, there were a couple of more accidents for the road— and new questions about the trustworthiness of Homeland Security. On August 17, a *Newsday* story reported that foot-and-mouth virus had briefly spread within the facility on two different occasions in June and July. The public learned of the incidents only after an anonymous tip reached the ears of a reporter; the incidents were then confirmed by DHS. A DHS spokesperson explained the secrecy thus: "It was within the laboratory environment, safely sealed in biocontainment. This was really an operational issue."[50]

The June incident occurred one day before officials and visitors converged on the island to celebrate the lab's fiftieth anniversary.[51] *Science* said the cases "have added to concerns that such accidents may become more common as biodefense research expands."[52] That those concerns were justified would be demonstrated by the 2007 Associated Press story on recent accidents and the 2007 GAO report and congressional hearing, documenting a reckless, uncontrolled explosion of new facilities and poorly-trained researchers.

CHAPTER TWO

Breaches of Containment
The Juggernaut Sputters in the Passing Lane

1

2007 was an embarrassing year for America's biodefense complex.

It had all started a bit earlier, with some routine carelessness at one of the Department of Homeland Security's new Centers of Excellence. A few fault lines in what a University of Georgia propagandist brazenly described as a "culture of safety." The carelessness would lead to a brucella infection (or brucellosis); the infection would lead to a cover-up; the outing of the cover-up would lead to a research shutdown and investigations—one by the regulatory agency of record, the CDC; the other by Congress's investigative arm, the Government Accountability Office. The investigation would culminate in a full-blown congressional hearing in October 2007, and the startling conclusion of the GAO's expert witnesses that a huge proliferation of biodefense research aimed at making Americans safer was actually putting them at risk.

Then, a year later, came the FBI's public confirmation that the anthrax terrorist who had scared us all half to death in 2001 and thereby helped fuel the current biodefense boom was not an Al Qaeda protege with a Wal-Mart chemistry set, but rather a biodefense insider, a researcher at America's oldest and biggest bioweapons research facility, supposedly now to be read "biodefense" facility. One of the select facilities which had helped the FBI with its investigations. "Now you have *me* to fight for you," Oedipus had told the Thebans in that earlier story of kingly murderer-turned-detective. And in our own dark times we had Fort Detrick to fight for us. There would be no blinded, humbled penitence here. No, Fort Detrick would become home to a bloated new biodefense "campus," home of new BSL-4 labs for the Army, Homeland Security, NIH, USDA, aimed at battling the same terrorism they had given birth to.

And Fort Detrick would not be a lonely Cyclops, as the Congressional hearing would reveal, with its testimony about a sevenfold increase in biodefense funding, a twelvefold increase in BSL-4 lab space, and the unquantified mushrooming of BSL-3 labs.

Of course, these are all state-of-the-art, high-tech facilities, we are told. Ultra-secure. Dedicated, meticulous scientists, imbued with a culture of safety, doing their bit to defend us from—well, dedicated and meticulous scientists, I suppose. "It's not just a matter of *if* bioterrorism will happen. It's a matter of *when*." If this includes the odd mad scientist demonstrating the effectiveness of his wares, of course, it's not even a question of when it will happen, but when it will happen *again*. Which raises the complicated question of how many more mad scientists we ought to hire to defend ourselves from other mad scientists.

Meanwhile, back at the OK Corral, alleged anthrax rustler Bruce Ivins conveniently does away with himself to spare the government the expense of a trial. Seems it was all connected with a sorority fetish or something. No conspiracy theories need apply. And Steven Hatfill, previous anthrax rustler of interest—the fresh marks of the noose still around his neck—rides off into the sunset with his $5.8 million settlement from the government.

2

In February 2006, back at the DHS Center of Excellence, the culture of safety nodded off at the wheel and ran into a ditch. Not for the first time, it would later be learned. And such bad timing, too. Texas

A&M was preparing to join twenty-eight other potential sites responding to DHS's January 19 Request for Expressions of Interest in NBAF. A&M was then under the leadership of Robert M. Gates, former CIA Director, future Secretary of Defense, and Bush family confidant. During Gates's A&M tenure, the university had already been awarded $18 million to establish its Center of Excellence, the National Center for Foreign Animal and Zoonotic Disease Defense.[1] The Center's mission resembled the proposed mission for NBAF, then described as "an integrated human, foreign animal, and zoonotic disease research, development and testing facility," and so A&M seemed to many people a prime candidate for NBAF. On March 21, Gates would announce A&M's intention to submit an Expression of Interest.

On February 9, 2006, however, the culture of safety was letting down both the health of the university's researchers and the prospects for its NBAF application. By the time fan-hitting time arrived, Gates would be gone, off to Bush Junior's cabinet to deal with the excrement spattered by Bush's Iraq invasion.

Several A&M researchers were in a BSL-3 lab on February 9, training with the Madison Aerosol Chamber (MAC), a device manufactured by the University of Wisconsin at Madison, used to expose the lungs of lab animals to researchers' germs of choice. The training was being supervised by A&M professor David McMurray, who had helped develop the MAC. Following a "hot run" with aerosolized brucella, a female student researcher not wearing proper protective gear "cleaned the unit by climbing partially into the chamber to disinfect it."[2] If that was A&M's "Standard Operating Procedure," it was not a good one.

The young researcher would pay for this less-than-excellent safety protocol with a long unexplained illness. By April 2006, she had "been home sick for several weeks." Apparently no one suspected brucellosis, despite the fact that brucellosis is what they were working with, and despite the fact that Principal Investigator Thomas Ficht was an experienced brucellosis researcher.[3] If you can't trust a brucellosis researcher to recognize the symptoms of brucellosis, who can you trust?

Eventually, the hapless researcher's personal physician ordered

blood tests and made the brucellosis diagnosis. The patient was eventually cured by a week's regimen of intravenous antibiotics, followed
by a 45-day course of two additional antibiotics.[4]

According to the CDC's Web site, brucellosis kills only about two
percent of its victims. Treatment can be difficult, however, and may
take up to several months. Its various symptoms and side effects can
be similar to those of the flu: fever, sweats, malaise, anorexia,
headache, myalgia, back pain, arthritis, and common fatigue syndrome. Severe infections of the central nervous system or lining of
the heart (endocarditis) can occur in five percent of cases, causing
severe long-term problems.[5]

The incident itself did not place the public in any particular
danger, because brucellosis is not contagious between humans. (Not
all bioweapons pathogens are so accommodating.) The university
covered the incident up, however, violating federal select agent
reporting rules, and the eventual outing of its deception was what
caused all the embarrassment. Not just for A&M, but for the biodefense industry in general. A few folks outside the biodefense industry
consider it bad form to say you're oh-so-safe when you're oh-so-not.

Perhaps the industry's ritual psalms of praise for its high-tech,
state-of-the-art equipment were misdirected, for instance. This was
not, after all, the first incident involving the Madison Aerosol
Chamber. In late 2003, a Seattle, Washington BSL-3 lab had begun
using the MAC to infect guinea pigs with tuberculosis, and had
processed several batches of guinea pig victims before trouble arose.
In March 2004, three employees who'd previously tested negative for
tuberculosis came back with positive tests, indicating they too had
been exposed. The State of Washington opened an investigation,
which concluded that a leaky airflow meter in the MAC was responsible. The investigation also determined that, even prior to the infection, the lab staff had encountered several other problems with
cracking seals and leaks from the MAC.

Before using the MAC, the Seattle staff had been trained by
McMurray in 2003. He'd told them the chamber was "foolproof" and
"so safe that there was no need to even locate it in a BSL-3 environment," and that "respirator use was not necessary." Famous last words.

As of 2005, over twenty MACs were in use at various locations, and the chamber had been promoted at a December 2003 NIAID biodefense workshop. The MAC itself was clearly at fault in the Washington incident, but at A&M the culprit was an "inappropriate safety protocol." McMurray had supervised the training at A&M, however, and his overconfidence about the machine may have contributed to the young researcher's carelessness. The lessons of the MAC are twofold: high-tech biodefense equipment is not necessarily foolproof; and even if it were, human carelessness can and regularly does defeat any state-of-the-art embodiment of technological perfection.

Had the disease involved in the A&M case been transmissible between humans, the two-month delay in diagnosis could have had more serious consequences. How confident can we be that biodefense and public health experts will act promptly and appropriately in the aftermath of future research accidents, if a mysteriously ill biodefense researcher can go undiagnosed for two months with no one suspecting her work might have caused the illness?

The most serious embarrassment, of course, was the coverup itself, which suggested that accidents were occurring and being concealed not just from the public, but even from regulatory authorities like the CDC, itself hardly the *National Enquirer* of biodefense incidents. When they learned of the brucellosis diagnosis in April, A&M officials apparently discussed the requirement to report the incident to the CDC but decided not to do so.

Nothing more happened until October, when a small watchdog organization, the Sunshine Project (SP), began requesting materials for a series of Transparency Reports on institutions bidding to host NBAF. Pursuant to the Texas Public Information Act, the SP's U.S. director, Edward Hammond, asked A&M for "all records on occupational exposures and/or laboratory-acquired infections with risk group 2 (RG2) or higher agents, from 1 January 2000 through the present."[6] A&M released one page total, reading as follows:

Occupational Exposure

Number of Incidents	Nature of Exposure	Required Treatment
1	Brucella sp.	antibiotics

The brevity roused Hammond's suspicions. He felt a brucella infection would have generated more documentation than what he received. A series of "tense E-mails" between Hammond and A&M's General Counsel followed, with Hammond threatening to report the incident to the Texas Attorney General. On April 10, 2007—over a year after the original incident—A&M E-mailed the CDC to inform them of the unreported incident, and that night the university mailed Hammond the requested documents. The next day, A&M filed—a year late—the APHIS/CDC Form 3 required under the Select Agent rules. On April 12, Hammond issued a press release about his discoveries, which was picked up by several Texas newspapers.[7]

The following week, CDC inspectors spent three days on the A&M campus investigating. On April 20, they suspended part of A&M's research program.[8]

On June 25, 2007, as Hammond's skirmishes with A&M over his freedom-of-information request continued, he received copies of a series of E-mails showing that in April 2006, about the same time as the brucella incident, three other A&M lab workers had tested positive for Q fever exposure. Once again, A&M had failed to report the incident to the CDC.[9]

Q fever is, like brucellosis, a zoonotic disease, one which normally spreads from animals to humans. It can be contracted through inhalation once the pathogen is airborne, however. According to the CDC Web site, only half those infected will develop such clinical symptoms as high fever, severe headache, general malaise, myalgia, chills/sweats, non-productive cough, nausea, vomiting, diarrhea, abdominal pain, and chest pain. Thirty to fifty percent of those who do exhibit symptoms, though, will eventually develop pneumonia. Five percent will succumb to chronic infections lasting anywhere from six months to twenty years. Possible complications other than pneumonia include weight loss, miscarriages, myocarditis, encephalitis, osteomyelitis, and hepatitis.[10]

The day after receiving the E-mails, Hammond issued a press release, elaborating on the sparsity of documentation about the Q-fever incident, and what he suspected had occurred:

What prompted the infected individuals to visit the hospital is not stated in the documents received by the Sunshine Project. Yet three individuals from the same lab visited the hospital at the same time and had the same tests for a very unusual pathogen performed. Circumstances strongly suggest a lab accident that led the researchers to suspect (correctly) that [they] had become infected. According to the A&M records, upon learning of the infections, the main action of the biosafety officer was to report the accident to the co-chairs of the Texas A&M Institutional Biosafety Committee, who include Thomas Ficht, the professor responsible for the researcher who contracted Brucella in February 2006. But no mention of a Q Fever accident appears in Texas A&M's biosafety committee meeting minutes.

In fact, Texas A&M has produced zero documentation, such as accident reports, lab paperwork, lessons learned, modified operating procedures, or anything else except a few sparse E-mails for either the Q-fever or the Brucella accident. This is despite open records requests for such paperwork.[11]

"If Texas A&M's replies under the Texas Public Information Act are to be believed," Hammond said, "then four people at the University have been infected with bioweapons agents without responsible A&M professors and other officials even bothering to file a simple incident report, much less alert the community or report to public health officials."[12]

A&M's provost told the *Dallas Morning News* on June 26 that the university's occupational health policy at the time didn't instruct the researchers to report the Q-fever exposures to the CDC, but that the university had changed its policy in April 2007, and reported the exposures at that time. The CDC said that as of June 26, however, it had not received such a report.[13]

On June 30, the CDC suspended all A&M's work with select agents and toxins while it conducted a "comprehensive review of all select agent and toxin activities at TAMU."[14]

The initial brucella incident had received some national coverage, including an article in *Science*, but national attention intensified after the Q-fever incident and the CDC's shutdown of A&M select agent research. *Nature* ran a July 11 editorial declaring that problems with new biodefense facilities needed to be addressed before further expansion. Referring to the Texas A&M incident, *Nature* said:

> [E]ven when Texas A&M made its report, neither the university nor the federal agencies found it [necessary] to share news of the infraction with the general population. It was only on 26 June that the problem became public, when the Sunshine Project, a small watchdog group based in Austin, Texas, revealed details of the infections at Texas A&M. A week later, the watchdog disclosed associated problems at nine other laboratories nationally.
>
> The federal government should not have to be prompted by activists into telling the American public the truth.[15]

Hammond's revelations about problems at nine other laboratories had come in a July 3 press release titled "Texas A&M Bioweapons Accidents More the Norm than an Exception." The new disclosures involved additional lab-acquired infections and exposures, unauthorized research, equipment malfunction, and disregard for safety protocols. They included an incident in which researchers at the University of California-Berkeley mistakenly handled dozens of samples of Rocky Mountain spotted fever as if they contained a different, harmless organism, then failed to tell the community about the mishap; an incident in which the University of Iowa performed unauthorized experiments to genetically engineer antibiotic resistance into the biowarfare agent tularemia; an incident in which lab workers at a University of Illinois at Chicago BSL-3 facility propped open lab doors—a major safety violation, completely undoing all those high-tech "containment" features like negative air pressure; failure of the exhaust fan in a University of North Carolina BSL-3 lab; a tuberculosis exposure at the Albert Einstein College of Medicine involving our old friend the Madison Aerosol Chamber; and failure of a steam valve (culprit in the Three Mile Island accident) in biological waste treatment tanks at the NIH campus in Bethesda, Maryland, causing major damage which forced the closing of Building 41A for repairs.[16]

Hammond pointed out that these disclosures came from biosafety committee minutes of **"institutions that actually record such incidents in records that are (at least nominally) available to the public. Often, this is not the case"** [emphasis in original]. As Hammond said, "There is no reason not to presume that many more similar accidents have occurred but have yet to come to light."[17]

The CDC sent eighteen inspectors to the A&M campus for four days in late July.[18] Shortly after their departure, the *Dallas Morning News* obtained personal injury and occupational safety reports from the six Texas universities conducting lab-based biodefense research. In addition to discovering references in the reports to "dozens of needle pricks, splashes, inhalations and exposure to other deadly, highly monitored diseases over the last several years," the *DMN* learned that the lab employee infected with brucella in February 2006 had not even been authorized to work with the agent.[19]

The CDC issued a twenty-one-page report on its investigation on September 4, 2007. In continuing the ban on select agent research at A&M, the agency cited a host of problems: workers conducting experiments without federal approval, in labs not authorized for the specific agents, and without proper medical monitoring; vials of brucella that were missing or unaccounted for; employees granted access to select agents without the proper approval; workers failing to wear proper protective gear, and wearing coats used for experiments outside the labs; aerosol experiments conducted with no barrier between the chamber and an adjacent research lab; failure to properly dispose of infected animal remains; failure to adequately maintain logbooks of who came into and out of the labs; "grossly inadequate" safety and security plans and staff training procedures.[20]

The CDC also faulted the university for failing to document corrections of deficiencies noted in previous inspections. This, ironically, reflected on the CDC itself, which had conducted annual inspections at A&M in 2004, 2005, and 2006, repeatedly finding a host of minor problems and not following up to verify that the problems had been dealt with. Inspections in both February 2006 and February 2007 had missed most of the major problems cited in the new CDC report, including the brucella and Q-fever exposures and the unauthorized research. Hammond said: "They caught almost nothing. . . . People take solace in the fact that we have a stringent permit and inspection program. In reality, it's pretty depressing." Hammond also pointed out that it was actually the public information act (and though he did not put it this way, his own diligence in using the act), not the inspections, which had alerted the government to the brucella and Q-fever incidents.[21]

On the day of the CDC report, A&M released a new batch of documents to the Sunshine Project. The documents revealed that there had been another unreported brucella exposure in 2007, and that the facility had experienced "major flooding" in February 2007. An inspection following the flood revealed significant damage—but the inspection had not been performed until almost two months later, four days after the Sunshine Project's news release on the brucella infection, and the same day a CDC inspection team arrived in College Station.[22]

Throughout the A&M events, A&M officials were spinning, trying to walk a tightrope: on the one hand expressing remorse for their failures and their eagerness to cooperate with the CDC investigation, on the other hand suggesting that the CDC was unfairly making them a scapegoat, that the situation at A&M was actually typical of what went on elsewhere, or that the whole mess was the result of a new, confusing, and increasingly strict regulatory environment.

"'Folks here, they're saying, Nobody died. That's not the point,' A&M Chancellor Mike McKinney said. 'That doesn't make it OK to make a mistake. Yes, we messed up, but we didn't mess up on purpose. There's that saying, never assume a conspiracy if it could just be incompetence.'"[23]

After the CDC's report was issued, McKinney would complain that "nobody else has a clean record either." TAMU microbiologist Vernon Tesh, one of the lab leaders singled out for safety lapses in the report, said: "If you were to apply an equivalent level of scrutiny at other institutions, I think you would find issues of concern." And interim TAMU president Eddie Davis likewise suggested that other institutions "under that same level of review would probably have findings that would be reportable to the CDC."[24]

A&M's vice president for research had already resigned; its safety director now did likewise.[25]

3

Even as the CDC was publicly chastising A&M, it was trying to spin an incident at its own facilities. On June 15, a lightning strike had knocked out power for about an hour at the agency's new $214 million Emerging Infectious Diseases building—which included six new "high-tech" BSL-4 labs, the "crown jewel of CDC's $1.5 billion construction plan." Backup power had not come on, which meant that negative air pressure, the high-tech safety feature routinely recited as reassurance for those worried about high-tech killer germs in their midst, had been lost.[26]

The BSL-4 suite was still unoccupied, but about 500 CDC scientists and staff had moved into other parts of the building containing BSL-3 labs.[27]

The *Atlanta Journal-Constitution* broke the story on July 7, 2007. According to CDC's building and safety director George Chandler, the lightning strike had caused a power surge, tripping breakers as it was supposed to. But the system hadn't alerted the CDC's new emergency generators that there was a power gap.[28] The CDC would later say that the lightning protection system had been damaged by construction at a nearby building.[29]

"In no way do we believe this incident in any way would have caused any risk to the public," CDC health and safety director Dr. Casey Chosewood said. He would cite the multiple, redundant safety features which remained operable.[30] (Of course, backup power was itself a "redundancy," and one that had failed.) And CDC spokesman Tom Skinner would declare, "We've worked in labs in Atlanta for over 60 years and not once have we had an environmental exposure that impacted the community."[31] Presumably, Texas A&M could also claim not "to have had an environmental exposure that impacted the community." But drunk driving and faulty brakes are still unsafe, even if a car hasn't yet crossed the median and hit another car head-on.

The *AJC* noted in its July 7 article that it had requested records relating to the power outage and the safety of the new BSL-4 labs under the federal Freedom of Information Act, but that "The CDC has denied the newspaper's request for expedited release of the information saying 'there is no urgency to inform the public.'" Almost a year later, in May 2008, the *AJC* reported that the CDC still had not released any records.[32]

Texas A&M had concealed its problems both from the public and from the CDC; now the nation's leading public health agency was itself showing a marked penchant for keeping the public out of the loop. On March 15, 2007, the Sunshine Project had filed a formal complaint with the NIH's Office of Biotechnology Activities against the CDC's Institutional Biosafety Committee, for violating the public access provisions of the NIH Guidelines regarding IBCs. Hammond charged that the CDC had completely ignored four separate requests for IBC minutes, had responded to the fifth request only after NIH sent CDC a letter, and that the minutes it had just provided were "complete and shameless junk" because "[t]hey provide no substantive information on the research under review or the committee's review of it."[33]

It was Hammond's review of similar documents from Texas A&M—not the CDC's regulation of A&M under the Select Agent Program—that had discovered the problems at A&M.

A July 24 article in the *AJC* suggested why the CDC preferred to keep its records to itself. The *AJC* obtained copies of internal documents dating back to 2001 showing that CDC engineers had warned higher-ups that plans to centralize backup power rather than locate individual systems in critical buildings would result in a vulnerable, unreliable backup power system. Mechanical engineer Johnnie West had said in a September 2002 E-mail: "I have very little confidence in the [backup] generators being able to operate as designed, due to the complexity of so many generators being connected in parallel, and the time it will take them to synchronize before they can provide power to the grid. I've gone way beyond the call of duty to point out all of these issues during the past year, and have paid a heavy price for doing so. After visiting NIH and Kings Bay, I'm even more convinced that if we want reliable power, our plans must include having a generator at the critical buildings."[34]

Not everyone agreed that the CDC power outage was so innocuous as the CDC was suggesting. Dr. Richard Ebright of Rutgers University pointed out that BSL-2 and BSL-3 labs in the building had been operational, and elaborated as follows:

> In BSL-2, one works in an air-filtered biocontainment cabinet. If the power goes out, the filter fails. In BSL-3, one works inside a cabinet inside a room with negative air pressure. Both cease when power fails. If there had been BSL-4 in effect, they would have lost the negative air pressure in that as well.
>
> The CDC isn't saying what they're working with, but it probably includes the 1918 influenza virus, H5N1 avian influenza, and certainly anthrax and plague. It would have gotten more play anywhere else, but the agency responsible for investigating was CDC itself.
>
> It's too early to know if there were infections. This would count as exposure. . . . This wouldn't be acceptable at a small company or university lab, and was astonishing for [a] lab working with 1918 influenza.[35]

4

A third embarrassment for America's biodefense complex came on August 3. An outbreak of foot-and-mouth disease was confirmed on a Surrey farm in the United Kingdom. The disease is not dangerous to humans, but is extremely contagious and can be devastating to the livestock industry. A 2001 U.K. outbreak had resulted in the slaughter of over 6 million animals. A study by the United Kingdom's National Audit Office had estimated a direct cost to the public sector of over $5.71 billion, and a cost to the private sector of over $9.51 billion.[36] The 2007 discovery triggered contingency plans developed after the 2001 outbreak. The herd on the outbreak farm was destroyed. An immediate national livestock movement ban was imposed; a three-kilometer protection zone and ten-kilometer surveillance zone were initiated; and a total export ban on meat and animal products was imposed by the European Union.[37]

Tests run the next day determined that the foot-and-mouth virus strain involved, O1 BFS 1860, also known as O1 BFS 67, was one which did not occur in nature and was held only at the nearby U.K.

Pirbright research complex.[38] Pirbright was shared by two "world class research facilities"—Merial Animal Health, the world's leading producer of foot-and-mouth vaccine, and the Institute of Animal Health, "the world's foremost reference laboratory for identifying and monitoring outbreaks of foot and mouth."[39]

The obvious conclusion was that the FMD virus had either escaped or been deliberately introduced from the "high-containment" facilities at Pirbright. Government-commissioned independent investigations by the Health and Safety Executive[40] and Professor Brian G. Spratt[41] would later conclude that most likely, waste water containing the live virus escaped from poorly maintained drainage pipe into the surrounding soil and was then carried off the site by construction vehicles. Excessive rainfall may have exacerbated the release from the drain.

HSE inspection of the drainage system had found evidence of "long-term damage and leakage, including cracked pipes, unsealed manholes and tree root ingress."[42] Spratt concluded that "[r]elease of infectious virus from Merial and consequent surface contamination from the drainage system, and mechanical spread to the outbreak farm, is . . . the most likely cause." He cautioned, however, that "[t]he cause of the escape of FMDV from Pirbright has still not been established, and may never be."[43] And the HSE report noted that while nucleotide sequencing indicated Pirbright as the source of the outbreak, the small differences between the strains used by IAH or Merial were not sufficient to establish one entity or the other as the ultimate source.[44]

Then on August 6 the disease was detected on a second farm. The herd on this farm was also "culled" or destroyed. No additional cases occurred for a month. On September 8, the Chief Veterinary Officer declared that the disease was over and that the remaining restrictions on animal movements would end. The EU had already lifted most of its export restrictions on August 23.[45] One of the independent investigators would later describe this as the first "confident" phase, noting wryly: "Formal control measures were confidently introduced on 3 August and just as confidently removed on 8 September with an unqualified announcement. On 12 September another case was confirmed."[46]

All the old restrictions went back into effect. Six new cases would be confirmed in all, the last on September 30.[47] 2160 animals, mostly cattle and pigs plus a few sheep and goats, would be slaughtered, with a total estimated cost to the government (as of March 2008) of 47 million pounds, and to the British livestock industry of 100 million pounds.[48] These figures were provided by a government regulatory agency with obvious incentives to keep the costs low. An August 29, 2007 editorial in the *Financial Times*—published before the second round of outbreaks that began on September 12—indicated that movement restrictions and the EU export ban had already cost Britain's meat and dairy industry hundreds of millions of pounds.[49]

The U.S. had long feared a foot-and-mouth disease outbreak. The last such outbreak on the mainland had occurred in 1929. For years a congressional law had prohibited introduction of live foot-and-mouth virus onto the U.S. mainland, and so such research had been confined to Plum Island, where there had been several foot-and-mouth outbreaks (situations in which the virus had escaped from "containment" areas and infected animals outside them). Now Homeland Security, which acquired control of Plum Island from USDA in 2003, had announced its intention to move Plum Island's research to a new facility—the National Bio and Agro-Defense Facility—on the mainland. It had just announced the five mainland finalists for the facility—sites in Texas, Kansas, North Carolina, Georgia, and Mississippi, ironically all significant livestock production states. (It had also just added as a "by the way" that it would also consider building the new facility at Plum Island, though neither the existing Plum Island facility nor the states of New York or Connecticut had solicited such consideration.) In the spring of 2008, the Government Accountability Office would conclude—and would testify at a congressional hearing—that DHS had not established the safety of moving foot-and-mouth research to the mainland.

The Pirbright situation raised troubling questions about DHS's decision. It also raised echoes of germ safety lapses in the U.S. There was the Oedipus factor, for instance. Scientists at Fort Detrick had helped investigate anthrax terrorism that the FBI would later indicate had been perpetrated from Detrick itself. Now, the world's "foremost reference laboratory" for foot-and-mouth disease—like Detrick, a high-containment operation—was investigating an outbreak that had originated from the facility itself. Defend us, O Lord, from our defenders.

There were also shades of Texas A&M. Once the cat was out of the bag, the memo out of the file folder, the germ out of containment—once "a problem" had been identified—investigators swarmed around the blood like mosquitoes at a Boy Scout jamboree. See, we do have regulators.

The HSE discovered the following "biosecurity lapses" at Pirbright: the compromised state of the site drainage system; "the practice employed by IAH of using bowsers and hoses in the intermediate site effluent drains to clear blockages without a standard operating procedure (SOP)"; the failure to record all human and vehicle movements via the IAH gatehouse to the site; and poor monitoring and control of access to restricted areas within IAH facilities.[50]

And there were shades of the CDC's problems with backup power. It turned out that a former construction and maintenance overseer for Pirbright had repeatedly warned superiors and agency administrators about a year earlier that the antiquated drainage system was "a disaster waiting to happen."[51]

The major difference between the U.K. incident and similar ones in America was the transparency of the U.K. investigation. The U.K. government commissioned four separate independent investigations of issues related to the outbreak,[52] and full interim and final reports were posted online.

5

On August 7, 2007, just as British investigators were announcing that the new foot-and-mouth outbreak there had probably originated at one of Pirbright's high-containment research facilities, the House Energy and Commerce Committee was announcing plans to conduct a hearing on safety at the U.S.'s biodefense labs. The hearing would be held on October 4 before the Oversight and Investigations subcommittee, chaired by Rep. Bart Stupak, D-MI, and would coincide with the release of the GAO's preliminary report. Part of the information the GAO had apparently obtained, but did not release as part of its report, were "confidential records" of incident reports submitted to the CDC over the past four years involving "select agents"—a list of 72 pathogens the government considers the most deadly and in need of regulation. On October 2, the Associated Press obtained these records and published a news story, "More than 100 Incidents Reported at Labs Handling Deadly Germs." A timeline and details were posted on the AP's Web site. The incidents involved a variety of accidents and missing shipments at 44 labs in 24 states. The CDC had refused to release them under the Freedom of Information Act, citing an anti-bioterrorism law.[53]

Thirty-six of the accidents had been reported in the first eight months of 2007, nearly double the number reported during all of 2004. Researchers were either getting more reckless, or they were getting more honest in the wake of the Texas A&M scandal.

NBAF 2

THE NBAF DATING GAME, BETA VERSION

ON AUGUST 22, 2005, DHS SAID IT WAS THINKING OF REPLACING "the important but aging facility" at Plum Island with a "next-generation biological and agricultural defense facility." President Bush's FY06 budget was requesting $23 million for needs assessment and design.[1] And on January 19, 2006, DHS published a Request for Expressions of Interest by Potential Sites in the Federal Register.

The Fact Sheet said the design study would explore three different NBAF options, one of which would keep the scope the same as the current Plum Island mission. The Federal Register notice, however, emphasized that the proposed facility would be "an integrated human, foreign animal, and zoonotic disease research, development and testing facility" with "the capability to address threats from human pathogens, high-consequence zoonotic disease agents, and foreign animal diseases"—something quite different and certainly scarier than Plum Island's current animal disease focus.[2]

Answers to Frequently Asked Questions posted on the DHS Web site in early 2006 also emphasized the expansive new mission of NBAF, and DHS's intention to build something quite different than a

Plum Island clone.[3] One respondent asked hopefully: "In other words, no human pathogens will be used for these BLS3 [sic] and BSL4 facilities, only for agricultural diseases?" The reason for the question is obvious: humans get sick from human pathogens, and we tend to think them—and the zoonotic diseases that spread from animals to humans—more dangerous than diseases affecting only animals or plants. And with BSL-4 diseases, we're talking about things *without* a vaccine or cure, highly likely to cause death.

In early 2006, DHS didn't provide the reassurance the respondent sought:

> The interrelated Homeland Security missions of the three Departments (DHS, HHS and USDA) all require new research and development infrastructure that can accommodate extensive testing with a variety of animal models. The proposed NBAF is envisioned to provide the nation with the first integrated agricultural, zoonotic disease and public health research, development, testing and evaluation facility with the capability to address threats from human pathogens, high-consequence zoonotic disease agents, and foreign animal disease.[4]

Over a year later, after encountering protests in Kentucky, California, Wisconsin, and Missouri, DHS would quietly and mysteriously post a stripped-down "Diseases of Interest" list on its Web site. (Whether the list's echoes of Steven Hatfield and the FBI were deliberate or simply obtuse is itself a mystery.) Then, as they visited the five finalist locations that fall (Texas, Kansas, Georgia, North Carolina, and Mississippi), DHS spokespeople would assure worried locals that no, those scary human diseases would not be studied at NBAF, not even those scary, well-known zoonotics they had all heard of, like anthrax and Ebola.

DHS would repeat this ad nauseam even though the fine print of the "Diseases of Interest" list flatly stated that "This list may change based upon continued threat assessments and risk assessments,"[5] and even though, according to a little-read December 2005 GAO report on Plum Island, "DHS officials emphasized that the dynamic nature of threat assessments makes it difficult to firmly commit to long-term priorities because information and research needs may change frequently depending on the nature of the threat."[6]

DHS wasn't committing to much that spring. In fact, it wasn't saying much, period, just that it needed and planned to build a shiny new $451 million, 500,000-square-foot high-security biolab to replace Plum Island.[7] Odd how it decided that this was what it needed, since the "needs assessment and design process" funded in the 2006 budget hadn't yet begun. (DHS stated in the Answers to Frequently Asked Questions that the conceptual design study spoken of in 2005 had not been completed yet and would not be initiated until late 2006.)

One of the problems that many would have (and still have) with NBAF was its control by DHS. We didn't trust a quasi-military organization, oriented toward security, to be anything but secretive and duplicitous. There was sufficient evidence, in the spring of 2006, that secrecy and duplicity were part of the NBAF future. Online life sciences magazine *The Scientist*, in its Feb. 6, 2006 article on DHS's plans, indicated that "DHS and USDA officials declined to respond to any of *The Scientist*'s questions about the lab." In the Answers to FAQ, responding to the question whether DHS was "willing to have a pre-application FAQ session with members of the consortium to answer questions related to the application process," DHS responded tersely, "There is not a pre-application meeting planned."

When asked "How will continual public transparency of NBAF operations be accomplished?" DHS appeared either not to know what the concept meant, or to be deliberately evading the question. DHS's response: that the public would be able to provide comments during the National Environmental Policy Act process! (One would assume that real "continual public transparency" would mean something like the public knowing precisely what the hell DHS and USDA were up to, after they got up to it.)

DHS similarly evaded the question of whether it would "declare that any research related to developing or preparing biological warfare agents will never be allowed." The question itself was naïve, because of course DHS wasn't going to declare any such thing, since it could only do so by lying. "Biological warfare agents" and "bioweapons agents" are terms commonly applied to all those pathogens that *might* be used for biowarfare or bioterrorism. The whole purpose of the biodefense enterprise is to research such agents.

The question was therefore naïve, and DHS treated it with the duplicitous contempt that it perhaps deserved, responding: "DHS conducts all of its biological agent research in full-compliance [*sic*] with the regulations governing such work in the Biological Weapons Convention (BWC) and other applicable laws and regulations. The BWC explicitly bans the production of biological agents for hostile purposes or armed conflict." The response is a double dose of skullduggery. First, it doesn't answer the question, which was not whether DHS plans to violate the BWC, engage in biowarfare, or even produce agents "for hostile purposes or armed conflict." It was simply whether DHS was going to do any research "related to developing or preparing biological warfare agents." And the answer is: you bet. That's what biodefense is all about, working with biological warfare agents—existing or new ones—to try to develop a vaccine or some other defense against them. The biodefense complex doesn't like the term "biowarfare agents" because it reminds us that these are deadly germs. But it's part of the basic biodefense two-step to scare us all to death by telling us how dangerous these germs would be in the hands of terrorists, but then to tell us how perfectly innocuous they are in our mushrooming network of state-of-the-art and not-so-state-of-the-art biodefense labs.

There's a second bit of skullduggery in the DHS answer. DHS knows very well that with respect to biowarfare agent research, the line between offense and defense is a thin and hotly debated one. Many feel that so-called threat assessment research, which creates threats, including new pathogens that did not exist, with the ultimate notion of then developing defenses against them, is a violation of the BWC. So what DHS should have said (assuming that even this is true) is that it conducts all its biological agent research in accordance with the BWC *as it interprets the BWC.*

Stunned by the August 2005 announcement, Senator Clinton and Representative Bishop had their own reasons for smelling skullduggery. Writing DHS Secretary Michael Chertoff, they said his predecessor Tom Ridge had promised them "full communication about the issues and developments that affect Plum Island and our common interests. Therefore, it is most disappointing to learn of your agency's plans for a facility of this significance by receiving a 'fact sheet.'"[8]

They also said Ridge had assured them that Plum Island would be updated, but would remain a Level 3 center to study foot-and-mouth disease. So they were unhappy when DHS said it would build a replacement facility, probably a BSL-4. In a meeting with Chertoff, they argued forcefully that Plum Island should be upgraded and kept open, but only at a BSL-3 level. They did so want to make the Plum Island thing work, if only Chertoff wouldn't insist on that BSL-4 thing. But they were adamant about that. They were not going to do the BSL-4. "It is just very hard," Clinton said, "to justify putting that kind of facility in one of the most populated areas."[9]

Chertoff listened to the two argumentative representatives from the opposing political party, yawned, and sent them on their way. He promised to keep them informed. He had "heard and addressed" their concerns. But he just wasn't satisfied with a BSL-3 any more. There were lots of eager consortiums out there who would gladly do the BSL-4 with a $500 million NBAF monster—or anything else he might want. Come on down, Miss Kansas!

CHAPTER THREE

The "Silent Proliferation" of High-Containment Germ Labs
Congress Takes a Look

In Which the Titanic's *Crew Denies the Existence of Icebergs*
The following statement from the GAO's October 2007 report should be posted at each of the many turn-offs on the road to biodefense heaven:

> According to the experts, there is a baseline risk associated with any high-containment [facility]. With expansion, the aggregate risks will increase. However, the associated safety and security risks will be greater for new labs with less experience. In addition, high-containment labs have health risks for individual lab workers as well as the surrounding community. According to a CDC official, the risks due to accidental exposure or release can never be completely eliminated, and even labs within sophisticated biological research programs— including those most extensively regulated—have had and will continue to have safety failures.[1]

Condensed version (*Germ Labs for Idiots?*), for those too busy chasing bioboodle to worry about safety: high-containment germ labs (all 1300-plus of them) are risky. The more labs, the more risk. The risks are even greater at new, inexperienced labs. But even labs with "sophisticated biological research programs . . . have had and will continue to have safety failures."

The report, and the subcommittee hearing that considered it, made clear that the actual, real-time risk goes far beyond the baseline risk attributable to human error and other failures. The new labs created by the recent sevenfold increase in biodefense funding—including a stunning twelvefold increase in BSL-4 lab space for studying no-vaccine, no-cure death-sentence diseases like Ebola—are all chasing after the same limited pool of trained, experienced researchers. And current regulation amounts to little more than self-policing: "Since the labs are largely overseeing themselves at this point," Keith Rhodes warned, "it is not the regulators but only the operators of these labs who can tell you whether the three recent incidents [Texas A&M, the CDC power loss, and the U.K. foot-and-mouth outbreak] are the tip of the iceberg or the iceberg itself."[2]

Yes, it's the wild, wild west in Germland these days:

No single federal agency, according to 12 agencies' responses to our survey, has the mission to track the overall number of BSL-3 and BSL-4 labs in the United States. Though several agencies have a need to know, no one agency knows the number and location of these labs in the United States. Consequently, no agency is responsible for determining the risks associated with the proliferation of these labs.[3]

Subcommittee chairman Stupak expressed his frustration about this biodefense "field of dreams":

No one can tell me how many labs that we have, the quantity of stuff we are looking at, the quality of stuff we are looking at that could be a threat to this country. It seems like if we put the money out there then germs will come, so we will build these labs. I mean, what has really changed since the fall [of 2001], other than anthrax, OK?[4]

And speaking of anthrax: one thing that made the Amerithrax investigation difficult from the start was that *no one* knew, in 2001, exactly how many labs handled anthrax in general, or the Ames strain in particular. Since the FBI determined within a few months that the killer anthrax came from the U.S.'s own facilities, one would have expected a rush to get a handle on which U.S. facilities had anthrax in general, the Ames strain in particular, and what they were doing with it. But in 2007, the U.S. still didn't know,[5] and the proliferation of new facilities wasn't helping matters. According to Rhodes:

> I mean, right now, we do not even know where they are and we do not know what is being done and we do not know who is doing it. And from my standpoint and my colleague's, as well as a lot of safety professionals and security professionals, including our own Federal Bureau of Investigation and our own Intelligence Community, that is a worrisome subject.[6]

Part of the reason for the ignorance was the CDC's decision to keep everything it knew about the labs it regulated, their locations and their mishaps, as hush-hush as possible. As of fall 2007, according to Dr. Sushil Sharma of the GAO, *the FBI itself* hadn't been able to obtain the listing of select agent labs registered with the CDC.[7]

Ed Hammond of the Sunshine Project testified at the hearing about his frustrations trying to get information from the CDC, which was behaving more like the Defense Department or the Soviet Union's old Biopreparat than like the land-of-the-free's leading public health agency. The CDC, he said, denies "absolutely all requests for anything," and thus there is no transparency whatsoever with respect to its oversight of select agents.[8]

Congressman Stupak had his own story about a needs assessment the CDC said existed but hadn't furnished to the committee.[9] That history led to the following interchange between Stupak and CDC representative Richard Besser:

> MR. STUPAK: Dr. Besser, on your page four of your testimony the statement is made that the NIAID estimated the new BSL-3 and BSL-4 facilities [that] would be required to accomplish the

research agenda. Our committee has asked for a copy of that assessment. When will you provide that copy of the assessment?

DR. BESSER: We will work with your committee to provide that as soon as we can.

MR. STUPAK: Yes, but we would have liked it before the hearing, that is why I am asking, so get it to us, OK?

DR. BESSER: Yes, sir.[10]

Stupak and Energy and Commerce Chairman Dingell (by way of written statement) both indicated displeasure at the Department of Homeland Security's refusal to send a witness. A few months later, continuing difficulties getting information out of DHS would lead to a heated interchange between Dingell and DHS Undersecretary Jay Cohen. With even the FBI and Congress having trouble getting the facts from America's biodefense behemoth, what hope did ordinary citizens have of keeping tabs on the monster?

Not all their congressmen, and certainly not all the afternoon's witnesses, were anxious to help them with the process. The hearing was lightly attended in the first place, with multiple simultaneous hearings competing for attention, but there was a decent showing for the testimony of GAO representatives Rhodes and Sharma. After that, it was just subcommittee chairman Stupak and the Texans. (Stupak noted wryly that he was "the only one here not representing Texas A&M."[11]) The Texans' primary concern was to run interference for A&M interim president Eddie Davis. Representing a state with two of the country's existing BSL-4 facilities, and actively pursuing NBAF, they also took frequent opportunities to raise doubts about the seriousness of the problem.

Congressman Joseph Barton (R-Texas) began the rehabilitation process by openly pulling the cooperative Davis's strings:

MR. BARTON: The individual that was infected was infected in doing a procedure which she did voluntarily and was not instructed to do so against protocol, is that correct?

. . . .

MR. BARTON: But in spite of that, this employee is currently cured and, so far as you know, has no complaint against the university, is that correct?

. . . .

MR. BARTON: Now currently, is the Texas A&M University system fully cooperating with the CDC in their investigation or re-examination of the facilities and procedures at Texas A&M?

. . . .

MR. BARTON: And so long as you are the acting president of Texas A&M, are you committed to doing everything within your power to make sure that A&M fully complies with the CDC directives and cooperates in every way to ensure the safety of these agents if this type of research is allowed to be commenced again?[12]

Congressman Gene Green tried to imply that A&M might not have reported the incidents to the CDC because they just didn't know they were supposed to.[13]

Michael Burgess tried repeatedly to get somebody to say that the status quo was actually not so bad. He failed with Rhodes, but by continually cutting off Rhodes's efforts at caveats, did obtain an "admission" that we might eventually arrive at an acceptable level of "managed" risk—ignoring the fact that currently the risk was not even being acknowledged, much less managed:

MR. BURGESS: Well, let me ask you this. Obviously, we have put some time and effort into protecting the homeland with the proliferation of labs, is it the opinion of the two individuals before us from the GAO that we have moved on that continuum of being more secure or are we stationary or are we less secure?

MR. RHODES: The fact that there is so much that is unknown at the moment, I would have to say we are at greater risk. Because as the number increases, the risk increases and it is not just the increase in the material, it is the increase in laboratories that have less experience than others.

MR. BURGESS: So the actual risk may be generated by the fact that we are studying to prepare for the risk?

MR. RHODES: Yes. It is a dilemma that we are in.

MR. BURGESS: But that is one of the prices you pay for doing the research, correct?

MR. SHARMA: That is correct.

MR. BURGESS: And you'll never get to a point of relative security if you are not willing to invest the time and effort and risk in doing the research, is that correct?

MR. RHODES: That is correct but doing—

MR. BURGESS: And we need to manage the risk.

MR. RHODES: Yes. We are not—

MR. BURGESS: So my question is, are we doing a good job of managing the risk. I would assume the answer to that question today is no.

MR. RHODES: No.

MR. BURGESS: But is it your opinion that we can get to that point of managed risk which now is acceptable?

MR. RHODES: Yes, we can. It could be done.

MR. BURGESS: Thank you.[14]

The CDC and NIH representatives were happy, later in the hearing, to provide Burgess with the answer he so desperately sought: that we were in better shape now than we had been in 2001. All because of their rigorous oversight, of course.

Green, whose daughter was on an infectious disease fellowship at the University of Texas Medical Branch at Galveston, home to one of Texas's two BSL-4s, suggested that high-containment risks mainly applied to the researchers themselves, and asked the CDC witnesses to confirm: "Generally, where are the risks in this line of work and would you say that they are primary to the lab worker or would it be to the broader community?" Dr. Casey Chosewood of the CDC rose to the occasion: "The vast majority of incidents that have occurred and given the vast amount of work that has gone on, we believe that the actual number of events is very small. But when those events have occurred, they have affected primarily the laboratory workers. . . . But the risk to the environment, in all those cases, has been non-existent in our opinion."[15]

The GAO had not been so sanguine in its report. With respect to the A&M incident, the GAO said it was fortunate that the disease involved, brucella, was one not easily spread from person to person.[16] The biodefense complex typically talks about laboratory-acquired infections (LAIs) as though their only risk is to the researchers themselves. The GAO noted—to the contrary—that many select agents are easily transmitted from person to person, and that:

> According to BMBL, the causative incident for most laboratory-acquired infections is often unknown. It can only be concluded that an exposure took place after a worker reports illness—with symptoms suggestive of a disease caused by the relevant agent—some time later. Since clinical symptoms can take weeks to become apparent, during which time an infected person may be contagious, it is important that exposure be identified as soon as possible and proper diagnosis and prompt medical treatment provided.[17]

Green also objected to use of the term "bio-weapons agents" by Ed Hammond of the Sunshine Project:

> MR. GREEN: Dr. [Robbin] Weyant, I am interested in clarifying the type of agents that are being researched in the BSL-3 and BSL-4 labs. Our witness on the fourth panel has referred to these agents as biological weapons agents. A term that certainly elicits strong reaction from the public. My understanding, however, is that these are not actually weapons agents by definition, rather they are infectious agents occurring naturally in nature. Is it fair to assume that the BSL-4 labs are necessarily working on biological weapons agents and can you clarify the distinction of the two?

DR. WEYANT: Well, it depends on usage. An agent such as bacillus anthracis, the agent that causes anthrax, exists in the environment. It exists in soil in many parts of the world. However, the agent can be grown up and purified and weaponized as was demonstrated in the events beginning October 4, so it is difficult to take a single organism on this list and say it is absolutely a weapons agent or it is absolutely a naturally-occurring agent. I would say it is fair to say that for the agents listed on this list, it is possible that they could be both.[18]

Ah yes, poor misunderstood bioweapons agents. All those decades of being abused by mad scientists in the U.S., Great Britain, and Russia, brutalized and trained to be CIA and KGB killers—anthrax, brucella, tularemia, smallpox, Ebola. Forcefully tattooed with swastikas; implanted with Bjorg control modules; forced to assassinate generations of monkeys and guinea pigs and pretend they enjoyed it. When all they really wanted to do was snuggle up in a dead cow leg somewhere in a Texas ditch and get back in touch with their inner cowboy.

And poor misunderstood biodefense complex, which is not in the scary old bioweapons agent business at all, but is just trying to do the humane thing and give these poor abandoned germ orphans some rehabilitation and stability in their lives. Poor misunderstood A&M. Poor misunderstood CDC, NIH, DHS. Poor misunderstood congressmen. Save the Germ Children! And isn't it wonderful how we were able to spend $60 billion plus to save the millions of Americans the Graham-Talent WMD Commission[19] predicts will die sometime in the next decade from excessive consumption of sick rabbits and the careless handling of cow carrion? Please, whatever you do, don't call the germs in biodefense laboratories bioweapons agents. Call them emerging pathogens, or Little Orphan Annies, or little boll weevils, jes-a-looking-for-a home.

I mean, the Texans had those little doggies all rustled up together and headed for the big germ roundup and biodefense barbecue in the sky, and here came this Hammond guy from the Sunshine Project, and started complaining about accidents, making everybody think our biodefense research was dangerous or something. Why, the CDC has been doing this stuff forever in downtown Atlanta, and they ain't killed nobody but a janitor or two. What does he mean hollering about "bioweapons agents," and screwing things up for our Aggies?

And it's true. Doubtful this hearing would ever have been held if it wasn't for the A&M revelations. And for sure the A&M revelations wouldn't have been made if it hadn't been for Ed Hammond and the Texas Public Information Act. The CDC didn't seem to have much of a clue, and what clues it had didn't concern it much. But Hammond just had to keep demanding those documents, and then he had to read them, and then he had to tell everybody about the accidents he found that hadn't been reported the way they were supposed to have been. And that embarrassed the hell out of everybody. Why, it made it look like the reason there were so few accidents is because the accidents were being hid. And it made it look like the CDC's oversight didn't amount to a hill of beans. It had been a nice little gentlemanly world in which we all just trusted each other to do the safe thing, because, after all, we're world-class scientists, and we wouldn't do nothing to hurt nobody. And Ed Hammond had to go and screw all that up, rude son of a bitch.

Enter Burgess (who at one point in the hearing had suggested we might need more lab space rather than less). Burgess opened by stating that he'd never heard of the Sunshine Project before (which, if so, would mean he had ignored several front-page stories in the *Fort Worth Star-Telegram* and the *Dallas Morning News*). Then he decided to inquire into the Sunshine Project's funding. Just curious, but oh so cleverly informed about an organization he'd never heard of before:

> MR. BURGESS: At the bottom, just before the table at the bottom of the first page, you reference the Sunshine Project, the Margaret Race of the SETI Institute. What does that acronym stand for?
>
> MR. HAMMOND: It is a NASA-funded institute that has to—
>
> MR. BURGESS: Is that the Search for Extra Terrestial—
>
> MR. HAMMOND: Yes, Extra Terrestial Intelligence.
>
> MR. BURGESS: OK.

MR. HAMMOND: If I may, the interest there is that—and it can be corrected if I misspeak but the interest there is that the Government, NASA, has a long-term interest in potentially constructing a level 4 laboratory in the event that they return samples from Mars and so, therefore, NASA is interested in— it has funded work at the SETI Institution to keep track of issues related to biosafety level 4 labs.

MR. BURGESS: As I recall, this group out of Berkeley was the one that connected personal computers across the country to evaluate whether there were meaningful signals coming from outer space. Do I remember that correctly?

MR. HAMMOND: Sir, I honestly do not know but I do not believe so.

MR. BURGESS: OK, I just had to ask.[20]

He just had to ask, because inquiring Texas congressmen want to know. They don't want to know if the huge explosion of biodefense research threatens the country, but they do want to know if there's any possible way they can discredit this David who singlehandedly wrecked the NBAF dreams and the research reputation of one of the Lone Star State's premier institutions. So if they can just connect him up with these "kooks"[21] who, perhaps disappointed at the scarcity of terrestial intelligence, look for it elsewhere, the job will be done. Of course it's a low blow, because Hammond has had to scramble for funds each of the Sunshine Project's seven years, and will close shop the following spring because every tiny financial spigot has been closed. In the midst of all this, he had just brought to the attention of the public in a way never accomplished before the threats posed by an out-of-control biodefense enterprise, and exposed a situation that the CDC, embarrassed by its own inattention, deemed so threatening that it suspended all A&M's select agent research, an unprecedented step.

Given the performance of the Texas congressional delegation this day, a cynic might welcome a search for intelligence in any far-flung extraterrestial parts whatsoever. A cynic might also add that looking for extraterrestial intelligence is not the same thing as believing in it.

An organization devoted to a Search for Intelligence in the Texas congressional delegation, or for "meaningful signals" from that delegation, wouldn't necessarily expect to find them.

One would hope to find intelligence—and perhaps a little honesty—in the halls of the CDC and NIH, however. But one of the things that happens so often with biodefense research is the removal of the truth gene. The GAO said of the June 2007 CDC power outage: "The incident showed that, even in the hands of experienced owners and operators, safety and security of high-containment labs can still be compromised. The incident also raises concerns about the security of other similar labs being built around the nation."[22]

The CDC, however, seemed to think the outage no big deal, even suggesting everything had worked as it was supposed to:

DR. BESSER: I think that the laboratories that are being built, these state of the art laboratories, are extremely safe. That does not mean that an error will not occur.

MR. STUPAK: Yes, your own CDC lab in Atlanta was supposed to have redundancy in the electricity event when the lightning struck. Everything shut down in a level 4 lab. You did not have the redundancy that was required and that is a brand new lab.

DR. BESSER: Dr. Chosewood?

DR. CHOSEWOOD: Sure. I would love to comment on that. In fact, we believe that the GAO findings about the lack of redundant power is absolutely incorrect.

MR. STUPAK: It is incorrect?

DR. CHOSEWOOD: Yes, sir.

MR. STUPAK: You should have just one power source at a level 4 lab, is that what you are saying?

DR. CHOSEWOOD: No, but in fact, that is not the case. The power outage in our building 18 laboratory occurred as a result of an error.

MR. STUPAK: A lightning strike, right?

DR. CHOSEWOOD: A lightning strike to the building.

MR. STUPAK: Yes.

DR. CHOSEWOOD: And unfortunately, the lightning protection system in that building had been interrupted by ongoing construction nearby. And so the power failure in that instance was completely appropriate. It was as if you were having a power surge in your own home.

MR. STUPAK: So you think——

DR. CHOSEWOOD: And if that were the case——

MR. STUPAK: Power outages at level 4 labs are certainly appropriate?

DR. CHOSEWOOD: No, I did not say that. One of the things that I think is important is to imagine a power surge in your own home and you have a breaker that trips appropriately. That is exactly what occurred in this situation. . . And that is what you would want.

MR. STUPAK: The backup system cable was cut, was it not?

DR. CHOSEWOOD: This was an interruption of the lightning protection system but not the backup cable for power.

MR. STUPAK: So then why did not the backup one come on then?

DR. CHOSEWOOD: Because it was not supposed to in an overload situation like a lightning strike. So basically at the time of the lab—we should tell you that we had no active work going on. The maximum containment labs in building 18 are actually not functional at this point. But even if they had been functional, there are multiple systems of safety in place to avoid escape of any dangerous pathogens.

MR. STUPAK: But if you do not have any power, those backup systems are not going to work.

DR. CHOSEWOOD: No, I would disagree because the facilities are designed to withstand higher levels of containment than the typical space. These are pressurized areas. If you have a power loss in a maximum containment laboratory, the actual air flow goes neutral, it does not become positive. You do not have the escape of that air in the lab to the outside.

MR. STUPAK: OK.

DR. CHOSEWOOD: Backup power is important. It is a critical thing but that was not the case here and our laboratories do have backup power.[23]

This was the first time anyone had suggested publicly that the CDC power failure was "appropriate." (The CDC, in classic batten-down-the-hatches-and-public-need-for-information-be-damned fashion, had ignored the *Atlanta Journal-Constitution*'s Freedom of Information request for details on the incident; the newspaper had to rely on confidential informants instead.) Apparently, according to Chosewood, it is appropriate for a high-containment germ lab to lose all power in the face of a power surge. Comparing a power loss at a high-tech facility inhabited by smallpox, Ebola, and pandemic flu pathogens to the flipping of a household breaker is richly ironic, given the way biodefense propagandists make the CDC a poster child for BSL-4 safety.

This may also have marked the first time anyone suggested that the absence of negative air pressure equals "neutral" air pressure, and that such an equivocal equilibrium suffices to keep state-of-the-art germ inmates from busting out.

A subsequent power outage occurred at occupied CDC BSL-4 labs in 2008. Reportedly, the CDC used duct tape to seal doors, apparently not confident that "neutral" air pressure and "multiple systems of safety" would do the trick.[24] Three cheers for Yankee (well, Southern) ingenuity and high-tech duct tape.

The GAO said the cracked and leaky drainage pipes blamed for the British foot-and-mouth outbreak indicated poor maintenance practice, and that the U.S. had reason to worry about maintenance at its own facilities:

High-containment labs are expensive to build and expensive to maintain. Adequate funding for each stage needs to be addressed. Typically, in large-scale construction projects, funding for initial construction comes from one source. But funding for ongoing operations and maintenance comes from somewhere else. For example, in the NIAID's recent funding of the 13 BSL-3 labs as RBLs and 2 BSL-4 labs as National Bio-containment Labs (NBL), the NIAID contributed to the initial costs for planning, design, construction, and commissioning. But the NIAID did not provide funding to support the operation of these facilities. In this case, the universities themselves are responsible for funding any maintenance costs after initial construction.[25]

Following the 2008 economic collapse, most of the universities constructing new biodefense facilities would be sharply slashing overall campus budgets, adding an unnoticed exclamation point to the GAO's concerns about maintenance.

The GAO said current regulation amounted to self-policing, and that the Texas A&M incident suggested the inadequacies of CDC select agent oversight. But the CDC and NIH representatives praised their own "oversight," "protocols," and even the fact they used specific checklists to guide their occasional inspections.[26] More relevant may have been the CDC's acknowledgement that so far it had announced all its inspections in advance.

Chairman Stupak questioned the NIH and CDC representatives closely about why and how they decided to build so many new labs. Auchincloss cited a Blue Ribbon Panel as basis for the NIH's decision to build 14 new BSL-3 facilities and 4 new BSL-4 facilities. Stupak's questioning revealed, however, that the panel only recommended a research agenda, and that the NIH itself determined the number of new facilities that would be "needed" to pursue the agenda. It gave no indication of how it had arrived at its numbers. Further questioning revealed that the NIH only focused on what it (and the scientists it funded) needed, and the CDC had done its own math on how many labs it "needed." A similar tunnel vision had prevailed at multiple federal agencies.[27]

Stupak's questioning certainly corroborated the GAO's conclusions that American biodefense construction had run, and was still running, amok:

> MR. STUPAK: Four more additional labs at CDC. So it looks like every agency is making their own assessment and doing their own thing basically, right?

> DR. BESSER: I think that there is room for a more comprehensive look at our national needs in both of these areas.

> MR. STUPAK: Well, if you got 15 different agencies, was your other testimony, you are up to four. If we did five, four at each one, four times 15 is 60. We would need 60 more level 4 labs if every agency did their own assessment. Is anyone in control [at all]?[28]

The Texas delegation, and some of the biodefense apologists, expressed enthusiasm about the concept of an anonymous, voluntary, nonpunitive reporting system suggested by Gigi Kwik Gronvall of the Center for Biosecurity at the University of Pittsburgh Medical Center :

> One possible model for high-containment laboratories to emulate is the reporting mechanism used for aviation incidents, wherein airlines can contribute operational experience without fear of regulatory action. Mistakes are analyzed and learned from, but they are not attributed to individuals (except when mistakes result from criminal actions, such as drunkenness). Institutional anonymity may also be required in order to get robust reporting from research institutions. Procedures would need to define thresholds and mechanisms for reporting if an accident poses a danger to the community surrounding the laboratory, however.[29]

The rationale for such a system, according to Gronvall:

> Generally, there is a disincentive to report acquired infections and other mishaps at research institutions. Infections lead to

negative publicity and scrutiny from the granting agency, adversely affecting future research funding. In addition, after a scientist acquires an infection in the laboratory, neither the scientist nor the laboratory wishes to advertise the mistake. These barriers need to be cleared so biosafety can be enhanced through shared learning from operational experiences, and also so the public may be reassured that accidents are being thoroughly examined and contained.[30]

Gronvall's recommendations drew on a 2007 article in the journal *Biosecurity and Bioterrorism*, reporting on a 2006 meeting at the UPMC's Center for Biosecurity, about high-containment biodefense research. The article itself reveals what Gronvall's testimony did not, that there was disagreement among the meeting participants as to whether the institutions reporting accidents could or should remain anonymous.[31]

Such a system proposes sparing errant individuals and institutions embarrassment in exchange for their sharing details of incidents. The hope would be that an institution like Texas A&M would reveal, anonymously, that an individual contracted brucellosis as a result of carelessness cleaning the Madison Aerosol Chamber, and that the brucellosis had not been diagnosed for some period of time afterward. One supposes other institutions might learn from such an incident that, indeed, they should follow "standard safety protocols" to the letter and wear proper protective gear when cleaning germ-infested equipment. They might also learn that when a biodefense researcher gets sick, she should consider that the disease she's researching might be the culprit. One would hope that a biodefense complex fond of using the words "world-class" to describe its researchers, institutions, procedures, and equipment could figure all this out without the remedial lessons of an anonymous reporting system.

What such an anonymous system *does* do is keep the biodefense complex's dirty laundry out of the public eye. The current system had already done a pretty good job of that, had it not been for the work of a soon-to-be-defunct watchdog organization, an occasional tip to the newspapers from within the complex itself, and the blatant "oops" of the British foot-and-mouth outbreak. Had A&M reported the brucella incident in the first place—despite the lack of anonymity vis-à-vis the CDC—the CDC would almost certainly have kept the university's secrets and probably withheld the regulatory hammer. Only the attention of the Sunshine Project discovered the problems at A&M and forced the CDC's hand. And if the CDC is to be believed, the situation there was serious enough—once the CDC began to look closely and not go through the motions of a routine inspection—to merit an extended shutdown of all select agent research at A&M.

One wonders how a voluntary, anonymous, nonpunitive reporting system would have discovered the A&M problems. And if Ed Hammond of the Sunshine Project is right and A&M represents the tip of the iceberg, one wonders how increasing anonymity and secrecy about accidents will encourage care and safety among biodefense researchers. Other institutions began acquiring a sudden interest in incident reports following A&M's embarrassment. The Gronvall proposal would apparently eliminate embarrassment and the application of penalties as incentives for enforcement. It would also conceal the problems from the public, a questionable result indeed.

Anonymous? Nonpunitive? You could read the drooling lips of the Texas congressional delegation and A&M's president.[32]

Gronvall did allude to the need for "thresholds and mechanisms for reporting if an accident poses a danger to the community surrounding the laboratory."[33] Presumably the determination of whether the requisite threshold had been met would be made by the facility itself or by a regulatory agency like the CDC. How high that threshold would be is suggested by the CDC's conclusions about the 105 incidents reported to the CDC from 2003 to September 25, 2007, incidents involving three confirmed "releases" of a select agent. The CDC considered none of these incidents—including, apparently, the ones at Texas A&M—to have presented a public health threat.[34] The biodefense complex's basic double standard is on display here. Bioterrorism by Al Qaeda or anyone with a college chemistry set is considered a threat, even if it hasn't occurred yet; accidents at high-containment germ facilities are only a "threat" if they actually produce illness in the public. Besser's testimony at the hearing also suggested considerable reluctance about sharing information with surrounding communities about the agents being researched, or about whether a lab shifts from BSL-3 to BSL-4 research: "I think that we have to weigh the issue of sharing information that could do harm to a community versus being open about what is being done."[35]

Gronvall talked about the need for "a more aggressive federal effort to standardize public engagement and transparency of operations for high-containment laboratories and to direct funds to this purpose."[36] She spoke of how "communities' concerns could be actively addressed both by HHS [the Department of Health and Human Services] and NIAID and by the institution sponsoring the laboratory."[37] Gronvall is calling not for transparency (what transparency can there be in secret, anonymous reporting of accidents?) but for a massive propaganda effort, funded by the federal government. "The community has a right to know that the people who are working in these high-containment laboratories are well trained, that if there is an accident, that it is being dealt with appropriately."[38] The public needs to be "reassured that accidents are being thoroughly examined and contained."[39] She does *not* say that the public needs to be told when an accident has occurred, so it can determine for itself whether an accident is being dealt with appropriately. No, the public only has a right to know that the biodefense wizards are doing the safety thing, and so the public will be told that this is the case until they believe it. Trust us to be safe, and if it turns out we haven't been, trust us anyway, or we won't tell anyone about our accidents.

Screening of Visitors? But It Hasn't Been Peer-Reviewed!

In September 2008, not long after the FBI's Ivins announcement, the GAO published a report on perimeter security controls at the nations's five existing BSL-4 labs. The review had been conducted at the request of the House Energy and Commerce Committee. What the GAO discovered was startling: two of those five labs, containing the nation's deadliest pathogens—those for which no cure or vaccine was available—lacked most of 15 security controls the GAO felt advisable:

> Although the presence of the security controls GAO assessed
> does not automatically ensure a secure perimeter, having

most controls provides increased assurance that a strong perimeter security system is in place and reduces the likelihood of unauthorized intrusion. For example, the two labs with fewer security controls lacked both visible deterrents and a means to respond to intrusion. One lab even had a window that looked directly into the room where BSL-4 agents were handled. In addition to creating the perception of vulnerability, the lack of key security controls means that security officials have fewer opportunities to stop an intruder or attacker.[40]

The fifteen security controls the GAO looked for were (1) an outer/tiered perimeter boundary; (2) a blast stand-off area (buffer zone) between the lab and perimeter barriers; (3) barriers to prevent vehicles from approaching the lab; (4) loading docks located outside the footprint of the main building; (5) no exterior windows providing direct access to the lab; (6) a command and control center; (7) closed-circuit television (CCTV) monitored by the command and control center; (8) an active intrusion detection system integrated with the CCTV; (9) camera coverage for all exterior lab building entrances; (10) perimeter lighting of the complex; (11) a visible armed guard presence at all public entrances to the lab; (12) roving armed guard patrols of the perimeter; (13) X-ray magnetometer machines in operation at the building entrances; (14) vehicle screening; (15) visitor screening.[41]

At the request of HHS, the GAO did not identify the labs by name, but journalists were able to identify the deficient labs as those at Georgia State in Atlanta and the Southwest Foundation for Medical Research in San Antonio.

All five labs were registered with the Select Agent Program at the CDC, and the registration process required each lab to develop a security plan, based on a site-specific risk assessment, "sufficient to safeguard against unauthorized theft, loss, or release of select agents." Inspections by the CDC were supposed to ensure that the labs met certain safety regulations. A recent report by HHS's Office of Inspector General (IG) had said labs regulated under the Select Agent program had weaknesses in such areas as access control and security plan implementation that could have compromised their ability to prevent accidental loss or theft.[42]

The GAO reported that CDC regulations do not mandate specific perimeter security controls. The CDC had told the GAO the differences in perimeter security were the result of specific risk-based planning, but hadn't commented on the specific vulnerabilities identified by the GAO and whether these should be addressed.[43]

More evidence of the current self-policing situation: the GAO discovered that the site-specific risk assessment required when a lab registers can be performed by officials of the lab itself.[44]

The three labs containing most of the recommended controls were all subject to additional federal security requirements beyond those required by the Select Agent Regulations. Security officials at the two "deficient" labs told the GAO that management and administration had little incentive to improve security because they already met Select Agent requirements. Some also cited budgetary restrictions.[45]

The CDC said specific security controls are not in place because Select Agent Regulations focus on performance objectives rather than specific methods of compliance.[46]

In its comments attached to the report, the CDC even challenged the GAO's specific list of controls, asking it "to provide references for the research that identified these 15 security controls as being appropriate for the assessment of the perimeter security of BSL-4 laboratories, identify the security experts that had been consulted in developing the list of security controls to use for the assessment, and indicate whether the use of this set of security controls . . . has ever been peer-reviewed."[47]

The GAO said it had used the survey tool for similar security assessments in the past, and that the 15 measures were not the only possible ones that might be considered, but that they represented a good baseline, and many, such as visitor screening and perimeter lighting, were obvious common-sense notions.[48]

The CDC also said the GAO's assessment was too "limited," that "GAO did not assess the security of the laboratories themselves or the threat of an insider attack" and suggested that the "final report include additional clarification of how perimeter security fits into overall select agent security."[49]

In fact, however, the GAO had already noted early in its report that "[p]erimeter security is just one aspect of overall security provisions under the Select Agent Regulations, which includes personnel training and inventory control."[50] The October 2007 GAO report and subcommittee hearing had already identified personnel training as a problem area; and the shutdown of USAMRIID in late 2008 while the facility tried to address inventory problems would suggest the problems in that area. One would think, if the CDC receives a report indicating that the security gate has a problem, it might be better to fix the security gate rather than fault the report for failing to consider the climate control and electrical systems as well.

The CDC did indicate it would begin seeking input "as to the need and advisability of requiring by federal regulation specific perimeter control(s) at each registered entity having a BSL-4 laboratory."[51] It undercut that concession, however, by declaring it would balance any BSL-4 security enhancements "against any impact on the important research being conducted by these laboratories."[52] After all, our regulators of record didn't want to intimidate scientists working with incurable diseases by using armed guards, cameras, perimeter lighting, visitor screening, and other security features to remind them how dangerous their work is.

NBAF 3

JUST HOW SAFE IS IT, DAN'L?

FILL IN THE BLANK QUIZ FOR SOUTHERN POLITICIANS: THE NBAF is as safe as _____.

Hal Rogers of Kentucky: Why, it's worth $500 million. Ergo, it's "about as safe as going to Wal-Mart."[1] Probably safer, when you think of the loose Wal-Mart shopping carts, and some of the places Wal-Mart locates Wal-Marts, compared to where Hal wants to put the NBAF, out in a rural part of Pulaski County. Nobody out here but us hoot owls. And if we need some quick training to run a biodefense lab, we can always do the online short course from Kentucky Tech.

Haley Barbour, governor of Mississippi, former chairman of the Republican National Committee: Why, it's worth $500 million. It will be as safe "as a submarine in a bank vault."[2] Maybe all those bank vaults for submarines explain why so many banks, and our government's whole optimistic, *expertly* regulated-not financial system, were struggling so much in 2008. And maybe business is so bad these days they do have a place to store your submarine for you, with a government "bailout" should the water start seeping through all those bank vault safety features.

While we're dealing with the metaphorical approach to biosafety, let me offer a comparison of my own: a Southern politician is about as trustworthy as a disbarred lawyer working as a used car salesman and moonlighting at a funeral home. And if you believe either Rogers or Barbour, let me hook you up with a few Manhattan, Kansas residents (none of them in elective office, or positions of authority at Kansas State) who'd be happy to offer you a considerably tarnished NBAF, sold "as is," no warranty, no returns.

Very interesting: in February 2006, nobody in Kentucky had yet raised a question about safety, and in March 2006, nobody in Mississippi had yet raised a question about safety; yet from the git-go, politicians in both states resorted to extravagant comparisons to assure us all how wonderfully safe it was playing around with death-sentence diseases lacking a vaccine or cure. It must be a hard life for a politician, having to flat-out deny what anyone with a lick of common sense can see for himself.

Fourteen states proposed twenty-nine sites for NBAF. Texas offered four sites, three in San Antonio and one at Texas A&M. Mississippi offered three sites in small towns near Jackson, the state capital. Georgia offered two different sites in Athens. Kentucky and Tennessee jointly offered up Pulaski County, not because of its central location, but because it was the alternate home for Hal Rogers, reigning chairman of the House Appropriations homeland security subcommittee, and for his co-owned local bank, Citizens National. A 2005 *Washington Post* article described how another of Rogers's government-funded ventures, the Southern Kentucky Rural Development Center, keeps millions of dollars on deposit at Citizens—from which he receives $100,000 and $1 million a year income on $5 million in bank shares. Clay Davis, president and chief executive of Citizens, was also founding director of the Center.[3]

In the spring of 2006, most of the proposals around the country received only a brief mention—if any at all—in the relevant local and state newspapers. The notable exceptions were Kentucky, Georgia, Mississippi, and Missouri.

Georgia's bid was covered on February 27 in the *Atlanta Business Chronicle*, on March 1, April 1, April 12, and June 4 in the *Athens Banner-Herald*, and on April 1 in the *Atlanta Journal-Constitution*. The stories also engendered the editorials. On March 9, the *AJC* exhorted state and local leaders to work up the same enthusiasm for NBAF that they'd shown in pursuing the NASCAR Hall of Fame.[4] The next day, the *ABH* urged local governments to cooperate in offering infrastructure.[5] On April 4, the *ABH* declared that the defense facility was a "natural fit," "a prize worth pursuing," and said it was "not too early for this community to begin coming together in support of getting the facility located here."[6] On May 14, editorial page editor Jim Thompson, declaring that "The future is now for this area's economic health," reiterated that "this community needs to put forth its best effort to get the proposed National Bio and Agro-Defense Facility."[7]

The Georgia truth-bending got started early. In an article ironically published on April Fool's Day, David Lee, UGA vice president for research, assured Athenians that the facility would be safe because "[t]he amount of pathogens being worked on would be miniscule [*sic*] and unable to survive even if they did escape the facility."[8] Apparently, Lee felt the harsh Georgia environment would simply overwhelm any escaped pathogens. Weakened by a diet of bread and water, shackled by heavy chains, and pursued across creek and swamp by bloodhounds and potbellied, gun-toting Georgia deputies, the anthrax and Ebola germs would of course beg to go back to their cramped, minuscule containers at NBAF.

Of course they wouldn't really all be in minuscule containers. Hundreds of livestock at a time might be functioning as breeder reactors for whatever minuscule amount of pathogen they might have started with. A single hog exhales about 500 million germs a day (for reasons of national security, the excretions attributable to said hog are classified information). The whole "minuscule amounts" claim is a classic example of the biodefense two-step. In the hands of terrorists, we are told, it takes only a minuscule amount of pathogen to kill thousands, hundreds of thousands of people. A Plum Island scientist once held a tiny vial of FMD pathogen and boasted that its contents could infect 10,000 trillion cattle with FMD—enough to destroy not only all the cattle now living on the planet, but all that had ever lived on it.[9] And the American team investigating the 1979 anthrax release in Sverdlovsk, which officially killed 64 people, concluded the deaths could have been caused by only a teaspoon or two of anthrax spores accidentally vented.

In Mississippi, Jackson's *Clarion-Ledger*, Biloxi's *Sun Herald*, and the *Madison County Journal* all ran NBAF stories on March 21-22. The *Clarion-Ledger*, which promoted the Mississippi efforts zealously, ran stories on March 21, 22, and 31 (the deadline for submitting expressions of interest), and lab-boosting editorials on March 22 and April 4. The March 22 story described the overwhelming community support for something that had first been announced to the public only the day before. ("Mississippi economic development officials think community support has put the state at the top of the list to land the National Bio and Agro-Defense Facility.") The "community support" apparently amounted to endorsements by the president of the Metro Jackson Chamber of Commerce and the executive director of the Hinds County Economic Development Association, accompanied by the Byram Business Association's vote to send a letter of support before the application deadline.[10]

Flora Mayor Scott Greaves, who would later characterize opponents as "a few idiots," told the *Madison County Journal* he'd received 25 phone calls from the "community," all positive, and that he personally was "tickled to death about the possibility."[11]

The Mississippi solicitation had some interesting angles. First, Ross Tucker of the Metro Jackson Economic Development Alliance said DHS had inspected eight possible sites in the tri-county area "and narrowed its choice" to three.[12] This sort of pre-application assistance seems unusual; no other site acknowledged having received anything like it. Also of interest, Mississippi proposed to have the facility managed by Battelle, a major laboratory and defense contractor (managing the new DHS facility at Detrick, for instance).[13] And Mississippi hired as a consultant Parney C. Albright, a former assistant secretary for science and technology for DHS.[14] Finally, Mississippi governor Haley Barbour, former Republican National Committee chairman—whom one might suspect of some inside knowledge—insisted the facility would not be developing bioweapons because "The Department of Defense does that, and the CIA."[15] Hard to figure which is the worse scenario: Barbour knew what he was talking about, and the Defense Department and CIA really are developing illegal bioweapons; or Barbour doesn't know diddly about the whole subject, including what sort of biodefense research is legal and what isn't.

The Missouri solicitation got attention because the University of Missouri's College of Veterinary Medicine had received a $12 million grant from NIAID in September 2003, for construction of a BSL-3 "regional biocontainment laboratory." Design for that facility was nearing completion, so UMC decided to schedule a public forum on both the pending BSL-3 facility and its current pursuit of NBAF. In marked contrast to most local newspapers around the country, both the *Columbia Daily Tribune* and *Columbia Missourian* reported on the upcoming forum more or less straightforwardly, and didn't straightway jump on the NBAF bandwagon.[16]

The two newspapers' stories following the March 23 forum described the concerns of about fifty local residents without denigrating them. ("At a public forum last night, on- and off-campus critics raised pointed questions."[17] "Safety of the facility's neighbors was a topic of heated discussion."[18] "MU officials sought to paint a rosy picture of the 500,000-square-foot facility."[19] Columbians were concerned about the proximity of homes and schools, and about the possibility of a terrorist attack. MU's facilities project manager said the inner fence would be about 250 feet from the facility, to ensure that the building wouldn't be damaged if an explosives-filled truck hit the fence. To which one resident retorted: "It may not destroy the fence, but what do you think it's going to do to my house?"[20]

Meanwhile, some 450 miles to the east of Columbia, the *Somerset Commonwealth-Journal* was making no pretense to objectivity, and no effort to be accurate with its steady stream of gushing adulation. Eventually Mississippi's *Clarion-Ledger*, Kansas's *Manhattan Mercury* and *Lawrence Journal-World*, Georgia's *Athens Banner-Herald*, and Texas's *San Antonio Express-News* would approach the *SCJ*'s level of Pollyanna-ish drool, but for now the *SCJ* was in a class to itself.

Some of the more notable headlines: "It Will Transform Us"; "National Lab: 'Safe as Going to Wal-Mart'"; "Lab's Impact: 'Mind-Boggling'"; "Hazmat Personnel Endorse New Lab"; "Picking Up Steam";

"Burnside [a small Pulaski County town on the shores of Lake Cumberland] Gives Thumbs Up to Lab"; "'Lab 257': Fact or Fiction"; "'It Will Be Safe': Past Plum Island Chief Says Fears Aren't Factual"; "Expert Proclaims Bio Lab 'Safe'"; "Science Hill [Middle] School OK with Lab." (One wonders why the Brownies, Girl Scouts, and a local pre-school or two weren't also brought in for endorsements.) February 21 and 22 alone saw four stories and sidebars, in addition to a long "Guest View" by Ewell Balltrip, director of the National Institute for Hometown Security and point man (after Rogers himself) for the Kentucky/Tennessee NBAF consortium. Presumably Balltrip had a lot of time on his hands due to the shortage of imminent hometown security threats in Somerset, and so could devote his attention to the consortium's current goal: "to fortify the region's role in the war on terrorism and expand a homeland security knowledge cluster."[21] *There's gold in that there homeland security, fellers.*

And who better to lead us all into the promised land of biodefense manna than U.S. Congressman Harold "Hal" Rogers, if not known for his groundbreaking contributions to U.S. public policy, at least famous for his ability to bring home the bacon, er, pork, and set it out there for his kinfolk and friends and neighbors to have at. (Of course, it didn't hurt that so much of the pork took the scenic route through Rogers's own bank.)

And who better to tell us why we all should expect big things from our Congressman than Leonard Lawson, the Kentucky road contractor indicted by a federal grand jury for bribery, conspiracy, and obstructing justice,[22] who shares with Rogers the dream of replacing an existing road, Kentucky 80—much of it already four-lane, and most of it lightly traveled—with a massive new east-to-west interstate, I-66? I-66 was envisioned in 1985 as a 3000-mile six-lane superhighway from southern California to Virginia, until a 1994 congressional study said it wasn't economically feasible, except for possibly a few sections in a few states. Rogers and state government officials decided Kentucky was one of those states that needed one of those sections, and Rogers managed to siphon $96 million of federal funds into the Kentucky project between 1998 and 2006.[23]

Rogers hopes to lay a ribbon of blacktop across southern Kentucky from Pikeville to Paducah, at a cost of $7 billion, cutting through the

Daniel Boone National Forest, and disrupting wildlife, the Rockcastle River, underground caves, and people in the path of and alongside the construction. Rogers probably never contemplated a road or building project he didn't love (or saw an existing road or building as anything but an obstacle to be bulldozed down), since each new bit of construction rains money on one or more campaign donors. And I-66 would pass smack dab through Rogers's district. Grateful road contractors might well push for the whole boondoggle to be named the Hal Rogers Memorial Highway or some such. Even a deceased Rogers probably wouldn't object to such an honor; a live one certainly wouldn't. Relevant example: sometime in the nineties Rogers calmly usurped Daniel Boone's place in highway history when the eastern segment of Ky. 80, the Daniel Boone Parkway, was renamed the Hal Rogers Parkway.

In August 2005, Rogers helped break ground for one I-66 portion, Somerset's Northern Bypass. The $33 million contract went to Bizzack, Inc., Leonard Lawson's company.[24] Nothing unusual about that: of $3.5 billion awarded for Kentucky road projects between 2000 and 2005, one-fourth went to Lawson companies.[25] Lawson's 2008 bid-rigging indictment suggested one way all this might have come about.

Lawson is one of Rogers's top campaign donors, and his admiration for Rogers is unblushing: "Hal Rogers is just incredible. . . . When you get someone like that on your side, you can't hardly tell him 'no.' He knows where the pockets of money are in Washington, all of 'em, and how to spend it."[26]

Oink.

The *Washington Post* and the *Lexington Herald-Leader* have each published multiple articles examining some of Rogers's specific pork achievements. (It was the *Herald-Leader* which dubbed him the "Prince" of that particular white meat.) They include such friend-larding examples as adding $500,000 to the Army Corps of Engineers' 2005 appropriation—money the Corps had not requested—to build a parking lot for a private resort (Lee's Ford Marina) owned by a campaign contributor. The contributor, J. D. Hamilton, explained, "That's the American system."[27] Rogers also set aside money for a no bid 511 service contract that went to Senture, owned by the family of Bill Deaton, another frequent campaign contributor. Deaton had recently been fined by the Securities and Exchange Commission for securities fraud; yet Rogers's son John had nonetheless gone to work for Senture. Rogers earmarked $7 million in federal funds to build a freight-transfer facility that would be leased to Norfolk Southern, another campaign donor, for $1 a year. Then there was the Southern Kentucky Rural Development Center, with its $15.5 million initial price tag, and its $18 million annual budget, most of it from federal funds.[28] The lobby contains a bust of Rogers, and wags have nicknamed the building the "Taj Ma-Hal."

Prince of Pork indeed. But nothing in the world tasted better than all that *homeland security* pork, which came Rogers's way through chairmanships first of the Appropriations transportation subcommittee, and then of its homeland security subcommittee. It was a fine thing to munch on crispy, homecured government millions *and* defend the homeland against terrorism all at the same time. Kind of like a staunch evangelist for faith-based marriage having a personal key to the Playboy mansion. It's such a deeply satisfying vocational mission that a number of other Congressmen have found their way to the same smokehouse.

In addition to the Taj Ma-Hal-based National Institute for Hometown Security and Kentucky Homeland Security University Consortium, one might note the interplay of government contracts and donations between Reveal Imaging Technologies Inc. and Rogers, explored by a 2005 *Washington Post* article. Reveal got a $463 million federal contract for baggage screening machines; Rogers got $122,211 in donations for his political action committee, HALPAC, and a pledge from the company to move $15 million worth of work to Rogers's district. As the *Post* put it: "Reveal's dealings with Rogers illuminate the intersection of politics, money and homeland security in the rush to make the nation safer since the Sept. 11, 2001 terrorist attacks."[29]

This was the Moses/Pharaoh who would lead Kentucky into the land of storied land of bioboodle. And we should follow him through the parted waters, according to the *Commonwealth-Journal*, because the Congressman "would not bring something that is dangerous or negative to Pulaski County. He has earned our trust and we have faith in his ventures."[30]

And the *CJ* would be at the congressman's side every step of the way, with its "things you need to know," its multiple lies about BSL-4 labs operating all round the country for over 50 years, with nary an incident of agent release or contamination, not even an accident, with its additional lie that BSL-4 labs in Geelong, Australia and Winnipeg, Canada had operated with "spotless safety records" and "without incident."[31]

A favorite tactic of the *CJ* was to ridicule opponents' concerns as just too farfetched to even contemplate. And it used the tactic from the very beginning, as a sort of precautionary measure, before opposition had even heard of the blessed lab. One of the three stories it ran on the first day, "How Safe Will National Lab Be?" (the others were "You Need to Know" and "It Will Transform Us"), opened as follows: "Before wild rumors begin about rampant bird flu in Shopville or anthrax in Faubush, Congressman Hal Rogers wants Pulaski Countians to know the truth about the proposed National Bio and Agro-Defense Facility." And the truth that was to set Pulaski Countians free from biolab worry?—"The truth is the lab is not a dangerous thing." Because, well, this lab would be the premier research laboratory in the world, with all those state-of-the-art safeguards—and because, in over fifty years of BSL-4 labs, *there had never been an accident of any type*.[32] In fact, there had been enough accidents to make the claim absurd.

Well, Moses had decided to stand in for God for the time being, and had spoken out of the burning bush himself, but what is one to do when Moses/God turns out to be a liar? The *CJ*'s three opening stories appeared on February 21. Over the next few days the *CJ* would continue to hammer home the extraordinary opportunity, the quantum leap, Pulaski Countians had before them. Dr. Michael Blackwell of the University of Tennessee would declare how happy he'd be to have the facility in his backyard.[33] (Since the facility would require about thirty acres, Blackwell should have had too much yard mowing to occupy himself with biolab propaganda.) Hazmat personnel expressed their confidence in their ability to handle anything the lab might throw at them: "What we deal with as a local team is a lot more dangerous than what I would have a problem with here." "More hazards go up and down the highways and byways of Pulaski

County."[34] And the mayor, county judge, and community college president all chimed in with their support.

As late as March 1, the *CJ* described the effort as "picking up steam." About 400 people attended a Chamber of Commerce forum and informational session at the Taj Ma-Hal. The congressman got a standing ovation. "There were numerous positive comments," the *Commonwwealth-Journal* reported, and only "a few expressions of concern" by nearby property owners. But their comments weren't negative at all, Greg Jones, the executive director of Southern Kentucky Economic Development Corporation, insisted. They were just run-of-the-mill property-owner questions.[35]

The congressman repeated the lie about there never having been "an accident or incident in the history of similar laboratories." (No wonder he thought the thing would be as safe as a Wal-Mart.) The *CJ*, and the congressman, told people not to get all steamed up with worry. This "was not a done deal." Once a site was selected, "there will be so many public hearings that you'll get blue in the face." The facility, the congressman said, would bring "tens of thousands" of new jobs and all that cold hard cash. The county judge said Pulaski Fiscal Court would vote on a resolution of support that afternoon. The City Council had just adopted one. The Burnside mayor said he'd ask Burnside City Council to follow suit next week.[36] A steamroller sure enough, if the *CJ* were believed.

Two days later, however, the local consortium held a different sort of meeting, at Pleasant View Baptist Church near the proposed site. Over a hundred people attended, and they had "dozens of questions" about "potential dangers, depressed land values and disruption of their quiet farming community." As one of the residents put it, "I don't want to knock the state out of such a prestigious thing. But at the same time, how does it treat us right next door? It's just a little scary." Not to worry, said Balltrip, and State Commissioner of Public Health William Hacker. There'd never been a release from a facility like this that caused harm to a community.[37] When queried about the 2004 SARS releases in Asia, Hacker said he hadn't heard of those particular incidents, despite their obvious relevance.[38]

About a week later, people began circulating a petition against the lab. By March 17, they had collected over 2800 signatures.[39] "If you don't want the thing, we'll just fold our hands and go away," Rogers had stated coyly at the Chamber of Commerce forum.[40] Now a large number of people were in fact saying they didn't want the thing, but Rogers made no move to fold his hands, and no one ever expected that he would. What the congressman wanted, the congressman usually got.

Lots of Pulaski Countians had considerably less faith in the *Commonwealth-Journal* than the *Commonwealth-Journal* had in the congressman. It was a rural area, and many possessed the farmer's natural skepticism about both the government and the claims made for state-of-the-art technology. They had ongoing experience with the contrariness of both machinery and people; their state-of-the-art tractors broke down just like the older tractors did.

Some of those raising questions—a doctor and a librarian among others—were trusted, educated members of the community. Michael Carroll's *Lab 257* came to their attention early, and they did more research on their own—something the lab's proponents never got around to. And up the road, one of the state's two chief newspapers, the *Lexington Herald-Leader*, acted like a newspaper instead of the congressman's private press agency. The *Herald-Leader* actually noticed that there was opposition to such facilities at other locations in the country, and consulted and quoted watchdog organizations like the Sunshine Project and the Council for Responsible Genetics. Whereas the *Commonwealth-Journal* had urged its readers to follow the Pied Piper's pipings in an orderly fashion, a February 26 *Herald-Leader* editorial said that while the proposed bio-lab was "worth exploring, Kentuckians must verify, not trust, [the] government on [the] facility's safety."[41]

Some of the Lexington TV stations paid attention, too. And there was a local Christian radio station, King of Kings Radio, with a large regional audience. King of Kings interviewed two members of the Pulaski County opposition during the very week the initial set of petitions was being circulated.

It was soon clear that the congressman's notion of eliciting massive public support with a blitzkrieg propaganda drive had failed. Elsewhere around the country, consortiums were keeping their initial NBAF efforts under the radar. It would work like a charm in Kansas, where opponents, facing the slickest and most well-heeled NBAF propaganda machine of all, didn't get clued in and organized until it was already too late. But the same strategy would backfire like crazy in North Carolina, where the home folks went on the warpath with a vengeance after they discovered what their elected officials had been up to.

Meanwhile, a sizable number of Kentucky locals were taking a good close look, and noticing that their bloated emperor really wasn't wearing any clothes. So far as they were concerned, it wasn't a pretty sight.

The Kentucky public relations battle was spiraling out of control, and the *Commonwealth-Journal* did its part to bring the unruly locals back into line. It started with a hatchet job on *Lab 257*. On March 18, the *CJ* ran an article titled "*'Lab 257': Fact or Fiction*," posing as an interview with Michael Carroll, except the *CJ* juxtaposed to Carroll's comments quotes from a critical review of the book. And after quoting Carroll on the questionableness of placing a BSL-4 livestock diseases lab in a livestock region (a concern the *Herald-Leader* had also raised in its February 26 editorial), the *CJ* added a little spontaneous editorializing: "But are these the words of a concerned investigative reporter—or an author trying to sell his book?"[42]

The following day, the *CJ* ran an interview with former Plum Island director Dr. Alfonse Torres: "'It Will Be Safe': Past Plum Island Chief Says Fears Aren't Factual." Torres ridiculed the "tales about aliens from outer space being kept there, along with this folklore about disease outbreak," and said *Lab 257* was such fiction that it didn't even deserve the label of investigative reporting.[43]

And two days later, the *CJ* ran an editorial, "We Should Support Effort to Bring Bio-Lab to Pulaski," decrying fear of the unknown, the "panic fallout" from *Lab 257*, and "people who simply don't want the facility here because they don't want things to change." After all, we should trust the congressman, welcome the "quantum leap" involved in taking a high-school-education-or-less region and bringing it onto "the world scientific landscape," and "do our part in our own rags-to-riches story."[44]

On April 5, 26-year-old staff writer Chris Harris ridiculed the notion that Somerset would ever be a target for terrorists—and maybe it wouldn't be, despite the 9/11 evacuation of Plum Island, and despite the frequent announcements of the ongoing danger of "agro-terrorism," which was one thing that certainly wasn't going to happen in New York City—but if Pulaski Countians would be immune from terrorism even with the presence of the country's second biggest biodefense facility, would someone please explain why the county had its very own National Institute of Hometown Security? If you're going to yank the terrorism chain every other day, you shouldn't be surprised when the sleeping dog actually barks.

Harris also ridiculed concerns about "The Attack of the 50-Foot Cockroach" and "three-headed mutant cows," and bemoaned Pulaski Countians' "frequent reluctance to accept progress and change." Because what we should be focusing on, he said, was money—because money is progress—even if the money comes from a bloated new bio-defense facility playing on those very terrorism fears Harris had just ridiculed.[45] The *Commonwealth-Journal* added its own conclusion that "Most of the outcry has come from residents living in the direct vicinity of where the proposed lab would be."[46] One wonders, by "outcry," does the *CJ* mean the hundred-plus people who showed up at the Pleasant View Baptist Church, or the 2,800-plus who signed the petition during one week of circulation? We might have a hundred hoot owls or so out in the direct vicinity, but we for sure didn't have 2,800.

On April 1, the *CJ* ran its interview with the University of Tennessee veterinary dean, "Expert Proclaims Bio Lab 'Safe.'" That article took another quick hatchet swipe at *Lab 257*, and opposed to the irresponsible claims of "fear-mongers" the dean's elegant statistical analysis: "The track record for these facilities around the nation is astounding. . . . In terms of safety you can't get any better."

And "from what I understand," the dean added, "the majority of the folks in Pulaski County are in favor of it because they understand the long-term benefits." (One would have liked to ask the dean what elegant statistical study he conducted—beyond talking to the consortium's inner circle—to arrive at this understanding.) "But for the people who are leery," he went on, "I would tell you that you have nothing to fear from this facility. It's completely safe."

And after reading the article, one would look up at the masthead and realize that it was indeed April Fool's Day.

ASK THEM NO QUESTIONS

I CRASH A PEP RALLY

IN LATE JULY 2006, A FRIEND THOUGHT HE'D HEARD A RADIO broadcast that the short list announcement would be made that weekend. Either his hearing or the radio station's sources were faulty, because no announcement was forthcoming, at least not that weekend. But that was par for the course: as DHS repeatedly missed target dates, those of us dreading bad news fell back on rumor. Someone tightly connected with Rogers was saying it was all a done deal: Kentucky had been chosen and DHS was just going through the formalities. We feared that the gossip might be true; at the same time, we weren't sure somebody wasn't spreading disinformation to discourage us.

On July 30, someone sent an E-mail saying that Ewell Balltrip would be the guest speaker for the Chamber of Commerce's monthly lunch meeting on August 1. The writer suggested it would be nice if someone attended and had some questions/remarks ready.

The Chamber holds such lunch meetings monthly in the Taj Ma-Hal. They're open to the public if payment for lunch is made. (When it voted its resolution of support for bringing NBAF to our county a few months earlier, however, the Chamber had met in closed session.)

Yours truly decided to attend the August meeting, with the hope of posing some questions. I suspect I was the only car with a "No Kentucky Biolab" sticker in the Taj Ma-Hal parking lot that day.

I shared my table with two women from Citizens National Bank, and Bill Mardis, the editor emeritus at the *CJ*.

We had our lunch, and then Balltrip got up to talk. He told us the short list decision might be made as early as next month. The consortium appreciated the 100 local businesses that had signed letters of support for the facility, and their support would be needed even more should Kentucky make the short list.

As evidence of safety, he cited the BSL-4 research being done in urban areas: the new NIH labs in Boston and Galveston; the NIH facility in Bethesda, Maryland; the CDC in Atlanta; and the regional biocontainment laboratory in Louisville. At the end of his talk, he said he could entertain a question or two.

I asked Balltrip if he were familiar with the facility at Detrick. He said yes. I started reading him my collection of Detrick headlines about anthrax leaks, broken safety rules, "sloppy methods," lost or stolen bacteria, government coverups, and "hazardous surprises." At some point in my reading someone behind me started booing. That was rude; I hadn't heckled Balltrip. I finished the list, and asked Balltrip if he thought those headlines should cause us concern.

Balltrip said no, because the headlines didn't indicate any catastrophic incidents or actual harm to the community.

After my first question, I began to ask Balltrip about a July 30, 2004 *Science* article, reporting on the CDC's efforts to artificially create a deadly variant of the bird flu virus that could be transmitted from person to person. The article voices concern about an accidental release causing a global epidemic. I wanted to know if we might anticipate this apocalyptic research being performed in south central Kentucky.

But the booing started up again, and I saw the Chamber president moving to the microphone. I assumed he was headed that way to say we were out of time—and maybe we were—so I didn't finish my question, just mumbled something to the effect of "So your answer basically is that you're not concerned because there weren't any catastrophes?"

It occurred to me later that the nice Chamber of Commerce moderator might have been coming to the microphone to admonish the rude heckler to let me ask my question. Maybe I should've stood my ground?

In the end, I don't think so. The moderator dismissed the meeting forthwith. He didn't ask if anybody else had a question.

Keenan Turner sent me an E-mail the next day saying he'd been to an Ag Council breakfast that morning and asked if anyone at his table had attended the Chamber luncheon. According to Keenan, "One of the fellows indicated he attended and 'someone spoke against the biolab and got cut off before he could really get on a roll.'"

Later, I would replay what had happened and think of what I should've said and done. I should have turned round and told the heckler that I hadn't heckled that mealy-mouthed Balltrip. I should have asked the whole luncheon group if they weren't a bit ashamed not even to listen to a few questions and answers about the sort of facility they had invited into their midst. I should have been more pointed in my questioning of Balltrip; I should have read back the *Commonwealth-Journal's* statement about there being no accidents in fifty years of BSL-4 research, and asked if those headlines didn't suggest the contrary.

I'm saying those things now, but I didn't say them then. I backed down because I was caught off guard and didn't want to seem too strident in the camp of the enemy. A final bit of comic irony arrived as the meeting dismissed. Bill Mardis, of the *CJ*, asked my name and occupation and contact information. I thought, *hey, he's disturbed by the stuff I brought out, and he wants to find out more. Bring on the truthseeker!*

I never heard from Mardis. In his article published the next day, he noted how the long-suffering Balltrip had "emphasized and re-emphasized that the laboratory will be safe." And if the site here went on the short list, it would "undergo a very comprehensive environmental impact study" and there would be many meetings for public input during this process[1] (in fact, there would only be two meetings—a scoping meeting, and a meeting for public comment on the Draft Environmental Impact Study).

And after noting that more than a hundred organizations and businesses, including the Chamber of Commerce, had endorsed locating the facility here, Mardis acknowledged that "there are opponents, including petition signers." My presence was described as follows:

> Kenneth King, a Somerset attorney and teacher at Western Kentucky University, voiced objections to the laboratory at Tuesday's chamber session.
> King read a list of what he identified as newspaper headlines that he said questioned the safety of similar laboratories.

I would call this damning with faint attribution. I alleged that I was reading a list of newspaper headlines, and I alleged that the headlines questioned the safety of similar laboratories. Mardis didn't know if I really was reading a list of newspaper headlines, and Mardis didn't know if they said what I said they did.

Actually, of course, I didn't say the headlines "questioned the safety of similar laboratories." I just read the headlines aloud; I thought they spoke for themselves. And if Mardis had wanted to "trust and verify," I'd have been happy to supply copies of the relevant articles.

But endorsing minds don't really want to know, do they?

CHAPTER FOUR

Growing Up in a Rough Neighborhood

The Troubling Antecedents of U.S. Biodefense Research

OKAY. THE BIODEFENSE FACILITY THAT SHOWS UP ON YOUR FRONT doorstep these days, wanting to sell you a lifetime subscription to the Pathogen-of-the-Month Club, will be an educated, smooth-talking son of a bitch. He'll tell you about the state-of-the-art technology, the meticulous containment protocols, the warm glow that will come to you from living on the edges of biodefense heaven. On the one hand, there is this deep inner satisfaction that comes from battling the bioterror bogeyman, that cunning faceless villain capable of con-cocting designer pathogens with a Wal-Mart chemistry set, and of dis-seminating them with his own private air force; on the other hand, wouldn't it be nice to get your hands on a big chunk of that bioterror loot while doing your bit for humanity? Wouldn't it be cool if your hometown became known around the world as a hub for all the inter-esting things people can do to germs these days while messing around to see what germs can do to people—make them resistant to anti-biotics; make them defeat existing vaccines; create new forms of the flu that don't exist yet, like a bird flu that spreads easily among people; bring the 1918 influenza back to life, splice some of those vigorous

1918 flu genes into our weak, anemic, twenty-first-century flus—and then—*after* creating all these awful things—work on a defense against them, because who knows, now that you've created them, the bioterror bogeyman might get his hands on them too, and doesn't that just scare the bejesus out of you?

Before the smooth talker has you signing on the dotted line, perhaps a small background check would be in order. If you discovered that the nice young man once worked for the dirty-tricks divisions of several intelligence agencies, specialized in learning how to kill people—large numbers of them—like, well, like a bioterrorist—**and** was buddies with several war criminals—**and** was widely known to be an uncontrollable, inveterate liar—perhaps you might call the police and ask to have him removed from your doorstep. Which may not be so easy if, as is likely, the nice young man (or woman) actually works for the Department of Homeland Security, the Defense Department, or even your local institution of higher learning.

Make no mistake about it, bioweapons were always cloak-and-dagger. As Jeanne Guillemin, author of *Biological Weapons*, probably the most informative history of the subject, explains, bioweapons programs in the United States, France, Britain, and Canada were launched in secret by "a small number of scientists, government officials, and military officers" and conducted secretly thereafter within "secure bureaucratic niches."[1]

Your U.S. biodefense researcher is a direct descendant of the U.S. bioweapons mafia, founded during World War II and concealed from the American public, Congress, and even most of the military. University scientists and pharmaceutical companies were part of the syndicate from the beginning. The first director of the U.S. bioweapons program, Ira Baldwin, came from chairing the bacteriology department at the University of Wisconsin. George Merck, president of Merck Pharmaceuticals, headed the War Research Service, formed in 1942 to make recommendations and initiate research projects in biological warfare. The WRS would pay biologists at more than two dozen universities and private laboratories to conduct biowarfare research. (Among them was even Harvard's environmental guru René Dubos.)[2]

It was necessary to avoid alarming the public, who from time to time succumb to the aberrant view that research involving killer germs is dangerous. So the very existence of the War Research Service was concealed by burying it within a larger civilian department, the Federal Security Agency, itself established "to promote social and economic security, advance educational opportunities and promote public health."[3]

In 1943, in green pastures surrounded by picturesque mountains, Camp Detrick became headquarters for the Chemical Warfare Service's new biowarfare division. Detrick took "elaborate precautions" to conceal its true purpose.[4]

The Special Projects Division at Detrick was a secret within a secret; even Detrick's commanding officer was kept out of the loop. Its technical director and staff reported directly to Washington or to Chemical Warfare Service officers elsewhere. And President Roosevelt may have been kept out of the loop, at least with respect to the seamier details, by Secretary of War Stimson.[5]

The idyllic rural outpost quickly grew into a metropolis of 250 buildings and living quarters for five thousand people, ringed by fences, towers, floodlights, and guards with machine guns.[6] By the end of the war, Detrick was engaged in more than 100 projects, producing everything from botulinum and anthrax to killer mosquitoes.[7]

A brief period of openness after the war brought more media attention than the Army wanted,[8] and in 1947 Army Chief of Staff Eisenhower banned further disclosures. A 1949 press statement by Secretary of Defense James Forrestal focused on defensive research and disguised the program's offensive goals.[9]

After the war, Detrick biowarfare researchers coveted the "state-of-the-art research" of their former Japanese enemies. During the war, the Japanese had killed thousands of Chinese civilians with anthrax, typhoid, and plague attacks[10] and performed gruesome experiments on prisoners of war. Some of them may have been Americans. John Powell, a scholar who obtained details of the research under the Freedom of Information Act in the late 1970s, estimated that at least 3,000 people were killed by the experiments alone.[11]

Even as the U.S. tried Nazi scientists in Nuremberg for similar coldbloodedness, American biological warfare authorities granted Japanese germ-war criminals immunity (and pay) for their data.[12]

Some dozen or so went on to become medical school professors and administrators or pursue successful careers in commercial research and industry.[13] The U.S. obtained 15,000 slides taken from more than 500 patients, out of 850 human corpses autopsied by the dedicated Japanese investigators.[14]

Dr. Edwin Hill, the chief of Basic Science at Detrick, appreciated the Japanese scientists' hard work and the bargain-basement price at which the U.S. obtained it:

> Evidence gathered in this investigation has greatly supplemented and amplified previous aspects of this field. It represents data which have been obtained by Japanese scientists at the expenditure of many millions of dollars and years of work. Information has accrued with respect to human susceptibility to those diseases as indicated by specific infectious doses of bacteria. Such information could not be obtained in our own laboratories because of scruples attached to human experimentation. These data were secured with a total outlay of 250,000 Yen to date, a mere pittance compared with the actual cost of the studies.[15]

Hill showed a finely honed empathy for his Japanese counterparts, hoping that "individuals who voluntarily contributed this information will be spared embarrassment because of it."[16] Embarrassment had presumably not been the foremost concern of the 850 victims. These "human subjects," read the bland report of another Detrick scientist, had been "used in exactly the same manner as other experimental animals"—i.e., tied to stakes and protected with helmets and body armor, in hopes that when attacked with anthrax bombs they might die from the disease itself rather than bomb fragments.[17]

In another "excellent study" (words of the American scientist), methods of infecting the victims with plague-infected fleas were explored. In one case six of ten subjects became infected, and four of those died; in another, eight of ten subjects received flea bites and became infected; six of these infected died.[18]

Whatever the death rate from the diseases themselves, "mortality in experimental cases was 100 percent due to the procedure of sacrificing experimental subjects." Kidney, spleen, and liver sections from the victims were mounted on microscopic slides, in proper scientific manner.[19]

Jeanne Guillemin says the immunity bargains stymied national and international public debate about bioweapons by concealing historical examples of "how a civilized, technologically advanced society could for years conduct inhumane biological weapons experiments and attacks."[20]

The U.S. military was always restless about the human experimentation scruples that confined it to animal studies. "The Air Force could be fairly accurate in predicting what a biological warfare attack would do to a city full of monkeys," said one document, "but what an attack would do to a city full of human beings remained the 'sixty-four-dollar' question." "Now we know what to do if we ever go to war against guinea pigs," complained the Air Force general who had bombed 11,628 guinea pigs with brucella over a period of two months.[21]

The Naval Research Unit located at U. Cal-Berkeley conducted the U.S.'s only wartime human experiments, injecting San Quentin "volunteers" with bubonic plague. Supposedly, none became seriously ill.[22]

After enlisted men at Detrick staged a sitdown strike over efforts to "volunteer" them for tularemia research,[23] the Army turned to the Seventh Day Adventists. Adventists reject killing, fighting, or bearing arms, though not such noncombatant service as caring for the sick and wounded. The Army told them what a fine contribution they would be making to national defense and to public health generally by exposing their bodies to tularemia, typhoid fever, Rocky Mountain spotted fever, Rift Valley fever, and encephalitis pathogens. The Army led them to believe the main goal of Project Whitecoat was defensive, to find the best protections against disease, when in fact the project's main goal was to determine the best doses of germs for an effective bomb.[24] Some 2200 Adventists volunteered over a twenty-year period.[25] The U.S. would also conduct at least one additional test on convict "volunteers" at the Ohio State Penitentiary.[26]

The most wide-ranging human tests in the U.S. didn't involve volunteers at all, however. For at least two decades, Army bioweaponeers secretly sprayed millions of Americans with huge clouds of supposedly harmless bacteria and chemical particles. Focused on measuring air currents and pathogen survivability, the Army took no steps even to monitor the health of the people exposed.[27]

Details of the tests emerged during 1977 Senate subcommittee hearings. Army spokesmen acknowledged that 239 populated areas around the country "had been blanketed with bacteria between 1949 and 1969": Alaska and Hawaii; San Francisco; Washington, D.C. (mock germ attacks at the Greyhound bus station and the National Airport); Key West and Panama City, Florida; and the Pennsylvania Turnpike and New York Subway system.[28]

The incidence of illnesses had suddenly increased in some areas near the tests, but Army witnesses said the Army hadn't monitored health effects in the tested areas because it "made an assumption of the innocence of these organisms."[29]

The San Francisco tests, conducted for a week in September 1950, involved spraying zinc cadmium sulfate as a fluorescent tracer with one of two bacteria, *Bacillus globigii* or *Serratia marcescens*. The Army's scientific data revealed that nearly everyone in San Francisco inhaled 5000 or more fluorescent particles per minute during the several hours the particles remained airborne, so that San Francisco residents unknowingly inhaled millions of bacteria and particles every day during the week of testing.[30]

Four days after the Army sprayed the San Francisco Bay area with Serratia, a patient at the Stanford University hospital was diagnosed with a Serratia infection, the first ever recorded at the hospital. During the next five months, ten more patients became infected, and one died.[31]

Doctors at the hospital considered the outbreak so unusual that they wrote a journal article about it. After learning of the epidemic, the Army secretly convened a committee of four not-so-objective scientists (they had previously worked with the Army's biological warfare program). These scientists helpfully concluded that the outbreak "appeared coincidental."[32] The Army continued to spray huge amounts of bacteria, including Serratia, and chemical agents over populated areas across the U.S. Increasing numbers of serratia infections led to identifying the bacteria as a significant infection threat by the end of the 1970s.[33]

Leonard Cole's book on the spraying, *Clouds of Secrecy*, cites several reasons for questioning the Army's safety claims. First, since the Army itself made no effort to monitor the health of affected populations, it had no basis for asserting that health wasn't compromised. (In fact, data gathered by non-military groups suggested the contrary.) Second, some of the documents the Army submitted for the 1977 hearings acknowledged, however subtly and tentatively, that some agents may not have been so innocuous. Third, scientific literature published as the testing proceeded suggested the dangers of several simulants. Finally, several scientists testified that any microorganism can be harmful under the right conditions.[34]

Sixty-three articles suggesting the health risks of Serratia appeared between 1950 and 1970;[35] fifteen studies suggested the dangers of cadmium, the heavy-metal component of zinc cadmium sulfate.[36] And standard medical reference works pointed to the health risks of two other simulants, *Aspergillus fumigatus* and *Bacillus globigii* (aka *Bacillus subtilis*).[37]

Near the Bahama Islands, the U.S., Canada, and Britain sprayed not just simulants, but "highly virulent" pathogens. The tests killed thousands of animals, and the germs might have come into contact with people had the wind shifted or other miscalculations occurred. In addition, after obtaining whooping-cough bacteria from Detrick in March 1955, the CIA field-tested unidentified bacterial agents along Florida's coast. The incidence of whooping cough in Florida tripled that year.[38]

General William M. Creasy was completely unapologetic for the secrecy of the tests:

> I would feel it completely impossible to conduct such a test trying to obtain informed consent. I could only conduct such a test without informing the citizens it was being conducted. I could not have hoped to prevent panic in the uninformed world in which we live in telling them that we were going to spread non-pathogenic particles over their community. 99 percent of the people wouldn't know what non-pathogenic

meant, nor do any words I know appear to be such that you could get it across to them.[39]

The true irony, of course, is that the bacteria were not "non-pathogenic," so that the ignorance of the testers, not the tested, was the problem. That ignorance, Cole suggests, was almost willful: "The officials seemed to have convinced themselves that certain facts were illusory or unimportant, because they did not want to believe them."[40]

The 1977 Senate testimony disclosed many previous secrets of the U.S. bioweapons/biodefense programs. Charles Piller's Oct. 3, 1988 article in *The Nation*, "Lethal Lies about Fatal Diseases," later showed that even in 1977 the Army had continued to deceive. (See Dugway chapter in this book.)

Between 1954 and 1956, the U.S. Defense Department quietly adopted a new policy, allowing it to consider first-strike use of chemical or biological weapons.[41] It accompanied the new policy with a wave of planted magazine articles extolling the humanity of chemical and biological weapons (CBW).[42] It drew up contingency plans that contemplated a biological strike as prelude to a Cuban invasion.[43]

The CIA involved itself in American biowarfare efforts in a number of ways, including efforts to assassinate Castro and other leaders with bioweapons.[44] Dr. Sidney Gottlieb, director of the CIA's Technical Support Division at Detrick, as part of his research on mind-control drugs, slipped LSD into the drinks of Special Operations Division personnel.[45] Probably the only thing scarier than messing with the minds of bioweapons researchers would have been trying the same fun and games with people in charge of our nuclear attack force. The psyches of bioweapons researchers were fragile enough without psychedelic drugs, apparently: "[I]t was not easy working with hot agents on a daily basis, and a number of Camp Detrick technical personnel ended up as 'psychoneurotic cases' and were dismissed. 'The nature of work on the post,' said the dry Army report, 'was not considered conducive to rehabilitation.'"[46]

One of Gottlieb's victims, Frank Olson, chief of the Special Operations Division's Plans and Assessment Branch, began feeling "mixed up" and said he wanted out of the germ warfare business.[47] That in itself might seem sane to most people, but Olson reportedly began experiencing delusions that people were out to get him. While in New York to see a psychiatrist, as his CIA handler slept, Olson—presumably to escape the people who were out to get him—jumped out a 175-foot-high hotel window.[48] The CIA hid the details from the police and the FBI with an elaborate cover story.[49] (In their book *Dead Silence: Fear and Terror on the Anthrax Trail*, Bob Coen and Eric Nadler explore the possibility that the alleged suicide was itself a cover story.)[50]

Gottlieb, the agent involved, was mildly reprimanded, then promoted six years later. He would be involved in U.S. plans to assassinate Patrice Lumumba with botulinum toxin,[51] and in 1999 would be charged with having slipped LSD into the drink of an American expatriate who'd offended him. Gottlieb died just before the trial began.[52]

After 1969, America's bioweapons program officially became a biodefense program. Of course, the bioweapons program had always partly been a biodefense program, and the biodefense program would partly be a bioweapons program, producing or simulating bioweapons so that defenses can be developed against them. In terms of public relations, bioweapons are the Mr. Hyde, and biodefense is the Dr. Jekyll. Dr. Jekyll now insists, loudly, that he has nothing to do with Mr. Hyde. But with Jekyll and Hyde having cohabited for so long, one has to figure Jekyll picked up a few of Hyde's habits. The national security mindset, for instance, with its penchant for secrecy and deceptions.

After all, Detrick, the site of the nation's lead biodefense facility after 1969, had been the site of its lead bioweapons facility *before* 1969. In 1989, Neal Levitt, a former Detrick researcher, told a Utah newspaper how the Army had purged from his reports his concerns about germ contamination and missing samples.[53] Other chapters in this book amply document deceptions at Detrick, Dugway, Plum Island, and multiple university research facilities.

Biodefense has grown up (or dawdled in extended adolescence) among a culture that sanctions deceiving the public about its activities and the risks of those activities, and that places a higher value on its experimental goals than on the safety of the public. Politicians, business interests, and research universities bring their own special prevarications to this culture of deception. Like their military fore-bears, academic researchers conceal problems at their facilities to avoid raising public concern about their activities. Boston University concealed the tularemia infections of three researchers as it pursued one of the NIH's new national biocontainment laboratories, and as it sought public consent or acquiescence for the new facility. Texas A&M researchers concealed problems there as the university competed for DHS's NBAF facility.

The A&M incident in particular suggested that a lot more accidents and safety lapses were occurring than were being reported. A sharp uptick in new accident reports occurred just after A&M got caught.

After the 2007 GAO investigation, the biodefense complex began pushing—as an alternative to more intrusive regulation—a voluntary, anonymous reporting system that would protect wayward facilities from embarrassment. Otherwise, seemed to be the suggestion, universities and other facilities would continue to conceal problems when they felt they could get away with it. An ironic admission for an industry urging the public to trust it.

DHS PULLS AN AUGUST SURPRISE

ON AUGUST 1, BALLTRIP TOLD THE CHAMBER OF COMMERCE THAT the DHS decision might come "as early as September." On the morning of Wednesday, August 9, we got a tip from the Lexington media. Kentucky had made the short list. "We Made the Cut," the *Commonwealth-Journal* proudly announced the next day. Rogers was elated, "because this lets us know we can run with the big dogs."[1]

The *CJ* took the opportunity of informing us again that we would become another Oak Ridge, Tennessee if chosen. Taj Ma-Hal resident and economic development director Greg Jones was quoted, yet again, on how "mind-boggling" it would all be, bringing little old Pulaski County all those 21st-century jobs. Germs were the growth industry of the future, apparently. We had welcomed the Japanese to the Toyota factory at Georgetown. We had people putting out the welcome mat for the Chinese. Now we were welcoming a more sinister sort of immigrant germs. And all so that our little Johnnies and Jills, with their newly minted online PhD's in microbiology, wouldn't have to leave the farm to find a suitable lab position.

"Of course," the *CJ* noted, "there are a few dissenting voices—most of them coming from the area 10 miles northeast of Somerset where the lab would be constructed."

Rogers said he could understand people with farms out there, who would be displaced—"but, for the most part, opposition has been very limited." And we shouldn't let a few backward types stand in the path of germ lab progress: "There's always going to be a few people who disagree. . . . People protested when we first put up gas street lights. They protested in eastern Kentucky when we first put up flood walls. But some of those same people would tell you now they were wrong."

If these backward types would just get with the program, and put their faith in Hal Rogers, as the *Commonwealth-Journal* had. . . .

Around the country, local newspapers that noticed the down-select at all felt the honor, sort of, of having made it to the DHS short list. It was a modest honor, to be sure, since 18 of the 29 original sites had survived to fight another day. "The short list is not as short as we had expected," said Georgia's David Lee, of minuscule-amounts-of-pathogen fame.[2] "We had not anticipated getting the list this early," said San Antonio's Harold Timboe. "We had thought it was going to be in September, a month from now, and that it would be a much smaller number."[3]

In fact, everyone had been expecting a short list of three to five sites, because that's what DHS had been telling everybody. Yet all the consortiums, bar none, would soon be telling their concerned publics how open and transparent NBAF would be. Yet, at a time when DHS might have been expected to put on its best-behavior face, it wouldn't even give the facility's cheerleaders advance notice of changes in the site selection process. What sort of openness could the public really expect from DHS after the facility opened?

Wary residents of Tracy, California, already knew what kind of transparency they could expect. The California bidder for NBAF, Lawrence Livermore Laboratory, was one of two facilities where all U.S. nuclear weapons were designed. Livermore, owned by the U.S. Department of Energy and managed by the University of California-Berkeley, ran two sites. The main site, established in 1952, was located at Livermore, a few miles to the east of San Francisco. Site 300, where UC-Berkeley proposed to operate NBAF, was established in 1955 a bit further to the east, in Tracy. Livermore had used Site 300 as a place to develop and detonate high explosives and to conduct mock nuclear bomb tests using depleted uranium.

Both sites had been named to the EPA's Superfund list. TriValley CARES (Communities Against a Radioactive Environment) had been formed in 1983 to keep an eye on Livermore and resist its more dangerous plans. The organization had about 4800 members, was governed by a board of directors, and had a paid staff, including an executive director and staff attorney.

At the time Livermore submitted its bid for NBAF, CARES was already fighting Livermore's plans for a new BSL-3 lab to study anthrax, plague, and several other diseases—research which Livermore acknowledged would involve genetic modification experiments and "aerosol challenges."[4] Livermore had gotten into the biodetector business on a small scale in the 1990s, but CARES argued that this facility would be "the first advanced biowarfare agent research facility inside a US nuclear weapons lab" and criticized the notion of mixing "bugs and bombs," of having the U.S. conduct biodefense research inside a classified nuclear weapons facility.[5] The Energy Department had made similar plans for both its nuclear weapons facilities, Livermore and Los Alamos, but had shelved the plans for Los Alamos (for further environmental assessment) after CARES and a sister group, Nuclear Watch of New Mexico, filed suit in 2003. Litigation about Livermore proceeded, and in 2004 a federal district court had ruled in favor of the Energy Department. CARES and Nuclear Watch had appealed, and on October 16, 2006, the Ninth Circuit had ruled in their favor, remanding the environmental review to the Energy Department for further analysis of terrorist risks.[6]

CARES had filed Freedom of Information lawsuits in 1998 and 2000, after DOE failed to respond to requests for unclassified information, and it would file another one in the fall of 2006 for "failure to provide responsive documents as required by law for as long as three years on five separate information requests."[7]

So California-Berkeley's refusal to provide a copy of its NBAF proposal[8] was nothing new. And after Livermore made it onto the DHS short list, CARES began focusing on the new facility, with a September workshop featuring staff attorney Loulena Miles, Sunshine

Project director Edward Hammond, and biologist Dr. Judith Flanagan.

Workshop attendees learned considerably more than they would learn at an October 4 joint meeting between the Tracy Tomorrow and Beyond Committee and Livermore and Cal-Berkeley representatives. At the October 4 meeting:

> The representatives told committee members that they didn't know what types of diseases would be researched, which routes would be used to transport pathogens into or out of the laboratory, or whether the public would be told about accidents.
>
> Lab spokeswoman Susan Houghton said the Department of Homeland Security had kept secret most information about the potential laboratory, which could be built by 2010. She said Lawrence Livermore risked its bid for the project if it tried to guess the answers.[9]

So ask them no questions, and they'll tell you no lies, because they're not asking any questions themselves.

The consortiums now believed—or had been told—that the real short list would be chosen by early 2007 and the final selection made by early 2008. Just after Thanksgiving 2006, the University of Wisconsin apparently decided to test the waters on its under-the-radar application. On November 29, several months after the university had actually submitted its expression of interest, the *Wisconsin State Journal* ran a story headlined "UW Applies for Federal Disease Lab."[10] UW proposed to put the facility on 49 acres of its Kegonsa Research Campus, in the town of Dunn fifteen miles to the south of Madison, and the story announced that an informational open house would be held the next day.[11] Supporters included Wisconsin Governor Jim Doyle and Dane County Executive Kathleen Falk.[12]

Over a hundred people attended the November 30 meeting. UW representatives faced "many passionate, sometimes harsh" questions. "So, are we talking about absolutely the worst, the most infectious, terrible diseases in the world coming right here to the heartland?" one resident asked.[13] If answers were given, they apparently didn't go beyond broad claims about the need for and safety of the research.

Chairman of Dunn's town board Ed Minihan expressed concern about things like runoff into the lake, traffic congestion, accidental releases, and attacks by homegrown terrorists. "It makes me very nervous," he said.[14]

UW representatives responded to the concerns with some of the standard reassurances: everything leaving the facility would be sterilized first; look at how "safely" similar facilities at the CDC and Detrick had been operated; and Dane County residents would have plenty of time to ask questions later, should the decision be made to build in Dunn.[15]

Among those asking questions was 46-year-old patent attorney George Corrigan. University officials' inability to provide answers made him dig further. By December 7, he had put up a Web site, www.stopnbafkegonsa.com. That same month, the three-member Dunn Town Board voted unanimously to oppose the facility.[16]

A SMALL SKIRMISH WITH THE ACADEMIC-MILITARY-INDUSTRIAL COMPLEX

ON NOVEMBER 22, 2006, JUST BEFORE THANKSGIVING, FACULTY members at Western Kentucky University received an E-mail from WKU political science professor and Honors Program director Dr. Craig Cobane, inviting us to a talk by one of his former students, the recipient of a Homeland Security undergraduate scholarship. He advised us that Homeland Security offered students undergraduate scholarships and graduate fellowships consisting of full tuition, handsome monthly stipends, and paid summer internships "in a national laboratory."

I was deeply troubled by this tea-and-cookies Honors Program approach when I learned what the undergraduate had been up to: "analyzed recruitment potential of university professionals into a biological agent program; created representations of terrorist relationships using network analysis software; and researched government and private facilities relevant to a possible biological program."

I sent a two-page Reply All response to the E-mail indicating that while I intended no criticism of Dr. Cobane or the student, I was "troubled both by some of the research the student performed over the summer, and by the cozy relationship which seems to be developing between our nation's universities (this one included) and the Department of Homeland Security." I then explained my involvement with the No Ky Biolab group, the problems that biodefense proliferation poses, and some of the specific problems the Livermore facility posed for nearby residents.

I received almost unanimously positive responses from a dozen or so faculty, with a bit of demurring from two conservatives about my criticisms of the Bush administration. However, I also received an E-mail from the university president stating that while diverse views are valuable, he and I "could not be farther apart" "on this one." How could our homeland securities and technologies be developed in the absence of the world's greatest minds? Furthermore,

> the opportunity to bring the biolab you describe to Kentucky represents the greatest scientific venture this state has ever pursued. . . . To have it in Kentucky would create numerous opportunities for our faculty and students to be directly engaged in meaning research and learning [sic]. Few things worthwhile are without risk, and this one certainly has risk. Infectious diseases bring risk, some of them by design by parties which want to bring harm to America and Americans. Aren't we glad our nation's government and our best universities are at work trying to address the risk of such threats? Without that, you and I would have little to be encouraged about.

By this time, I'd had a close look at what our government and our "nation's greatest minds" were up to, and I obviously wasn't encouraged at all. Since the president cc'ed Dr. Cobane on the E-mail, I suspected the real purpose of his communication was to give Dr. Cobane a pat on the back and encourage him to continue pursuing such lucrative opportunities for Western's budding great minds.

I wrote back to the two conservative faculty members—I agreed with some of their rejoinders but not all—but I ignored the president's E-mail. One needs some sort of serious effort at thought—something more than a collection of platitudes about our "nation's greatest minds" and "best universities"—if one is to try grappling with it. The slippery slope of pious platitudes repels all efforts at engagement.

CHAPTER FIVE

Tales of Anthrax
Your Biodefense Dollars at Work

We dance round in a ring and suppose,
But the Secret sits in the middle and knows.
　　　　　　　—Robert Frost,"The Secret Sits"

"I think a lot of good has come from it [the anthrax
attacks]. From a biological or medical standpoint,
we've now five people who have died, but we've put
about $6 billion in our budget into defending against
bioterrorism."
　　　　　　　—Col. David Franz, former Commander,
　　　　　　　　USAMRIID, ABC News, April 4, 2002

Fear Factor

Twenty-two people were sickened and five died in the most serious bioterrorism incident on U.S. soil, the anthrax letter attacks of 2001.

　　The attacks came in two waves, involving seven or eight envelopes, four recovered by investigators. The envelopes included crudely written notes, poor spelling, and references to Allah, like this:

09-11-01. THIS IS NEXT. TAKE PENACILIN NOW. DEATH
TO AMERICA. DEATH TO ISRAEL. ALLAH IS GREAT.

Or this:

*We have anthrax. You die now. Are you afraid? Death to
America. Death to Israel. Allah is great.*

In the days after 9/11, someone clearly wanted Muslims suspected.
The first wave went to selected media outlets shortly after September 11; three weeks later, letters went to Senate Majority Leader
Tom Daschle and Judiciary Committee Chairman Patrick Leahy. Both
were Democratic senators wrestling with the Bush administration
over proposed details of the Patriot Act.

John Ezzell, the USAMRIID scientist who first examined the
Daschle anthrax under the microscope, called it "weaponized"; Major
General John Parker, the commanding officer at USAMRIID, then
decided "professionally done" and "energetic" were better terms for
mass consumption. Considerable debate would occur from that time
forth over whether the anthrax was technically "weaponized."

DNA testing eventually narrowed the short list of potential source
laboratories to four or five—three of them (Dugway, Battelle, and Fort
Detrick) important parts of the U.S. biodefense effort. In early 2002,
Dugway admitted it had been producing small amounts of
"weaponized anthrax" for biodefense testing.

All three biodefense facilities tested samples and otherwise assisted
the FBI with its investigation—even as the FBI administered polygraphs to their scientists. Equally ironic, all three facilities took part,
in a big way, in the subsequent biodefense expansion.

The timing of the attacks—and the genealogy of the anthrax—
caused some to suspect a conspiracy from within the U.S. government, including Francis Boyle, the University of Illinois law professor
who had drafted the Biological Weapons Anti-Terrorism Act of 1989
for the first Bush administration. In a 2005 book, *Biowarfare and Terrorism*, Boyle charged:

I believe the FBI knows exactly who was behind these terrorist
anthrax attacks upon the United States Congress in the Fall of

2001, and that the culprits were U.S. government-related scientists involved in a criminal U.S. government biowarfare program that violated both the BWC and U.S. domestic legislation
implementing the same. For that reason, the FBI is not going
to apprehend and have indicted the culprits.[1]

A little-noticed *New York Times* article that appeared just before
September 11, revealing that U.S. government labs were involved in
some highly questionable research, including developing genetically
modified, deadlier strains of anthrax, helped establish some basis for
such suspicions. In addition, Dugway had been developing
"weaponized" anthrax—small amounts, it said—for some time. In
the midst of the Amerithrax investigation, Dugway "acknowledged
producing small quantities of dry anthrax and shipping it in paste
form to a few unspecified locations."[2]

Soon after the FBI made its allegations about Ivins, Glenn Greenwald began blogging at Salon.com about the unanswered questions
surrounding the attacks and the FBI's investigation. He reminded
readers just how significant the attacks had been:

> After 9/11 itself, the anthrax attacks were probably the most
> consequential event of the Bush presidency. One could make
> a persuasive case that they were actually more consequential.
> The 9/11 attacks were obviously traumatic for the country,
> but in the absence of the anthrax attacks, 9/11 could easily
> have been perceived as a single, isolated event. It was really
> the anthrax letters—with the first one sent on September 18,
> just one week after 9/11—that severely ratcheted up the fear
> levels and created the climate that would dominate in this
> country for the next several years after. It was anthrax—sent
> directly into the heart of the country's elite political and
> media institutions, to then-Senate Majority Leader Tom
> Daschle (D-SD), Sen. Pat Leahy (D-VT), NBC News Anchor
> Tom Brokaw, and other leading media outlets—that created
> the impression that social order itself was genuinely threat-
> ened by Islamic radicalism.[3]

In fact, however, the attacks were an "autoimmune disease," launched from within our own biodefense establishment.

Trust the Government, Not: The Spin

The government's early responses to the anthrax attacks should trouble all of us confronted by current reassurances about the safety of biodefense research. At crucial points during September and October 2001, the government withheld information from the public to prevent or reduce panic. And "experts" at the CDC and elsewhere—with their confident assertions that postal workers were not endangered by the anthrax-laden envelopes that passed through their facilities—showed how experts sometimes get it wrong, and assume they know things they don't, such as the degree of danger that a situation presents.

From the very beginning, a greater value was placed on avoiding panic than on fully informing the public. After the first anthrax death—of Globe Communications photo editor Bob Stevens—and the discovery of anthrax spores in the AMI building where he worked, "Federal and local officials told the public that the anthrax release was limited to a single building and that the strain of anthrax involved in Stevens's death appeared to be a naturally occurring one. Behind the scenes, however, Popovic's [CDC] laboratory had identified the strain as Ames, the one used in U.S. military labs."[4]

In Florida, "confusion reigned" as multiple agencies—federal, state, and local—"scrambled to decide who was in charge and just how much the public needed to know."[5] Florida public health officials wanted to get as much information to the public as possible, including the name of the strain which had been identified. Then they got a call from Florida Governor Jeb Bush, citing national security interests.[6]

The CDC, also, was under an early gag order. Lack of information helped to increase fear. A former past president of the American Society of Microbiology, Abigail Salyers, blaming the gag order for the fear, thought: "It's the U.S. government."[7] Her thoughts had a deeper relevance than she then realized.

The experts, being experts, knew what they knew. They mistakenly believed anthrax couldn't re-aerosolize, for instance.[8] The CDC told the USPS and state health authorities that the chances of anthrax escaping a sealed envelope were next to none. "It never dawned on the CDC officials that refined anthrax could simply seep through the pores of a standard envelope."[9]

At crucial points, CDC experts were skeptical of the tentative anthrax diagnoses made by treating physicians, including those for Bob Stevens, Ernesto Blanco,[10] and Leroy Richmond.[11] Dr. Elsie Jones had trouble getting someone to listen to her fears that Teresa Heller had anthrax.[12]

The CDC's diagnostic skepticism probably did not significantly affect any of the ultimate outcomes. The same could not be said for the skepticism of some treating physicians. The doctor who first saw postal worker Thomas Morris dismissed his concerns he might have anthrax, and sent him home. Three days later Morris called 911. He died later that evening.[13]

Joseph Curseen was diagnosed with gastroenteritis and also sent home. His wife called 911 the next day, but he, too, died the same day he was brought into the hospital.[14]

One bit of expert advice that would have helped—a Canadian study showing the ease with which anthrax might spread from an envelope —sat in an unopened E-mail.[15] So while congressmen and their staffers underwent nasal swabs and were put on Cipro, postal workers were treated to a public relations event. To demonstrate that there was nothing to fear, no reason to lose billions of dollars by shutting down facilities—and to show his trust in the CDC's conclusion that 2400 Brentwood employees didn't need preventive antibiotics, Postmaster General John E. Potter held a press conference at Brentwood. A few days later, after the deaths of Brentwood workers Thomas Morris and Joseph Curseen, he joined other postal employees in the Cipro line.[16]

After the postal workers' deaths, President Bush met with postal and union representatives at the White House, telling them they had been drafted into "an unprecedented war on terrorism."

Is This Weapon "Weaponized?"

Early on, the refinement and characteristics of the Daschle/Leahy anthrax caused astonished scientists to call it "weaponized." In her book *The Killer Strain*, Marilyn Thompson offered multiple examples of what weaponized anthrax was like. There was former USAMRIID bioweaponeer Bill Patrick's weaponized-anthrax simulant, for instance: "a white powder finer than confectioner's sugar, so airy that with each slight turn of the bottle it burst into movement and rebounded off the sides of the jar."[17]

Before he received the Daschle letter for testing, current USAMRIID employee John Ezzell "had yet to see *B. anthracis* in its most deadly and scientifically exciting form, the highly dispersible free-floating weaponized powder produced by Fort Detrick's bioweaponeers before Nixon pulled the plug on germ warfare."[18] Now, when Ezzell opened the Daschle letter, "Out burst a spore powder so pure that it evaporated in midair."

In a politically incorrect rush of frankness or alarm, Ezzell used the term "weaponized anthrax" in speaking to the White House about the "astoundingly fine aerosol material—more easily dispersed than any he had ever seen." In a later meeting, Peter Jahrling would choose his words more carefully, describing the letter anthrax as "professionally done" and "energetic."[19] USAMRIID's commanding officer, Major General John Parker, would use the more carefully chosen terms in hearings on Capitol Hill.[20]

Other early evidence of "weaponization" was the presence of silica, an additive used in past U.S. biological warfare programs to make the anthrax disperse better, fly about in the air so it would be more likely to be inhaled.[21]

The semantic dispute over whether the anthrax was "weaponized" had implications for who or what could have produced the anthrax. The high refinement typical of weaponization, and the addition of silica, both implied a state-run program involving more than one person. Since microbial forensics had already, more or less reliably, associated the Amerithrax anthrax with U.S. government laboratories, weaponization and silica strongly suggested some sort of U.S. government conspiracy.

An FBI researcher would publish a 2006 article explaining why the anthrax was *not* weaponized.[22] Other scientists would challenge him to publish his data, saying he could not reliably reach his conclusions based only on the data he had just provided.[23]

At the "science" press conference, scheduled by the FBI after the first Ivins press conference left huge numbers of skeptics, the FBI said the silica found in the Daschle/Leahy letters had been naturally absorbed by the spores. The finely milled powder everyone had marveled at, and called high-grade or weaponized, was simply the result of the envelopes' passage through the sorting machines.

Peter Jahrling, the senior USAMRIID scientist who had called attention to the silica additive in an October 2001 White House meeting, now recanted and said he had been mistaken.[24]

U.S. Rep. Roscoe Bartlett, who represents Frederick, Maryland in Congress (and holds a Ph.D. in physiology), called the mail sorter explanation "patently ridiculous." He also said he was skeptical that Ivins had sent the anthrax.[25]

At a House committee hearing in the fall of 2008, Representative Jerrold Nadler asked FBI director Robert Mueller to provide the relative weight of the silica in the attack powder. Mueller "didn't know" the answer to that question; he'd have to get back to Nadler on that.[26]

Nadler said "only a handful of laboratories" could achieve a silica content of over 1 percent. Had the FBI investigated all of these labs—and how had it ruled them out as potential sources?[27]

Mueller said he'd have to get back to Nadler on the second part of that question as well.[28]

One thing was clear: this was not your mother's anthrax.

Unless your mother worked in a government lab.

Nor was this I-dug-up-this-here-dead-cow-leg-and-refined-it-in-my-barn-lab anthrax.

By focusing on the semantics of whether the anthrax had been "weaponized," says *Killer Strain* author Marilyn Thompson, the Bush administration missed the most important consideration: the anthrax's dangerousness.

Sorry, postal workers. You've been drafted into an "unprecedented war on terrorism."

Trust the Government, Not, 2: Blaming It on You Know Who
In August 2008, Glenn Greenwald reminded readers of a pivotal 2001 ABC story.[29]

During the last week of October 2001, ABC News in general, Brian

Ross in particular, ran repeated stories that government tests showed bentonite in the Daschle anthrax, and that this pointed to Iraq as the source. As Greenwald pointed out: "ABC News' claim—which they said came at first from 'three well-placed but separate sources,' followed by 'four well-placed and separate sources'—was completely false from the beginning. . . . **No tests ever found or even suggested the presence of bentonite**. The claim was just concocted from the start" [emphasis in original].

The "four sources" obviously fed ABC News completely false information—and since the sources said the alleged tests had been conducted at Fort Detrick, those sources likely also had some connection with Detrick. "It's extremely possible," Greenwald said—"one could say highly likely—that the same people responsible for perpetrating the attacks were the ones who fed the false reports to the public, through ABC News, that Saddam was behind them."

A number of other media outlets followed ABC in pushing the notion that Iraq was behind the attacks. The *Washington Post's* Richard Cohen, who initially supported the invasion of Iraq but later regretted it, said in 2004 that anthrax had played a major role in his 2002 thinking: in his mind, Saddam Hussein was linked with anthrax, and, by way of the anthrax, to September 11.

Greenwald said this sort of Iraq-anthrax-September 11 linkage was a common one in the months leading up to the Bush administration's Iraq invasion. ABC News never retracted its story, just continued to note that the White House denied the bentonite reports. (Everyone knew the White House had to officially deny them, of course; the real story was in the behind-the-scenes "facts" that had been ferreted out by the country's diligent TV journalists, with the aid of helpful government insiders.) "And thus, the linkage between Saddam and the anthrax attacks—**every bit as false as the linkage between Saddam and the 9/11 attacks**—persisted" [emphasis in original].

"Surely," Greenwald insisted, "the question of who generated those false Iraq-anthrax reports is one of the most significant and explosive stories of the last decade." And ABC knew, he said, who concocted the false bentonite reports in an effort to link the anthrax to Iraq and conceal the true culprit(s), yet the network continued to conceal this information. Greenwald reacted scornfully to the notion of "source" protection under these circumstances: "The people who fed them the bentonite story aren't 'sources.' They're fabricators and liars who purposely used ABC News to disseminate to the American public an extremely consequential and damaging falsehood. But by protecting the wrongdoers, ABC News has made itself complicit in this fraud perpetrated on the public, rather than a news organization uncovering such frauds."

On October 18, 2001, on the very heels of the attacks, John McCain appeared on the David Letterman Show and said Iraq was "the second phase" after Afghanistan, because: "There is some indication, and I don't have the conclusions, but some of this anthrax may—and I emphasize may—have come from Iraq."

Three days later, McCain and Joe Lieberman both suggested on *Meet the Press* that we would have to attack Iraq. Lieberman said of the anthrax: "There's either a significant amount of money behind this, or this is state-sponsored, or this is stuff that was stolen from the former Soviet program."

Two months after the ABC reports, in his January 2002 "Axis of Evil" State of the Union address, President Bush continued the Iraqi-anthrax linkage by declaring: "The Iraqi regime has plotted to develop anthrax, and nerve gas, and nuclear weapons for over a decade."

In February 2003, Secretary of State Colin Powell put the final touches on the Iraqi-anthrax "fiction" when he appeared before the UN Security Council to condemn Saddam Hussein's alleged cache of biological weapons. Powell, who had argued against the Iraqi invasion within the administration but played the obedient soldier in public, had been seriously misled by the CIA, who assured him that "intelligence" about Iraqi WMDs was credible. Powell now assured the United Nations that: "We have firsthand descriptions of biological weapons factories on wheels and on rails."[30] He held up a glass vial of powder and told the audience that less than a teaspoon of dried anthrax had shut down the United States Senate.[31]

Long after the invasion, multiple sources would reveal that intelligence professing to document Iraq's continued possession of WMDs had been deliberately "cooked"—and that the deception extended from the president downward.

In August 2008, the *New York Daily News* reported that, in the immediate aftermath of the attacks, the White House had pressured FBI Director Mueller to "prove" the attacks were a second-wave assault by Al Qaeda.[32]

Ron Suskind's book *The Way of the World* quoted intelligence officials who said the White House had first ignored—several months before the Iraq invasion—intelligence that Saddam Hussein no longer possessed weapons of mass destruction, hadn't possessed them since 1996—and then, after the Iraq invasion proved the intelligence to be inaccurate, concocted a fake, backdated letter from the head of Iraqi intelligence to Hussein, stating that 9/11 ringleader Mohammed Atta had actually trained in Iraq.[33]

When I learned in 2006 that the 2001 anthrax had been traced to a small number of U.S. military labs (something I'd not previously known because media that happily speculated about the possible role of Iraq went on vacation once the anthrax's domestic origins became clear), I immediately wondered whether those attacks might have been some sort of sinister planned "inside job." By 2006, some of the Bush administration's fabrications about Iraq had become known. Taking us to war under false pretenses had resulted in the unnecessary deaths of thousands of U.S. soldiers and many more Iraqis. I believed an administration capable of that sort of deceit might willingly sacrifice a few innocent civilians to bolster its "unprecedented war on terrorism."

Trust the Government, Not, 3: The Amerithrax Investigation
When the FBI declared that dead suspect Bruce Ivins was the sole person responsible for the anthrax attacks—and suggested it was preparing to close a case that it would, regretfully (it said), not be able to try in a court of law—more than a few people were suspicious.

Whether you regarded the FBI as the Keystone Kops or the KGB, you couldn't help but notice their prior false steps. Depending on your inclination, you could doubt their competence, their veracity, or both.

For six years, the Bureau had tried the wrong man—or so it now said—in absentia: not in the absence of the man, but in the absence of a trial, using media leaks, ongoing harassment, and the subtle innuendo of the phrase "person of interest."

Just two months earlier, the U.S. government had agreed to a $5.8 million settlement with Steven Hatfill, the only announced person of interest during those six years. Yet in reaching the settlement, the government acknowledged no wrongdoing, no violation of the Privacy Act, nothing.[34]

After all, pressuring suspects was a time-honored method of criminal investigation. In this instance, it resembled the ancient trial by ordeal. Steven Hatfill had been thrown into the water with a millstone about his neck, and had floated back up to the top: ergo, not guilty. Ivins had drowned under the same treatment: ergo, guilty.

And once again the FBI substituted leaks and innuendo for an actual trial. By the time the Bureau got around to releasing some of its evidence, a series of piecemeal leaks had outed Ivins's possible motives, miscellaneous eccentricities, and evidences of mental shakiness, indicting him in the public eye, if not before a grand jury, as the sort of unstable "mad scientist" who might have committed the attacks on his own:

> Ivins at one time maintained a mailbox under an assumed name where he received pornographic magazines. He had once been "obsessed" with a Princeton sorority because of a failed college romance, and the Princeton mailbox where one of the letters originated was located within 100 yards of a storage facility used by the sorority—in a location Ivins apparently last visited 27 years ago. He drank. He made homicidal statements to a mental-health support group. He wrote rambling letters to the editor of his local paper.[35]

Scott Shane's January 4, 2009 *New York Times* article, representing itself as "the deepest look so far at the investigation," lavished attention on Ivins's "troubled life" even as it concluded that "unless new evidence were to surface, the enormous public investment in the case would appear to have yielded nothing more persuasive than a strong hunch, based on a pattern of damning circumstances, that Dr. Ivins was the perpetrator."[36]

Shane's article cited both a former graduate school colleague and the FBI as suspecting Ivins because of safety breaches. The former colleague noticed—in a picture Ivins said showed him examining anthrax samples—that he wasn't wearing gloves—"a safety breach she thought showed an unnerving 'hubris.'"[37]

Then in December 2001 Ivins had swabbed for spores outside a containment area and wiped the areas with bleach, and failed to report the suspected spill to superiors. Apparently this "flagrant breach of biosafety standards" raised the FBI's interest in Ivins—or at least it did so several years later, after Steven Hatfill's value as an earlier stalking horse had diminished.[38]

Unfortunately, as this book demonstrates, flagrant breaches of biosafety standards—or "hubris" about safety matters—are much more frequent than we would like. Even at USAMRIID, apparently. And that certainly applies to anthrax matters: see "Bill and Ted's Excellent Secret Adventures with Anthrax," below.

Moreover, as interested as the FBI professed to be in the containment breach, it never examined samples from that contamination, a decision anthrax sleuther Paul Keim called "weird."[39]

Neglecting obvious promising evidence had been a hallmark of the FBI's early investigations. The *Hartford Courant* noted the FBI's failure to follow up on one particularly suggestive matter. Someone had sent the FBI an anonymous letter identifying Ayaad Assaad, an Egyptian-born former USAMRIID biologist, as a member of a terrorist cell possibly linked to the anthrax attacks. Assaad was quickly exonerated, but the FBI apparently made no effort to track down the letter's author.

The letter had been "sent prior to the arrival of the anthrax letters, implying foreknowledge of the attacks, and its language was similar to that of the deadly mail. Moreover, it displayed an intimate knowledge of USAMRIID operations, suggesting it came from within the limited ranks of Fort Detrick researchers."

Assaad had named two people he thought likely to carry a grudge against him. One had been videotaped making after-hours trips to the labs. But there is no indication that the FBI investigated either person.

The FBI refused to make a copy of the letter publicly available, or to give one to Assaad.[40]

Then there was the Iowa State incident. At the time of the attacks, Iowa State (located in Ames, Iowa) maintained a comprehensive collection of some 100 vials of the Ames strain, amassed since 1928. When ISU researchers, fearful of a terrorist theft, offered to destroy the cultures, the FBI did not object. According to one critic of the action, "It should have been preserved as evidence. This was a roadmap of everybody and anybody that had gotten access [directly from Iowa State?] to develop the [strain] that hit Leahy and Daschle."[41]

In March 2008, FOX News reported (and they did this during the Bush presidency, bit of a small miracle) that the FBI had narrowed its focus to "about four" suspects—and that among the pool were three scientists linked to USAMRIID: a former deputy commander, a leading anthrax scientist, and a microbiologist.[42]

If that weren't enough public embarrassment for USAMRIID, there was more. In December 2001, USAMRIID's commanding general had tried to dispel the idea of a Detrick connection by saying that the Institute used only liquid anthrax, not powder, for its experiments. But FOX said it had obtained an E-mail in which USAMRIID scientists discussed the fact that the powder being examined was nearly identical to a colleague's:

> Then he said he had to look at a lot of samples that the FBI had prepared . . . to duplicate the letter material. Then the bombshell. He said that the best duplication of the material was the stuff made by [name redacted]. He said that it was almost exactly the same . . . his knees got shaky and he sputtered, "But I told the General we didn't make the spore powder!"[43]

When the FBI made the Ivins allegations four months later, some wondered how the other suspects had been eliminated so quickly.

Then there was the curious fact that many Washington insiders were being told to take Cipro soon after September 11. At Salon.com, Greenwald pointed out this significant revelation in a March 2008 article by the *Washington Post's* Richard Cohen: "The attacks were not entirely unexpected. **I had been told soon after Sept. 11 to secure Cipro, the antidote to anthrax. The tip had come in a roundabout way from a high government official, and I immediately acted on it.** I was carrying Cipro way before most people had ever heard of it" [emphasis in Greenwald].

Greenwald said many well-connected journalists were popping Cipro at the time because government sources had told them they should. "Leave aside," he said, "the ethical questions about the fact that these journalists kept those warnings to themselves. Wouldn't the most basic journalistic instincts lead them now—in light of the claims by our government that the attacks came from a Government scientist—to wonder why and how their Government sources were warning about an anthrax attack?"[44]

Though the major TV networks paid little attention to the Ivins revelations beyond briefly reporting (and accepting without question) the FBI's statements that the case was closed, several print publications editorialized on the weakness of the FBI's evidence and the need for independent review: *Washington Post* (8-7-08); *New York Times* (8-7-08), (8-19-08); *Nature* (8-21-08).

After first floating bits and pieces of its case through leaks, the FBI gave a quick "Amerithrax Press Conference for Dummies." That left way too many skeptics, so the FBI tried a "scientific" press conference. Some major congressmen weren't satisfied, so FBI Director Mueller appeared before the House and Senate Judiciary Committees, where he professed not to know the answers to a lot of crucial questions. All these events raised skepticism and questions among multiple critics. The FBI then said it would have its "science" reviewed by the National Academy of Sciences.

The problems with the NAS review were threefold: (1) the FBI was controlling the questions it wanted the NAS to look at; (2) the FBI had already co-opted (and sworn to confidentiality) 60 experts for its earlier investigations, leaving concern about who was still available, and how expert or independent they would be; and (3) the NAS inquiry would only look into scientific questions. One very crucial—and apparently nonscientific—question for many skeptics was how the FBI eliminated the "more than 100 other people" who had access to the alleged telltale flask.

One expert called the FBI's NAS "review" "a nice little jujitsu move" to deflect attention from nonscientific questions. Alan Pearson of the Center for Arms Control and Non-Proliferation said of one FBI question, that it wasn't relevant to ask whether "*Bacillus anthracis* samples dried with a rudimentary methodology can pose an inhalation hazard resulting in pulmonary anthrax." [Duh, even infected goat skin or drum heads can do that.] Of course they [the samples] can. The question is "whether [this method] can produce anthrax like that found in the letter."[45]

There were multiple holes and gaps in the FBI's case. Investigators had failed to place Ivins in New Jersey on the September and October 2001 dates when the letters were supposed to have been mailed from Princeton. They had swabbed his "residence, locker, several cars, the tools in his laboratory, and his office space, but found no trace of anthrax that genetically matched the bacteria in the letters."[46]

The *New York Times* reported that laboratory records revealed that the RMR-1029 anthrax supply had been stored in a building other than Ivins's during part of the time between 1997 and 2001. This reportedly doubled or tripled the number of people at USAMRIID alone with access to the "killer strain."

At the Senate Judiciary hearing, Senator Leahy said if Ivins had indeed been involved in the letter attacks, he didn't believe he'd acted alone.

In his August 3 blog, Greenwald listed about a dozen "unanswered questions" that needed to be explored by a full-scale independent investigation, including these:

> Why were White House aides given Cipro weeks before the anthrax attacks, and why "on the night of the Sept. 11 attacks, [did] the White House Medical Office dispense Cipro to staff accompanying Vice President Dick Cheney as he was secreted off to the safety of Camp David"?

Which "high government official" told Richard Cohen to take Cipro prior to the anthrax attacks?

Did the FBI meaningfully investigate who sent an anonymous letter to the FBI after the anthrax letters were sent, **but before they were made public**, accusing a former Fort Detrick scientist—the Arab-American Ayaad Assaad—of being a "potential biological terrorist," after Assaad was forced out of Fort Detrick by a group of USAMRIID bioweapons researchers who had exhibited extreme anti-Arab animus? [Emphasis added.]

Why did the FBI give its consent in October, 2001 for the remaining samples of the Ames anthrax strain to be destroyed, thereby losing crucial "genetic clues valuable to the criminal inquiry?"

If—as was publicly disclosed as early as 2004—Bruce Ivins' behavior in 2001 and 2002 in conducting unauthorized tests on anthrax residue was so suspicious, why was he allowed to remain with access to the nation's most dangerous toxins for many years after, and why wasn't he a top suspect much earlier?

If it's really the case—as principal Ivins antagonist Jean Duley claims—that Ivins, as far back as 2000, had "actually attempted to murder several other people, [including] through poisoning" and had threatened to kill his co-workers at his Fort Detrick lab, then why did he continue to maintain clearance to work on biological weapons, and why are his co-workers and friends, with virtual unanimity, insisting that he never displayed any behavior suggestive of being the anthrax attacker?

What was John McCain referencing when he went on national television in October, 2001 and claimed "there is some indication, and I don't have the conclusions, but some of this anthrax may—and I emphasize may—have come from Iraq"?

What was Joe Lieberman's basis for stating on national television, three days after McCain's Letterman appearance and
in the midst of advocating a U.S. attack on Iraq, that the
anthrax was so complex and potent that "there's either a significant amount of money behind this, or this is state-sponsored, or this is stuff that was stolen from the former Soviet
program"?

What did Pat Leahy mean when he said the following in a
September, 2007 interview:

> I don't think it's somebody insane. I'd accept everything
> else you said. But I don't think it's somebody insane. **And
> I think there are people within our government—cer-
> tainly from the source of it—who know where it came
> from. [Taps the table to let that settle in] And these
> people may not have had anything to do with it, but
> they certainly know where it came from.** (Emphasis by
> Greenwald)

Who were the "four separate and well-placed sources" who
told ABC News, falsely, that tests conducted at Fort Detrick
had found the presence of bentonite in the anthrax sent to
Tom Daschle, causing ABC News to aggressively link the
attacks to Iraq for five straight days in October, 2001?

Who was responsible for the numerous leaks even before
the ABC News bentonite reports linking the anthrax attacks
to Iraq?[47]

In September 2008, Representative Rush Holt, D-NJ, introduced a
bill (re-introduced in 2009) to create an independent commission to
review the anthrax investigation.[48] Some of his congressional colleagues questioned the need for an independent investigation, suggesting that a congressional investigation would suffice. Greenwald
questioned whether Congress was up to the task itself.

Trust the Government, Not, Russian-Style:
A 1979 Anthrax Chernobyl[49]

Standard Operating Procedure. Turn on the filtering switch. Then turn on the ventilation switch. On the afternoon of April 2, 1979, someone at Compound 19 forgot to follow the Standard Operating Procedure.[50]

A plume of vented air, containing a gram or less of anthrax spores, drifted southeast, borne along by fifteen-kilometer-per-hour winds. There was nothing to see, or smell, except the odd leaf still drifting after winter, or the random bit of paper lifted from the street.[51]

Why should this day be different? Why should I not breathe in, breathe out? Why should I not welcome the season into my lungs, as I had on each of my thousands of days of living?

The plume drifted through the windows of the nearby ceramics factory, drifted through the yards of Chkalovsky, then ambled over the countryside, over flocks of sheep, herds of cattle. The cattle and sheep breathed from the plume, and went on grazing.[52]

All governments find it easier if their people act like cattle (the placid, cud-chewing kind, not the half-crazed longhorns) and sheep. Leave it to the experts. Don't interfere with things you know nothing about. We will protect you. There is nothing to fear. Find your bit of pasture and browse it. Eat, drink, breathe, procreate. Make more cattle. Make more sheep.

The sheep began dying first, three days after the accident, in villages south of Sverdlovsk. This made it easier, when the people started dying, for authorities to blame it on anthrax-infected meat.

On April 4, two days after the accident, some of the people were sick. But they either stayed away from doctors hoping to get better on their own, or the doctors told them they had the flu and sent them home.[53]

Dr. Yakov Klipnitzer, the head of Hospital 20 in Chkalovsky, was a patient there himself, with nephritis, when one of his physicians warned him about the new arrivals and sudden deaths at Hospital 20, Hospital 24, and Hospital 1.[54]

Dr. Margarita Ilyenko was the Director of Hospital 24, a smaller hospital which often referred patients to Hospital 20. Klipnitzer called to tell her two of her referrals were dead. Three more patients at Hospital 20 died suddenly the next day. On the third day, three more patients died at Hospital 24.[55]

On April 11, an autopsy of one of the bodies confirmed the cause of death as anthrax.[56] Over a hundred autopsies would be conducted over the next two months. Sixty-four anthrax deaths would be confirmed.

The local health authorities contacted Moscow, and Dr. Vladimir Nikiforov, Director of the Infectious Diseases Department at Moscow's Central Postgraduate Institute at the Botkin Hospital, was sent to take charge. He arrived in Sverdlovsk April 12, launched a massive vaccination program, and told the local physicians that infected meat was the outbreak's cause.[57]

The authorities at Compound 19 said nothing, gave no clue. Though they claimed to be working on an improved vaccine for anthrax, they offered no help. They gave no warning, claimed no responsibility.[58]

Investigators and public health authorities descended on the village of Abramovo, where several sheep and a cow had died of anthrax between April 5 and April 10. One investigator approached the owners of the first dead sheep, demanding to know who had sent them "the bacteria." Sheds were ripped up and carted away; houses and yards disinfected; antibiotics dispensed. No one was allowed to leave the village for two weeks.[59]

The investigators treated the "ignorant" villagers roughly, as though they were responsible for the outbreak.

An anonymous article published in the Russian dissident broadside *Posev* in January 1980 first brought the incident to the attention of the West. The article said there had been a tremendous explosion of anthrax in Sverdlovsk, with 30 to 40 people dying each day for over a month, and causing over a thousand deaths in all.[60]

The Soviets first denied there had been any outbreak at all. Then they said an epidemic had occurred, but it had been caused by infected meat. In 1986, Matthew Meselson, a Harvard biologist and one of the experts the U.S. government had consulted about the incident, was invited to Moscow for a briefing by the three public health officials who had taken charge of the outbreak. They presented their evidence for the tainted meat explanation to Meselson, and later to a delegation from the U.S. National Academy of Sciences. In 1988, two of the physicians came to the U.S. and spoke at the National Academy of Sciences, at the Johns Hopkins School of Public Health, and at the American Academy of Arts and Sciences. "Their detailed presentation of the infected-meat scenario was judged plausible, even persuasive, although it lacked substantial clinical and epidemiological evidence."[61]

In 1992, an American team led by Meselson was allowed inside Russia in an effort to determine the cause of the outbreak for certain. General A. T. Kharechko, the current commander at Compound 19, had continued to deny that the facility was involved.[62] Major-General Valentin Yevstigneyev, who directed Compound 19 for a period after 1984, also doubted the facility's involvement, citing the small amounts of anthrax used and the same high-tech safety features that American biodefense propagandists repeatedly insist are 99.999% foolproof:

> [A]erosol chambers [Yevstigneyev claimed] were used there to test velvet monkeys and baboons injected with STI anthrax vaccine. The maximum used in such a test, he insists, was no more than forty milligrams (0.04 grams, around 40 billion spores), and he believes that no such challenge tests were done, at least according to records, in late March or early April 1979. Besides, as in the United States, the laboratory was maintained at a P3 safety level. Its rooms were hermetically sealed, with negative pressure recorded by manometers; a special filter and ventilation system controlled the release of any aerosols from the chamber. The filtered air, Matthew is told, was released at roof level, three to four meters above ground.[63]

During the outbreak, the KGB, doing its part for "community relations," had come to town and taken away most of the relevant documents. In Moscow, Boris Yeltsin's Councilor of the Environment and Health, Alexey Yablokov, told the American investigators that his research assistant, investigating the outbreak on the orders of Yeltsin, had discovered only an empty KGB folder on Sverdlovsk, labeled

Order to Confiscate all Documents Connected with Military Activity. The contents had apparently been destroyed in December 1990 by a top secret order of the Council of Ministers.[64]

Dr. Faina Abramova, however, who directed the Sverdlovsk autopsies, had hid the pathological specimens away some place the KGB wouldn't look. Examining the specimens showed the Americans that the anthrax was inhalatory; confirming and mapping most of the deaths, and checking the wind speed and direction for that day, showed that the anthrax spores had escaped from Compound 19 through the air and spread southeast. Painstaking tabulation of people's whereabouts on April 2 enabled Meselson's team to pin down the time of release to a three-hour period on the afternoon of April 2.

Extrapolating from Fort Detrick monkey experiments, and the fact that only a small portion of the people exposed had actually succumbed to anthrax, the American team estimated—hesitantly—that while 8000 spores per person would be required to infect half the exposed population, only 9 spores per person would be required to infect 2% of the population. The team concluded that the Sverdlovsk deaths were probably caused by a release ranging from a low of 2 to 4 milligrams to a high of slightly more than half a gram.[65]

While they were waiting at the Helsinki airport for their flight back to the U.S., the Americans learned from an abandoned copy of the *International Herald Tribune* that Boris Yeltsin had publicly stated that the military were responsible for the 1979 outbreak. Unknown to the Americans, *Pravda* had carried an interview with Yeltsin just before their trip (and probably prompted by it), acknowledging that there had been biological weapons development in Sverdlovsk and that the military had been responsible for the 1979 outbreak.[66]

Bill and Ted's Excellent Secret Adventures with Anthrax

Long after the 2001 attacks perpetrated from one of the U.S.'s own anthrax facilities, the 350 or so other U.S. facilities registered to handle live anthrax continued to offer their host communities a wide range of unique thrills. In many cases, however, communities remained (and remain) unaware of the excitements in their midst. Most of the details are courtesy of a 2007 AP story, using information provided (leaked) by an unidentified source. The CDC, which receives official reports of "select agent" accidents when accidents are actually reported, routinely conceals them from the public.

2004: a whoops that did make headlines. The Southern Research Institute shipped several test tubes of supposedly "killed" anthrax— via a customary method, Federal Express—to the Children's Hospital Oakland Research Institute. After 49 mice injected with the anthrax died, culturing of the sample and the dead mice showed the anthrax was live. Seven Oakland researchers were treated with antibiotics for anthrax exposure. Not to worry, though, the Institute said: the anthrax was never airborne.[67]

The Oakland Institute behaved with unusual integrity and caution for a biodefense research institution, going public with the information and suspending the research.

"Southern Research spokeswoman Rhonda Jung said the company has never before had such a problem and does not yet know what went wrong.

"'We have well-seasoned and experienced researchers who are accustomed to working with this agent,' she said. 'It's my understanding we followed the standard procedure that has been the gold standard for working with this bacteria for 20 years.'"[68]

That's what can happen with the gold standard. Many new institutions lack "well-seasoned and experienced researchers." Silver standard? Bronze standard?

About a fourth of the CDC accident reports obtained by the Associated Press in fall 2007 involved anthrax. Most involved shipping or inventory malfunctions: "anthrax reported as missing due to poor record keeping"; "inventory discrepancy involving anthrax bacteria"; "poor record keeping blamed for empty vial of anthrax spores"; discrepancies between shipping invoices and amounts actually shipped.[69] These discrepancies may arise from poor record-keeping, but they could also, obviously, indicate theft or diversion. The great irony of biodefense gone amok is this: a diversion from America's own facilities prompted an expansion of those facilities, which has multiplied opportunities for similar diversions—and for such accidents as a graduate student breaking a flask of anthrax in the University of Mississippi Medical Center's BSL-3 lab.[70] Anthrax research at a medical center, conducted by a graduate student—not exactly suggesting a gold standard for anthrax security.

According to the 2007 AP story, several anthrax shipments have involved leaking packages.

Someone at Detrick turned the (negative air pressure?) fan off while working with anthrax; "worker exposure confirmed."[71] (Such protocol lapses perhaps account for the June 2007 appearance of anthrax bacteria outside the appropriate containment area at Detrick, on the freezer handle, light switch, and shoes in the changing room—places from which, presumably, anthrax spores might be carried into the outside world—and we all know what a fearsome thing Detrick anthrax spores can be.)

At the Midwest Research Institute in Kansas City, two lab workers opened a package with leaking tubes of anthrax on an open bench. The incident was noted in the 2007 AP summary; no other details are available.

October 2001: the Los Alamos National Laboratory, one of the nation's nuclear weapons laboratories, received an unauthorized shipment of virulent anthrax from Northern Arizona University in Flagstaff. The Project on Government Oversight, which discovered the incident, called it "the latest indicator of security problems at the nation's nuclear labs. The security breach also illustrates the ease with which scientific researchers can obtain the potentially lethal material."[72]

The *Albuquerque Journal* wrote that "LANL Owes An Explanation."[73] LANL, which only had a BSL-2 lab, was seeking a BSL-3 facility, probably for aerosolization experiments. Critics had pointed to the dangers of earthquakes, and to the suspicions aroused by conducting biodefense research in a nuclear weapons lab.

In 2005, the Department of Energy's Office of the Inspector General issued a report on a 2003 incident at Oak Ridge National Laboratory involving a guest researcher's work with "killed" anthrax spores. The Inspector General, investigating an anonymous complaint, found that the Guest Researcher had not been authorized to work on the specific anthrax spore project; the Guest Researcher had worked on the project in a laboratory not authorized for the project; the anthrax spores were not adequately controlled and secured (they were routinely left on a countertop or in an unlocked refrigerator, instead of being "*inventoried, controlled,* and *secured* in locked storage when not in use"); "key laboratory personnel typically involved with biological select agent projects were not *aware* of the anthrax spore project."[74]

Also in 2005, the *New Scientist* reported on Ed Hammond's discovery of a series of contracts at Dugway, asking companies to supply bulk quantities of a non-virulent *Sterne* strain of anthrax, and for fermentation equipment for producing 3000-liter batches of an unspecified biological agent. Alan Pearson said, "It raises a serious question over how the US is going to demonstrate its compliance with obligations under the Biological Weapons Convention if it brings these tanks online." The tanks could be used to grow the lethal Ames strain. Some feared the anthrax might be used for such "threat assessment research" as trying to determine how effectively anthrax is dispersed when released from crop-spraying aircraft or bombs. Dugway refused to tell the *New Scientist* what the anthrax would be used for.[75]

Anthrax Emergency! Bring on the Stimulus Money!

In October 2008, just as a CDC advisory panel was on the verge of recommending that state and local public health officials administer anthrax vaccines to as many as 3 million first responders nationwide, U.S. Health and Human Services Secretary Michael Leavitt declared a seven-year "public health emergency" due to the risk of bioterrorism. The declaration established legal immunity for public and private health officials who oversee the production or distribution of the anthrax vaccine, which has been controversial because of reports of serious adverse effects, including deaths, among recipients.[76]

John Michels, an attorney in litigation involving the Pentagon's forced inoculation program, "said commercial interests appear to be playing a role in the legal immunity issue. He questioned whether there had been any bona fide escalation in the anthrax threat sufficient to justify the declaration of an emergency." Dr. Meryl Nass, a critic of the government's mandatory vaccination program, "accused Leavitt of taking more interest in protecting bureaucrats from legal action than in protecting the public from health threats."[77]

Universal Detection Technology, however, a maker of anthrax detection devices, applauded the government's recognition of the ongoing "anthrax emergency" and of the huge dangers posed by anthrax, "the only agent used against the US."[78] (In the interest of full disclosure, UDT should have said "the only agent used against the US by the US." After all, anthrax, like guns, doesn't kill; people do.) Perhaps, in the not too distant and profitable future, anthrax detection devices might be as ubiquitous as cell phones.

Nobody Knows, or More Where This Came From

At a November 6, 2001, hearing, Senator Dianne Feinstein asked FBI spokesperson James Caruso how many American labs handled anthrax:

CARUSO: We do not know at this time.

FEINSTEIN: You don't know that.

CARUSO: No, we do not. We're pressing hard to determine . . .

FEINSTEIN: Could you possibly tell me why you do not know that?

CARUSO: The research capabilities of thousands of researchers is something that we're just continuing to run down. I know it's an unsatisfactory answer, and unsatisfying to us as well.[79]

The FBI said it also didn't know how many labs had access to the Ames strain.

In 2007, GAO representative Sushil Sharma testified that the FBI had been asking the CDC—unsuccessfully—for a list of labs handling anthrax. The CDC, in its new role as protector of government secrets, had denied the FBI the information. Sharma said he understood the impasse was in the process of being worked out.

What we do know is that since 2001, the number of high-containment germ labs in the U.S. has risen from fewer than 200 to more than 1300, all in the name of protecting ourselves from the sort of attack perpetrated from one of those facilities—either by a lone mad scientist or by something more sinister.

I LOSE A REASON FOR CARING

I DROVE TO LOUISVILLE AND BACK THAT THURSDAY MORNING. I had registered for the Kentucky Law Update, the two-day continuing-education event offered by the Kentucky Bar Association in different locations. I'd decided the night before to skip the training this year because I was busy, and had enough carry-over CLE to keep me legal for another year. But I drove up to get the materials nonetheless.

I ate on the way back at a Cracker Barrel. I stopped by the Bowling Green Public Library and checked out DVDs for the weekend. I unloaded my car.

During one of my trips outside, Joslynn Newman, one of my and Betty's oldest friends, called and left a message on my voice mail. She left a number. She said I needed to call the number as soon as possible.

I called the number. I didn't get an answer. I kept trying; I kept getting no answer. I was worried. I feared the call had something to do with either Betty or Joslynn's husband Ira. I called Betty's number, and left a message. I told her of Joslynn's call, said I was worried about her, and asked her, if she came back and got the message, to call and let me know she was okay.

Joslynn reached me eventually. She asked if there was someone who could be with me. I said no, what was the matter. She said Betty had died of a heart attack that evening. I said no again, screaming it this time. I said it again. And again.

I would pick up Muffin, our cat, on Saturday. I would go to a service in Somerset on Sunday. Ira would meet me as I came into the funeral home, tell me that Betty's brothers and sisters didn't want to speak with me. They asked that I sit at the back of the chapel. I said no (no as in this can't be happening) again, but I did what they asked.

I made it through finals the next week. The following week, I went to the ASPI (Appalachian Science in the Public Interest) office in Mt. Vernon on Monday night and helped prepare the first issue of *Downwind*, our new newsletter, for mailing. Then, before Christmas, my parents and brother-in-law helped me move my remaining possessions from Bowling Green back to the house where Betty and I had lived for twenty-two years. I would take a leave of absence from my job in the spring, and sit there in the house week after week, grief gathered round me like a thick fog.

It was hard to care about hypothetical threats when the person I'd loved and cared about since 1974 had died at the age of 51. During that year after the divorce, as we struggled to live without each other, I'd not allowed myself to acknowledge I still loved her. Now, when it could change nothing, I acknowledged it: what I had felt, what I had lost, what would not be coming back.

I'd known I would not stay in the house forever without her. At the end, as we were about to sign the separation agreement, I'd wanted her to keep the house. I'd given her another chance to change her mind even after the divorce decree was entered. She'd wavered, then decided to keep things as they were. But she'd advised me not to be in a rush to sell it. I'd suspected she, like me, was still not sure we were finished.

Now I'd never find out. And I'd lost any lingering thought I would still be in the house by the time they put the last brick on NBAF.

I'd never expected to still be around anyway, of course. (And now "still being around" had acquired a whole different meaning.) I'd gotten involved in this battle because of my anger at the lies, because of my residue of concern for this place where we'd planted gardens, raised animals, entertained friends. So I expected to keep fighting.

It was harder for me to care, though, about NBAF or anything else. It was hard surrounding myself with stories of death: painful to read about the dead victims of Sverdlovsk, how their lives, too, had been suddenly and irrevocably changed by events beyond their control. It was painful to read about the U.S. anthrax victims, about Ebola and smallpox deaths, and to imagine the agonies of the animals sacrificed to biowarfare agent research.

It was harder for me to be patient with my fellow NBAF opponents, with those who I felt weren't pulling their share of the load, or were letting ego get in the way of our mutual efforts.

If I could have taken a leave of absence from this part of my life, I would have.

CHAPTER SIX

Culture of Denial
Accidents in the World of High-Tech Germ Research

Beyond the history of incidents at Plum Island, we found no evidence that the study considered the history of accidents in or releases from biocontainment facilities generally. Had the study considered this history, it would have shown that no facility for handling dangerous pathogens can ever be completely safe and that no technology can be totally relied on to ensure safety.

—Government Accountability Office,
High-Containment Biosafety Laboratories:
DHS Lacks Evidence to Conclude That
Foot-and-Mouth-Disease Research Can Be
Done Safely on the U.S. Mainland, May 22, 2008

"We're getting as close to fail-safe as possible . . . as fail-safe as the space shuttle."

—Jim Orzechowski, chief executive
of Smith Carter, then the world's
leading producer of BSL-4 labs, 2003.

In January 2009, the Associated Press reported that the University of Wisconsin-Madison had "quietly decided to stop manufacturing" the Madison Aerosol Chamber (MAC), used to infect the lungs of lab animals with everything from tuberculosis to anthrax and implicated in the Texas A&M brucella incident.[1]

Concern about huge liability costs prompted the decision. An internal audit that raised the liability concerns acknowledged the risks of biodefense research, and of the current proliferation, in a way the biodefense complex rarely does in public. "Like any mechanical thing, it has seals and gaskets," the auditor explained to AP reporter Ryan Foley. "Those seals and gaskets can fail and then that would release those toxic agents to human exposure. That's the risk."[2]

At the time of Foley's statement, the $40,000 device was being used by some 25 universities and companies from the Bronx to Bangalore. Demand was growing because of the surge in biodefense funding, and NIH had encouraged the MAC's use in biodefense workshops.[3]

Increased use meant increased liability risk, because of the "increased opportunity and risk of field failures." UW's Engineering College, which produced the MACs, was selling them without a warranty agreement which might have limited liability by requiring buyers to do routine maintenance. In one case, UW had filled a purchase order requiring it to accept liability for any problems.[4] In addition, "University of Wisconsin officials tended to fix problems as they popped up rather than develop routine maintenance plans to prevent them."[5]

Such was the "state of the art" with respect to aerosol exposure chambers. Even more startling, David McMurray of Texas A&M complained that the only privately manufactured alternative was even less efficient and safe than the risky MAC.[6]

Biodefense propaganda asserts that state-of-the-art equals safety. A producer of one state-of-the-art device—the MAC—says state-of-the-art equals risk. And that's also what Government Accountability Office technologists say. In critiquing a 2002 study DHS relied on for its NBAF plans, the GAO rejected the study's blanket statement that "biocontainment technology allows safe research." Technology is only one part of a comprehensive biosafety program, the GAO said, and human error accounts for the majority of accidents in high containment laboratories. "This risk persists, even in the most modern facilities and with the latest technology."[7]

At the October 2007 subcommittee hearing, the GAO had already warned that there is a "baseline risk" associated with any high-containment facility. The current expansion increases the aggregate risks (the same conclusion drawn by the University of Wisconsin's auditor), and the risk is even *greater* for new facilities with less experience. There are a lot of those new facilities, and a lot of new researchers to go with them. "According to a CDC official, the risks due to accidental exposure or release can never be completely eliminated, and even labs within sophisticated biological research programs—including those most extensively regulated—have had and will continue to have safety failures."[8]

This is basic common sense for any and *all* industries, and a working assumption of concerned biodefense neighbors everywhere. The biodefense industry opposes this common sense with a sort of pseudoscientific, magical thinking. Catastrophe is not worth thinking about, we are told, because the pathogens have been enchanted with a series of spells to make them safe: biosafety cabinets; negative air pressure; changing rooms, air locks, and multiple sterilizations. The very pathogens which are supposed to frighten us because of their potential use by bad-witch bioterrorists become innocuous in the hands of white-magician biodefense researchers. Or so we are asked to believe.

The industry's propaganda trump card is that so far no member of the general public has died from an *accident* at a high-containment U.S. germ lab. It would be inaccurate to say that no member of the general public has been harmed or killed by *activity* at a U.S. high-containment germ lab, of course. By 2002, the anthrax used in the 2001 letter attacks had been traced to a small number of U.S. biodefense labs and would eventually be attributed by the FBI to the elder statesman of those facilities, USAMRIID at Fort Detrick. Multiple institutes and centers for the study of bioterrorism mushroomed after the attacks, but as yet there is no Center for the Study of Bioterrorism Committed by Centers for the Study of Bioterrorism. But that's what America's biodefense behemoth adds up to: a country on alert, and an industry in full production mode, because of a self-administered wake-up call.

The other documented instances of tangible harm to the public also came from deliberate acts of the U.S. bioweapons/biodefense complex: the secret spraying of pathogens and toxic tracer compounds on U.S. cities during the 1950s. If those responsible for the projects truly believed the spraying was harmless, then the resulting harm, though not the spraying itself, was accidental. Leonard Cole's book *Clouds of Secrecy*, however, establishes that enough research was out there to suggest the harmfulness of the stimulants and tracers, so that the spraying involved either knowing or reckless infliction of harm.

The biodefense complex denies—to the public and to itself—that its activities pose any significant risk. This denial requires asserting the flawlessness and impermeability of its facilities, equipment, and "standard protocols." Accidents obviously show that the whole state-of-the-art ensemble isn't so flawless after all. And so the biodefense complex withholds information about accidents from the public whenever possible, in order to keep clinging to the myth. When news of an incident does get out, the industry will spin the events by denying that the public was ever at any risk.

A typical example occurred at the end of 2008. The frozen remains of two plague-infested dead mice went missing from the Public Health Research Institute, a Biosafety Level 3 facility at the University of Medicine and Dentistry of New Jersey in Newark. The facility notified the CDC, FBI, and state health authorities, but withheld information from the public until *The Star-Ledger* began asking questions several weeks later.[9] The university then sent a mass E-mail about the incident to the university community.[10]

The university defended its decision to keep the matter confidential by denying that there was any risk to the public, and asserting that it was therefore inappropriate to alarm the public. A senior public relations vice-president said: "Plague sounds like it is Black Death. It would have been irresponsible to raise concerns."[11]

Yes, plague sounds like Black Death because it *is* Black Death. Mortality in untreated cases is either fifty percent or one hundred percent, depending on whether the pathogen has been inhaled (primary pneumonic) or transmitted by an insect bite (bubonic).

The desire not "to raise concerns" illustrates the double-standard fear-factor at work in the current proliferation. We, the public, are asked to be terribly alarmed that some foreign terrorist might unleash deadly pathogens upon us. Otherwise we won't fund the important work of the biodefense industry. The terrorism experts are constantly alarmed by the threat of bioterrorism, or at least by the thought that the biodefense milk cow will dry up. In the New Jersey situation, "[t]he mice were part [of] ongoing work in a federally-funded bio-defense program to find vaccinations that could be effective for the plague, which terrorism experts fear could be used as a biological weapon."[12] Yet we, the public, should understand that when the same pathogen which so frightens the bioterror experts goes missing from a biodefense facility, there's absolutely no risk to the public and nothing at all to be concerned about, even though the only significant bioterror incident to date involved small quantities of *Bacillus anthracis* diverted from a U.S. biodefense facility into the U.S. Postal Service.

After the Southern Research Institute inadvertently shipped live anthrax samples to the Children's Hospital Oakland Research Institute in 2004,[13] *Science* noted that the incident nipped at the heels of other recent miscues—three accidental SARS releases in Asia; an Ebola exposure in the U.S.; and an Ebola death in Russia—and would likely damage public trust further. C. J. Peters, the biodefense guru at UTMB-Galveston, complained about all the negative attention: "We're facing a frenzy of concern, and it isn't always rational."[14]

Several scientific magazines had carried articles about the 2003-2004 SARS releases, the first time since the 1979 Sverdlovsk incident that a high-containment laboratory lapse had infected and killed members of the general public. *Science* would cite the SARS incidents in a July 30, 2004 article, "Tiptoeing around Pandora's Box," describing efforts to genetically "tweak" the bird flu virus so it could be transmitted from person to person. *Science* would say: "Such experiments can give the world a better handle on the risks, but they could also create dangerous new viruses that would have to be destroyed or locked up forever in a scientific high-security prison. An accidental release—not so far-fetched a scenario given that the severe acute respiratory syndrome (SARS) virus managed to escape from three Asian labs in the past year—could lead to global disaster."[15]

It's hard to see what Peters would find irrational in the public's (or

Science's) concern about any of these incidents. Live anthrax bacteria are considerably more dangerous than dead anthrax bacteria: we don't want scientists in a children's hospital unknowingly handling live anthrax; and we don't want the institutes and companies that supply *Bacillus anthracis* to the 350 or so U.S. facilities[16] registered to handle it (as of 2004) to be shipping it willy-nilly, dead or alive.

It only takes a little loose anthrax to wreak bioterror havoc, we've been told repeatedly (and shown once, just in case we didn't believe it). Yet more U.S. facilities now possess anthrax than at any time in our history—partly because there are more facilities period, and partly because research funding heavily favors work on anthrax and other bioweapons agents, instead of tuberculosis or MRSA (multi-resistant *Staphylococcus aureus*).

Shipping and inventory discrepancies are common for bioweapons agents, even at the nation's Select Agent overseer, the CDC, where in October 2005, *Coxiella burnetii*, the pathogen which causes Q fever, was reported missing, though the CDC report suggested in typical hopeful fashion that poor bookkeeping was suspected. Other missing agents identified in the accident reports obtained by the AP include anthrax, brucella, glanders, plague, and tularemia pathogens, and botulinum and staphylococcal enterotoxin. The reports generally assume that nothing escaped; the public was at no risk; everything was done properly except the bookkeeping. An "individual failed to note that botulinum neurotoxin was destroyed." (If the individual failed to note it, how can one be confident that it was in fact destroyed?) "Three Brucella bacteria specimens were missing; worker failed to note their destruction." "Poor record keeping blamed for empty vial of Burkholderia mallei that was recorded as including the agent."[17]

Infected mice, both live and dead ones, occasionally go missing. Optimism prevails in this situation also. The Newark, New Jersey facility that lost dead plague-infected mice in 2008 had lost live plague-infected mice in 2005, but "decided they were eaten by other mice or incinerated." (For all UMDNJ knew, the mice could have been merrily spreading the plague in the city's sewers.) In Stillwater, Oklahoma, a mouse inoculated with tularemia pathogen went missing; "the lab believes it was incinerated but not recorded."[18]

In 2007 a request by the *Honolulu Advertiser* for information on microorganism imports brought a twofold discovery: University of Hawaii researchers were not submitting required paperwork "on numerous imported bugs including dengue and West Nile viruses"; and the state Department of Agriculture had lost track of hundreds of viruses and microbes imported to Hawaii under 356 state permits dating as far back as fifty years.[19]

As a result, the Agriculture Department had prohibited the University from seeking new or renewed import permits for five months. UH's permit paperwork problems were eventually resolved, but six months later the Agriculture Department itself was still unable "to produce a list of restricted microorganisms that [could] be imported into Hawaii under previously granted permits. The department also could not produce a complete list of microorganism permits that were found not in compliance with reporting requirements."[20]

Dr. Lorrin Pang, a Maui physician who had served as a World Health Organization consultant, said the lack of appropriate records could affect public health if someone caught one of the diseases in question. Knowing that an Hawaii facility held the relevant microorganism would allow investigators to consider the facility as a possible source; not knowing, investigators would of course be clueless.[21]

Of course, UH and state officials argued that "there's little risk to the public of exposure to these viruses because access to the bugs is tightly controlled." Failure to keep the appropriate records obviously makes such a blanket statement questionable.[22]

The permit problems occurred even as the UH was ramping up virus imports in an effort "to specialize in infectious disease detection and drug discovery." The university already had 177 microorganism import permits covering avian flu, dengue fever, West Nile virus, SARS, and several encephalitis-causing viruses.[23]

Throughout the country, germs are generally shipped by common carrier. In March 2003, a package containing the West Nile virus exploded at a Federal Express facility near the Columbus, Ohio International Airport, forcing the evacuation of about fifty people.[24]

About half the incidents reported by the Associated Press involved an actual exposure or illness. Certain patterns tend to recur. Clumsy fingers will drop plates of plague. Knives and needles will slip and inflict cuts, in the midst of work with plague or anthrax. Pathogen packages leak, or somebody fails to properly decontaminate them. Guinea pigs bite; livestock breeds of pig bite; macaques and ferrets bite; monkeys scratch (and bite, and spit).

People forget to follow "standard operating procedures." At the CDC Division of Vector-Borne Infectious Diseases, in Fort Collins, Colorado, a lab worker comes across old containers of Russian spring/summer virus, and apparently does something with the containers—moves or opens them—without wearing the stipulated high-tech pressure suit. Someone at Detrick turns the (negative air pressure?) fan off while working with anthrax; "worker exposure confirmed."[25] At the Midwest Research Institute in Kansas City, two lab workers open a package with leaking tubes of anthrax on an open bench. At Washington University in St. Louis, eight employees enter a lab—presumably not wearing protective garb— because they're told the room contains no select agents, when in fact it houses "sealed" culture plates of plague. At a lab in New Mexico, a notebook—presumed contaminated—is removed from a lab working with monkeypox virus.[26]

In 2005 the influenza research world was rocked by a double error. The first error occurred when a Cincinnati-based test-kit maker producing sample vials for the College of American Pathologists mistakenly shipped the wrong influenza strain to about 4,000 laboratories in eight countries. The vials were labeled as being of the current A/Shanghai variety; actually, however, they were of the H2N2 influenza strain, which had killed about four million people in 1957, and had not circulated in the world since 1968.[27] No one born since that time would have any immunity.

A second miscue in a Vancouver hospital laboratory brought the erroneous shipment to light. A patient specimen indicated that a patient had influenza, when she did not in fact have clinical symptoms of influenza. Testing at the National Microbiology Laboratory in Winnipeg revealed that the sample had been cross-contaminated with H2N2.[28]

The Vancouver laboratory determined that, on the same day the patient's specimen was prepared for shipping, the hospital had also been conducting proficiency testing with the samples provided by the Cincinnati manufacturer. The lab didn't know how the cross-contamination had occurred, but human error, the usual culprit, was presumed to be the villain here as well. According to Dr. Danuta Skowronski of the B.C. Centre for Disease Control, "In a perfect world, cross-contamination would not occur and it is indeed very rare. But obviously, humans are fallible. [Duh!] Those who work in laboratories are taught to assume that every agent is potentially infectious, whether they know what it is or not."[29]

In 2009, Baxter International Inc. in Austria sent three neighboring countries lab samples contaminated with the bird flu virus. The error was discovered when ferrets at a Czech laboratory died after being inoculated with vaccine made from the samples.[30]

Equipment failures are also documented in the AP materials, suggesting that human error, while the chief danger, isn't the only one. A power outage occurs at a San Diego lab working with the fungus that causes valley fever. The report notes that this "occurred after" a necropsy of infected mice—as opposed, one presumes, "during" the necropsy, when a sudden absence of light and electrical power and negative air pressure would be not only inconvenient but downright risky.[31] (And might still be risky after a necropsy if the room and equipment hadn't yet been decontaminated.)

Other equipment failures in the AP disclosures include a centrifuge leak, a water supply leak, and leaky wastewater pipes. The wastewater pipe leak occurred at the National Animal Disease Center in Ames, Iowa.[32] Leaking wastewater pipes are thought to be the source of the 2007 British foot-and-mouth outbreak, an event that raised concern about DHS's plans to move foot-and-mouth research to the U.S. mainland. Ironically, at the Congressional subcommittee hearing on those plans, a couple of witnesses suggested American facilities like the National Animal Disease Center were safer than the British Pirbright facility.

For the American biodefense complex, absence of full-out catastrophe constitutes safety . By this standard, a drunk driver who hasn't hit anybody yet is driving perfectly safely, thank you.

The assumption that state-of-the-art technology assures safety is a dangerous assumption, breeding an overconfidence that can foster catastrophe. This is a point made over and over by scholarly studies of catastrophic and near-catastrophic accidents involving advanced technologies.

These studies show, first, that accidents will happen, indeed. As Lloyd Dumas points out in *Lethal Arrogance: Human Fallibility and Dangerous Technologies*:

> Because all systems human beings design, build and operate are flawed and subject to error, accidents are not bizarre aberrations, they are a normal part of system life. . . . Accidents and failures of dangerous technological systems differ from accidents and failures of other technological systems only in their consequences, not in their essence.[33]

Yale organizational sociologist Charles Perrow coined the term "normal accidents" to describe the sort of "system failures" that are inevitable in complex technological systems. Perrow says system failures typically begin as component failures, which occur because "nothing is perfect—neither designs, equipment, operating procedures, operators, materials, and supplies, nor the environment." Complex systems require designs that entail multiple invisible interactions.

> [If] the complex interactions defeat designed-in safety devices or go around them, there will be failures that are unexpected and incomprehensible. If the system is also tightly coupled, leaving little time for recovery from failure, little slack in resources or fortuitous safety devices, then the failure cannot be limited to parts or units, but will bring down subsystems or systems.[34]

Perrow, Dumas, and James R. Chiles, the author of *Inviting Disaster: Lessons from the Edge of Technology*, all agree that overconfidence is a serious problem for those working in risky areas. Chiles critiques the commonly held notion that "if nothing very bad has happened so far, it probably won't happen in the future either."[35] Dumas points out:

> When faced with the possibility of disaster, it is common to seek refuge in the belief that disaster is very unlikely. Once we

are convinced (or convince ourselves) that it is only a remote possibility, we often go one step farther and translate that into "it won't happen." Since there is no point in wasting time worrying over things that will never happen, when we think about them at all, we think about them only in passing, shudder, and then move on.[36]

Both Dumas and Perrow question the confidence which "experts" and "risk assessors" bring to their evaluations of risky technologies, and explore the reasons for what Dumas calls "lethal arrogance." Dumas first draws a distinction between "risk" and "uncertainty." "Risk" entails situations when "we do not know exactly what will happen, but we do know all of the possible outcomes, as well as the likelihood that any particular outcome will occur."[37] Uncertainty, on the other hand, "refers to a situation in which we don't know all possible outcomes and/or we don't know the probability of every possible outcome. We are lacking some critical information that we have in a comparable case of risk."[38]

Dumas finds a general human penchant "toward wanting to believe that there is a higher degree of predictability in situations than actually exists."[39] And he notes the particular psychological pressure on analysts and decision-makers to characterize a situation as more predictable than it really is:

> For analysts or decision makers to admit that a situation is one of uncertainty rather than risk is tantamount to admitting that they are unable to see all the relevant outcomes and/or evaluate their probabilities. It is an admission of ignorance or incapability. . . . Those we call experts are even more likely to fall into this trap.[40]

Dumas says that when experts are hired as consultants, clients expect to be told, if not what will happen, "at least what could happen and just how likely it is. They are not interested in being told that the situation is uncertain and therefore unpredictable. A report like that does not do wonders for the reputation or the future flow of consulting work."[41]

Perrow says the role of risk assessors is not just to "inform and advise" the operators of dangerous technologies, but also, "should the risk be taken, to legitimate it and reassure the subjects."[42]

Perrow says risk assessment in general is too narrowly focused, tending to monetarize social goods, and assuming that everything can be bought, including a life. Whatever cannot be bought does not enter into the "sophisticated calculations."[43] Risk assessment also implies, he says, that the public should either be excluded from discussions that affect them or "educated" and brought over to the experts' view.[44] This book's chapters on specific biodefense projects provide multiple illustrations of confident "experts" reassuring, ridiculing, or otherwise "educating" worried publics.

Over and over, risk assessors and decision-makers have seriously miscalculated the likelihood of catastrophic accidents. The "probabilistic risk analysis" used by NASA before the Challenger disaster concluded that the odds of disaster was only one in a hundred thousand flights.[45] In fact, designers of the space shuttle booster rockets hadn't even equipped them with sensors that might have warned of trouble—because they believed the boosters were "not susceptible to failure." The tragic explosion of the Challenger's right-side booster in 1986 showed otherwise.[46]

Within the nine-month period surrounding the Challenger disaster, four other failures of "highly reliable" launch vehicles occurred: two explosions of the Titan 34D rocket; a misfiring of a Nike-Orion rocket; and a misfiring of a Delta rocket. Perrow, who examined several catstrophic and near-catastrophic incidents in the space program,[47] said the program illustrated "how the best talent and organizational resources, while they certainly help, cannot overcome system accident potentials."[48]

Before Three Mile Island, the Nuclear Regulatory Commission had deemed a "loss of coolant" accident (LOCA) at full power so unlikely as not to require consideration in safety plans.[49] The LOCA which did occur brought the reactor core within a half hour of a complete meltdown.[50]

McDonnell-Douglas engineering studies had calculated the probability of a combined loss of engine power and slat control damage during aircraft takeoff as less than one in a billion. Yet 273 people were killed when this happened at Chicago's O'Hare on May 25, 1979. A similar accident had occurred in Pakistan in 1977; two more occurred in 1981—four cases of "one in a billion" in five years.[51]

A partial core meltdown of the Fermi reactor in a nuclear power plant near Detroit occurred in October 1966. A classified report by the Atomic Energy Commission, completed before the accident, had estimated that a severe accident combined with unfavorable wind conditions would expose 133,000 people to high doses of radiation, and that half of those would quickly die. Another 181,000 people would receive significant 150-rad exposures. Nobel laureate physicist and nuclear power advocate Hans Bethe had confidently predicted that a core meltdown could not occur with this type of reactor. Another expert had concluded that at worst, only one subassembly would melt. After this very close shave (described in an article titled "We Almost Lost Detroit"), investigators would eventually learn that four subassemblies had been damaged, and two of them were stuck together. It took four months to learn that the subassemblies were damaged, five more months to remove them, three years to remove some of the poisonous materials from the plant and to store radioactive sodium in steel drums on site.[52]

Perrow identifies six areas in which failures can arise in any complex technological system: design; equipment; procedures; operators; supplies and materials; environment.[53] Most "system accidents" involve multiple failures interacting in unexpected ways.

Both Perrow[54] and Dumas identify the complacency of operators as a significant problem. Says Dumas: "No matter how expensive or dangerous the systems might be, if things go well and all is calm for a long time, most people begin to assume that nothing will go wrong. The cutting edge of their vigilance begins to dull. Even if familiarity does not breed contempt, it does breed sloppiness."[55]

Biodefense research is subject to failures in any of the six system areas, and the complacency of operators is a particular danger. Accidents are frequent enough, but because no catastrophes have yet resulted from a U.S. biodefense "accident," biodefense promoters argue, and perhaps believe, that state-of-the-art technological design and equipment, coupled with "redundancy," will prevent catastrophe.

The appropriate reply to this "lethal arrogance" is suggested by

Dumas: "Anything that has ever happened, happened once for the first time. Before that time, it too had never happened. And for some of these catastrophes, we cannot afford even a first time."[56]

The biodefense complex clearly suffers from the mindset described by Dumas: first convincing itself disaster is unlikely, then translating that into "it won't happen." One finds frequent comments of the sort made by the director of a proposed BSL-4 project in Toronto, that the chances of infecting anyone on the outside "are zero. They are nil. This laboratory is totally safe."[57]

In 2003, the director of a different Canadian BSL-4 facility, the one at Winnipeg, boasted that no researchers had ever been infected by SARS in a lab.[58] Within a few months of that statement, however, SARS escaped—by way of infected researchers—from labs in Taiwan, Singapore, and China. In one instance, a researcher took the train from Beijing to Anhui province 400 miles away, and infected seven others. One person died.[59]

Then there is the irony of the 2002 SAIC (Science Applications International Corporation) study, holding up the UK Pirbright facility as an example of animal disease research proceeding safely amidst large animal populations: in 2007 a release from that facility caused a significant foot-and-mouth outbreak and concerned both potential NBAF neighbors and American livestock producers aware of DHS's plans to move FMD research onshore.[60]

We can't be sure how much of its own propaganda the biodefense complex actually believes. We can be sure of this: the biodefenders very much prefer to keep their "incidents" out of the public eye. Most people, after all, understand that drunk driving is hazardous even if it hasn't yet resulted in fatalities.

So the pattern is this: don't alarm the public; don't embarrass the institution. As it is seeking approvals from the city for its NIH BSL-4 facility, Boston University conceals the tularemia infections of three researchers and fails to update its previous submission to the NIH indicating no laboratory-acquired infections of researchers. Texas A&M conceals its brucella infection, elevated Q fever titers, etc., as it launches its NBAF bid. The UMDNJ in Newark "responsibly" protects the public from unnecessary alarm about its disposition of plague-infested mice.

In the fall of 2008, the University of Georgia, then an NBAF finalist, followed a similar pattern. Twice, on September 24 and October 2, floor drains in a large-animal lab at the university's "new" Animal Health Research Center overflowed, and a mixture of water and animal waste leaked into the basement (and thus outside the high-containment area). The first flood involved about 15,000 gallons, the second some 5,000 gallons. Cracks in the lab floor allowed some of this to leak into the basement.[61]

The AHRC, "conceived as one of the most technologically advanced facilities on a university campus," was first opened in 1999, amid lots of hoopla. Early inspections, however, revealed problems with heating and ventilation systems and the wastewater system. The building was eventually gutted and rebuilt from the inside, increasing the cost from $21 million to $63 million. The high-containment lab implicated here had just reopened in the spring of 2008.[62]

The College of Veterinary Medicine, which operated the lab, didn't inform a community liaison committee, or the university's public relations office, about the problems until the pro-NBAF *Athens Banner-Herald*—apparently deciding to play newspaper-for-a-day— began asking questions. The chair of the community liaison committee—also the University's Director of Community Relations—just managed to release the news to the committee on October 6, one day before the newspaper ran its story.[63]

The committee's mission was "to establish and maintain trust and confidence among the scientists associated with the Animal Health Research Center, the University of Georgia community, and the citizens of Athens-Clarke County," according to the lab's Web site. It was, in other words, a public relations tool for the university.[64]

In an October 7 memo to administration at the AHRC, the liaison committee chair complained that the public relations tool wouldn't be credible if the public relations people themselves were kept in the dark:

> I am concerned that I was not able to notify the committee about the recent water issues at the AHRC in what I believe was a timely fashion. While the messages I received yesterday from Mike (2:23 PM) and David (2:27 PM) did allow me to

get something to them before they read it in the newspaper this morning, they can easily conclude that we decided to advise them only after the media discovered it, and had it not been for the media, they might not have known at all. Frankly, I draw the same conclusion. I fear we may lose credibility with the group.[65]

Tom Jackson, the university's Vice President for Public Affairs, had already written that same morning to express his dismay at learning

so late about these two incidents at the AHRC . . . the first of which happened on Sept. 23. I learned about them yesterday—Oct. 6—after someone had gone to Lee Shearer with the Banner-Herald. We need to determine who knew about these incidents, who they informed when, and who made the decision not to inform my office. This does not bode well for building trust with the community about the way we're going to operate that building. In this atmosphere, and particularly with the NBAF application pending, there are no "minor incidents."[66]

When one of the AHRC administrators suggested that "it comes down to a question as to what events merit being conveyed to the committee and which do not,"[67] Jackson responded: "It will be a hard protocol to write, but one answer is, 'Any event that may generate media coverage.' And that is not always clear at the outset."

The main concern of the university memos seemed to be the public relations black eye, not the safety of its students, faculty, or town. The university spun the situation as well as it could, telling the *Atlanta Journal-Constitution* that "the recent incidents at the large animal lab posed no threat to workers, students, or the public; the water that leaked into the basement was not contaminated. They said the lab is experiencing the customary problems of a new facility, and they stressed that it has never had a breach of dangerous material."[68]

Richard Ebright told the *AJC* that UGA should have installed water-level alarms after the first flood, that the basement leak did in fact constitute a breach of containment, that such breaches are a "potentially serious hazard," and that it appeared protective measures were not put into place after the first flood.[69]

Jackson told the *ABH* that no one "made a decision" to keep the committee in the dark. Just no one thought to inform the committee, or anyone outside the College of Veterinary Medicine, apparently. Harry Dickerson, Dean of the College of Veterinary Medicine and a prime Georgia NBAF propagandist, argued that the committee would eventually have learned of the incidents at a January 2009 meeting, and that "the leaks didn't seem serious enough to warrant notifying the committee right away because the water didn't escape the building and didn't threaten human or animal health."[70]

This is invariably the sort of reason given for concealing accidents. It would seem that overflowing drains, and a breach of containment (at least within the building itself), would "threaten" human health. Apparently, only when human health is demonstrably affected will the biodefense complex acknowledge that one of its own lapses represents a threat.

The problem was eventually traced to a pump at the bulk autoclave (or sterilizer). The pump, which ran water to produce a vacuum to operate the doors, failed to shut off, overwhelming the effluent drainage system.

A memo from Steris Corporation, supplier of the autoclave, indicated that "the bulk sterilizer at the UGA-AHRC has had a number of problems since late July of this year. Those problems range from leaks on door exhaust heat exchangers to a defective CPU Board and contaminated program. Recently, there have been door seal/unseal problems."[71]

A UGA memo indicated that, as of October 3, "Steris is having trouble getting this thing operational again."[72] On October 30, Steris was still working on the various problems.[73]

AHRC Operations Committee minutes for October 9 showed that the sterilizer wasn't the only lab equipment with problems: "The incinerator has been cleaned out and put back together. It was not recommended to make the modification of removing the flange and replacing it with a metal plate. So the incinerator was drained and the flange replaced and tightened." The minutes for September 25 said that "Biosafety suggests suspending necropsy showers until directional air flow is resolved"—suggesting a problem with the negative air pressure system as well.[74]

The Georgia documents suggest that state-of-the-art biodefense equipment breaks down about as often as any other machinery. Problems were occurring simultaneously with the AHRC's autoclave, sterilizer, and negative air pressure system. Such multiple failures are generally behind the catastrophic failures of high-tech hazardous systems. "Redundancy" helps, but may not be enough when hell breaks loose all over. The Georgia documents also illustrate how the "experts" can be puzzled for lengthy periods of time about what is going wrong and why. Shades of Three Mile Island.

Meanwhile, the floor cracks that allowed water to leak into the basement also had to be repaired. The question raised by the AHRC situation is: what do you do when state-of-the-art equipment and facilities (this one is contemplating, among other things, research on SARS, West Nile, and avian flu) aren't so state of the art?

Yet another warning about the dangers of germ lab overconfidence came in the fall of 2009. A University of Chicago researcher apparently died from a supposedly weakened strain of the plague (*Yersinia pestis*) he was studying. The strain was supposed to be one that had most of the harmful components removed, and "[b]ecause this form of the bacteria is not known to cause problems in healthy people, special safety procedures are not required to handle it."[75] Investigators were exploring the possibility that there might have been a mutation in the bacteria that might have made it dangerous, or that the researcher had some sort of pre-existing condition—perhaps an abnormality in his iron metabolism—that made him more susceptible. His family, however, said he was very health-conscious and had no chronic medical conditions they knew of. Since the university and local health officials did not know what there was about either the plague or the researcher that caused him to die from it, their assurances that there was no threat to the public seemed premature.[76]

HOMELAND SECURITY SHOWS UP LATE FOR A DATE, AGAIN

DHS ORIGINALLY SAID IT WOULD CHOOSE THE NBAF FINALISTS in August 2006, then select the site itself a year later. When August 2006 came around, however, DHS had peeled off only 11 of the initial 29 sites, and pushed finalist selection back to early 2007. Or so it said.

December and January went by, and we found ourselves still waiting for the short list. Some thought the reversion of Congress to the Democrats had turned the existing political calculus on its head, that the site selection universe was different now that Democrats chaired the important committees. We were encouraged that Hal Rogers no longer chaired the Homeland Security Appropriations Subcommittee.

We also wondered, however, if the calendar changes were DHS's way of keeping NBAF opponents everywhere off balance, wearing us down by lengthening the process and disrupting our planning.

On February 16, we learned from a newspaper article distributed through the Sunshine Project's listserv that DHS was changing its timeline once again. According to the article, published in the

Lawrence Journal-World (Kansas), each of the fourteen semifinalists (representing eighteen sites) had a February 16 deadline to submit another round of information to DHS—and finalist selection would now be delayed until June.[1]

Christopher Kelly, DHS's spokesperson on this occasion, said: "It's simply another milestone in the whole process of selecting the National Bio and Agro-Defense Facility." For us, though, it was just another example of a felony suspect not showing up during a stakeout.

We mailed our second newsletter at the end of February. DHS's strategy—if it was one—was working, in Kentucky at least. Our local steering group didn't meet for another six weeks. What happened in April would catch us off guard.

At many locations, people had never been "on guard" because there had been little news coverage, even of the bioboosting kind. This gradually began to change. In Kansas, for instance, the *Manhattan Mercury* and, especially, the *Lawrence Journal-World* began beating the biolab tom-toms on a regular basis.

It didn't hurt that one of the nine directors of the Kansas Bioscience Authority, the organization that launched the Kansas efforts, was Dolph Simons, publisher of the *Lawrence Journal-World*.[2]

The state legislature had created the Kansas Bioscience Authority in 2004, to take income tax withholdings from employees in the biosciences sector and reinvest them in an effort to attract or develop new bioscience businesses. It was expected to disburse an estimated $588 million in public funds over 15 years. Its initial revenues rose from $20 million to $50 million per year, and by 2009 it was projecting a budget of $75 million for 2010.

In October 2006 it would hire as its CEO Tom Thornton, previously the executive director of the Illinois Technology Development Alliance. Thornton, a former aide to Rep. Dennis Hastert of Illinois, would prove astute at the sort of political wheeling and dealing and public relations hoopla needed to reel in a massive pork project like NBAF. In a June 2007 editorial, Simons, smarting over the recent ouster of first KBA chairman Clay Blair, would accuse Thornton of not liking to share the spotlight, and characterized him as "an able individual with considerable experience in the Washington, D.C., game of lobbying and back-stabbing."[3] Simons would also accuse Kansas Governor Sebelius, the Kansas Technology Enterprise Corporation, and the University of Kansas of sullying the pure bioscience dreams that had prompted establishment of the KBA.[4] Eighteen months later, this sniping would all be quietly deleted from the final version of *The NBAF in Wonderland, a.k.a. Kansas.*

Kansas would mimic the other consortiums in pulling all the political wires it could while loudly protesting that it expected to win on the merits—and if it didn't, politics had to be the reason. While Kansas strutted its pure unsullied germ lab wiggle up and down the NBAF walkway, it acted like any other hardboiled beauty queen backstage. The KBA decided early on to spend $250,000 on a lobbyist and consultants for the ongoing NBAF seduction.[5]

Kansas was promoting two alternative sites: one just west of the U.S. Disciplinary Barracks at Fort Leavenworth,[6] and the other on the campus of Kansas State in Manhattan. Kansas political leaders of both parties were pulling all the political strings they could. Republican Senator Pat Roberts, engaging in a little remedial education of USDA and DHS officials "on the benefits of housing the new . . . NBAF . . . in Kansas," brought USDA officials on a tour of K-State's new Biosecurity Research Institute in January 2007.[7]

A March 2007 Associated Press article announced that the $50 million BRI—to be housed in, what else, Pat Roberts Hall—would open in April.[8] A series of stories would announce over the next two years that the BRI was just getting ready to open, but there it was in mid-2009, still largely unused—even as the Kansas delegation, worried about Congressional funding for NBAF, screamed loudly how desperately urgent this animal disease research was. Al Qaeda was coming to get our cattle, and our hogs, and our sheep, and our goats. And if they didn't, we don't know why not, because we keep telling them how they could bring the American economy to a standstill, if they would just let a little foot-and-mouth loose in one of our feedlots.

In a December 2008 article in *Nature*, a KSU spokesperson would let slip that the real motivating force "behind the Biosecurity Research Institute was 'build it and they would come' . . . University officials had hoped that biotech businesses or outside researchers would use the labs on a fee-for-service basis. Officials now hope that some Plum

Island labs may move their research there before the NBAF is completed in 2014."[9]

So biodefense research had suddenly reached the realm of real estate entrepreneurship, with universities building biodefense apartment complexes in the hopes of leasing space to the country's bright-eyed new horde of biodefense researchers.

Of course, it didn't hurt Kansas's NBAF chances, either, telling DHS that, by the way, we have this shiny new little ol' lab we'd just be tickled to death for you folks to take off our hands.

A BSL-3-Ag facility devoted to agricultural and food safety biosecurity, the BRI was expected to focus on research similar to that of NBAF. Randy Phebus, a K-State professor and first director of BRI, said the facility would initially focus on studying anthrax, staphylococcus, botulinum, and avian (bird) flu.[10] (Notice how NBAF, Jr. planned, like 350 or more U.S. facilities, to research anthrax as a more or less routine matter. To calm NBAF neighbor-(not)-wannabes, though, DHS would later assure them that NBAF, Sr. would not research anthrax under any circumstances, no sir, couldn't imagine where they even got such an idea.)

NBAF propagandists around the country were fond of saying that NBAF would only handle minuscule quantities of pathogens— teaspoon-size or less—ignoring the fact that hundreds of large animals, hardly teaspoon size, would be routinely infected with the pathogens. In March 2007, at least, Phebus was making no small-amount claims for the BRI, a fifth the size of NBAF. "In the old lab," he said, "we might have a small vial of bacteria and be doing tests on small quantities of foods. . . . In here, we're talking about being able to spray the entire carcass of a cow with anthrax or botulinum." The laboratory would have "a live animal holding area, a slaughter floor, meat processing equipment and the capability to bring in objects as large as an airplane cabin for contamination and de-contamination experiments."[11]

Democratic Governor Kathleen Sebelius indicated in her January 10 State of the State address that attracting NBAF would be one of the state's top priorities. Two weeks later, she named a 43-person task force made up of representatives from "industry, academia, public officials, agricultural organizations and economic development groups." The announced purpose of the task force was to "coordinate community outreach and advocacy and serve as a focal point for public input on their proposals."[12] That was all fancy lingo for propaganda. The best time for "public input" on proposals would have been before the proposals were submitted—but that had happened back in March 2006, before anyone outside the Kansas pro-NBAF consortium knew what was in the wind. For several months, the proposals had barely been mentioned in the Kansas media, and Kansas was one of several states that refused to publicly release its "expressions of interest."

Appropriately for a huge propaganda effort, one of the task force's co-chairs, former Clinton Agriculture Secretary Dan Glickman, was now president of the Motion Picture Association of America. Over the next two years, the real-time feel-good movie *The NBAF in Wonderland: The Kansas Story* would put to shame *The Wizard of Oz* and the soppiest flicks Disney ever made. Stay tuned.

On February 6, Roberts addressed a joint session of the Kansas legislature to encourage state support for the project.[13] Shortly afterward, the legislature passed bills allowing the Kansas Board of Regents to transfer 60 acres of land to the federal government should it build in Manhattan, and establishing an interagency group to help with an environmental impact study of Kansas sites. Local governments in Manhattan and Leavenworth each offered $5 million incentives.[14]

Later that month, as he was comparing the NBAF effort to an NCAA Final Four, Roberts revealed that the ultimate NBAF decision-maker, DHS undersecretary Jay Cohen, had accepted his invitation to visit Kansas in May.[15]

In March, Lieutenant Governor Mark Parkinson, co-chair of the NBAF Task Force, led a delegation of local and state officials to Washington to lobby for the Kansas solicitation.[16]

And of course, the *Lawrence Journal-World* and *Manhattan Mercury* were backing the project full bore, with never a word about any considerations other than the exciting economic opportunities and Kansas's chances of winning. In a February 18 editorial, the *Lawrence Journal-World* exulted that the NBAF solicitation showed how Kansas could "compete with the 'big boys.'"[17] Until NBAF came along, one never realized how many rural states suffered from pecker-order inferiority complexes. And the *Manhattan Mercury* editorialized on February 6 that Roberts's challenge was "one the entire state—and certainly the Manhattan area—should enthusiastically take up."[18]

A couple of days later, Walter Dodds, an aquatic ecologist in K-State's Department of Biology, published a letter to the editor expressing his surprise at the Mercury's editorial, since, he said, "We have virtually no information on this facility or any potential hazards associated with having such a facility in our city limits, but we have been asked to provide $5 million to bring it in."[19]

Dodds noted the 2003 GAO report about Plum Island and went on to lay out some basic questions which he felt needed answering:

> **Why is it OK to have such a facility inside the city limits of the Little Apple when there is worry over the existing facility on an island 12 miles from New Haven, Conn.?** Are any of the diseases to be housed in the facility infectious to humans, such as West Nile virus? Why is the Department of Homeland Security concerned about the level of local support for such a facility? Will the facility be converted to a Level 4 facility in the future? What are the plans for safely transporting hazardous materials to and from the facility? **Will our local agricultural economy be in danger if diseases escape from this facility, either inadvertently or as a result of sabotage?**
>
> I, for one, am uncertain if we should encourage such a facility to locate in Manhattan. **Perhaps the editorial staff of the Mercury has access to more information which it used to form its position in support of the facility. Some investigative articles would be helpful in providing the citizens of Manhattan more information than simply how much money will come into the local economy if we build the disease center here.** Given the potential stakes, the citizens of Manhattan and the surrounding area need more information and an opportunity to express their support or disapproval of housing such a facility in our community [emphasis added].[20]

Dodd's letter might have had more impact had he identified himself as a member of KSU's Biology Department—though such an identification might have been harmful to his status at KSU—and, judging from what I was told privately on my visit to Manhattan, and from Manhattan comments about the Draft Environmental Impact Statement, speaking out at all was courageous.

Dodd's admonishment to the *Mercury* would apply equally well to hometown newspapers in any of the NBAF states. With the sole exception of the *Lexington Herald-Leader*, however, almost the only examples of investigative articles came from alternative or weekly publications—the *Independent Weekly* of Raleigh, North Carolina; the *Current* of San Antonio; and the *Flagpole* of Athens, Georgia. Elsewhere, citizens needing information were on their own.

What the citizens of Manhattan got from the *Mercury* were articles like "What the City Gets with NBAF," in which Chamber of Commerce president Lyle Butler talked about the economic impact as "off the chart," and said, with respect to safety, "based on the briefing he has sat in on, he has heard 'really good assurances that this will be safe facilities.'" Then, with an inaccuracy straight out of the *Somerset Commonwealth-Journal* playbook, Butler noted that "there are already close to a dozen level 4 research facilities in various locales, 'some literally in downtowns of major metropolitan areas,' and those facilities operate without incident."[21]

Unlike their counterparts in Leavenworth (or Kentucky, Missouri, and Wisconsin), the Manhattan economic development folks did avoid the perils of holding an informational meeting for the general public. An event hosted by the Leavenworth County Development Corporation on February 28, presumably for the purpose of touting the marvels of NBAF Wonderland, brought forth dozens of skeptical residents. "If you have a problem over there, it's going to be a big problem. That's the way I'm seeing it," one resident said, after learning what sort of diseases a BSL-4 lab handled. Another complained about all the secrecy: "Nobody came around and mentioned that this was under consideration. I don't know what you're asking for. . . . You've given me no reason, not one specific reason, to support it."[22]

If the Leavenworth folks weren't up to the full propaganda

regimen, a famous figure from K-State was there to help them out. Jerry Jaax, now associate vice provost for research at K-State, and a prominent character in Richard Preston's Ebola thriller *The Hot Zone*, didn't bat an eyelash when asked if he'd take his own child through the lab. (Jaax was one of three USAMRIID researchers K-State had brought in to buttress its biolab credentials; the other were Jaax's wife Nancy and former USAMRIID director David Franz. All would propagandize shamelessly for the Kansas NBAF efforts.) "Absolutely," Jaax said.

This was the same Jaax whom Preston portrayed in *The Hot Zone* as, well, losing it after Nancy told him about a near-miss in an Ebola experiment:

> He was appalled. "God-damn it, Nancy! I told you not to get involved with that Ebola virus! That *fucking* Ebola!" And he went into a ten-minute diatribe about the dangers of doing hot work in a space suit, especially with Ebola.[23]

Since that time, Jaax had probably undergone some of that remedial education biolab proponents are always proposing for worried home folks.

Meanwhile, down in Dixie, the only remedial education the *Jackson Clarion-Ledger* understood was the sledgehammer kind. Residents near one of Mississippi's proposed sites, the one in Hinds County, had started posting No Bio-Lab signs, holding weekly meetings, and distributing petitions. District 5 supervisor George Smith, who represented the affected territory, attributed the opposition to the reputation of Plum Island: "I think people in the area are aware of that situation and they don't want Hinds County to become Plum Island."[24]

A week later, the *Clarion-Ledger* reacted indignantly, pointing out that DHS was getting ready to pay a visit, but that "already, some families in Hinds County are posting signs that read 'No Bio-Lab'— rejecting the idea." Landing the project would require local support, so "People in the metro area should get to know the facts and be comfortable with the plans. There is nothing to fear, and much to be gained. Does the metro area want this? Heck, yeah! We don't need signs saying 'Go away!'"[25] And then the *Clarion-Ledger* repeated Heck, Yeah two more times, once in capital letters, as if shouting Heck, yeah! often enough would make the bad old Hinds County hicks go away. It's a standard biolab propaganda technique. Repeat after us: safe safe safe.

The bad old Hinds County folks went away, all right. Just before the DHS site visit to Mississippi, Mississippi discreetly withdrew the Hinds County site from further consideration.[26]

A few weeks later, the *Clarion-Ledger* would run an article praising Flora—Mississippi's eventual NBAF finalist—as "the little town that could."[27] Seems folks there knew better than to question what the state capitol crowd wanted to bring their way—and the mayor knew what to do with ignorant yahoos—either "edyacate em" or call them idiots and be damned to them.

The *Clarion-Ledger* would still be fuming about Hinds County even after Flora chugged its way into the Final Five. "It's unfortunate that some communities let fear and ignorance curtail their consideration. . . . It was embarrassing earlier this year that some residents around Hinds County's J.C. 'Sonny' McDonald Industrial Center near Terry protested to have that site removed as a potential location."[28]

And how had they done this? Why, they had maliciously destroyed the reputation of an upright, God-fearing germ lab: "Fanning fears, local residents gossiped that the soil could be contaminated, or the water supply, or their livestock could die off, or that terrorists might target the area." When, if they had allowed themselves to be properly "educated," they would understand that none of these terrible things could possibly occur, because, because the facility would be "state of the art"—and because someone had even called it an "animal CDC."[29] Who's calling who ignorant here?

Down in Texas, it was also emphasize-the-need-for-community-support time, as the *San Antonio Express-News* threw out a brief flurry of biolab fluff just prior to the DHS February 16 deadline: "Alamo City a Perfect Fit for Biological Laboratory"[30] (editorial); "Biodefense Lab Needs Support of Community"[31] (article); "S.A. Well-Positioned to Win Federal Vaccine Lab"[32] (article); "City's Future Looks Incredibly Good"[33] (article).

Georgia was revving up local propaganda efforts, launching a Web site and a series of public meetings at organizations like the Athens Kiwanis Club. As part of its propaganda efforts, Georgia was hyping the presence of its new BSL-3-Ag facility, the $42 million Animal Health Research Center, and of the Southeastern Poultry Laboratory, which housed the main U.S. research program on avian flu. The *Athens Banner-Herald* described the lab "as one of the most biosecure buildings in the United States, designed and built so that none of the potentially dangerous viruses and bacteria used in the lab can ever escape to wreak havoc in the outside world." Interestingly, however, this technological marvel was being simultaneously characterized as "outdated" and "antiquated"—so that a $160 million new building could be constructed.[34] Biodefense two-step? Heck, yeah.

Out in Tracy, California, biolab propaganda fell on deaf ears. Residents there—like the neighbors of Plum Island—knew better than to trust government safety assurances and going-through-the-motions environmental impact statements. They had lived in the shadow of WMD research too long for that. They were already battling one BSL-3 facility with visions of deadly germ aerosols dancing in their heads. In January 2007 (two months after my WKU E-mail expressing concern about the WKU student's research at Livermore), the Tracy City Council voted 3-1 to oppose placing NBAF at Livermore's Site 300. By April, Tri-Valley CARES had gathered more than 7000 petition signatures, E-mails, and telephone lettergrams (carried by Working Assets for a small fee) against the facility.[35]

In Wisconsin, patent lawyer George Corrigan was continuing his crusade against NBAF there. In addition to the Web site he'd started in December, Corrigan would send out 1800 letters and distribute more than 800 yard signs.[36] His efforts fell on receptive ears. In Wisconsin, the opposition was not all—or even primarily—about safety; Dunn was known nationwide for its efforts to protect farmland and open space and, since 1996, residents had been paying several hundred dollars of extra taxes each year to fund a program buying development rights from farmers. Before they submitted their proposal, university officials had met with Dunn's town chairman, Ed Minihan, and land-use manager, Renee Lauber. The two had told UW bluntly that the facility would be at odds with

Dunn's land-use plan: "Our vision for that land is land, not an 11-acre block of cement."[37]

The facility would also require a conditional use permit from the Zoning and Land Regulation Committee of the Dane County Board of Supervisors, and on April 25, that committee followed two other County Board subcommittees in approving a resolution opposing UW's plan to build NBAF in Dunn.[38] The writing was clearly on the wall for the Wisconsin NBAF efforts.

CHAPTER SEVEN

Culture of Recklessness
Building the Frankenstein Germs

Brewing the Super Flus, and Swine Flu Paranoia

Early media reports, quoting CDC officials, described the swine flu circulating in spring 2009 as "having a unique combination of gene segments not seen in people or pigs before"—a mixture of "human virus, avian virus from North America and pig viruses from North America, Europe and Asia." Mixes of bird, pig, and human flu had been documented before, but not "such an intercontinental combination with more than one pig virus in the mix."[1]

The standard explanation for such hybrids is that pigs readily accept flu viruses from humans and birds, and serve, Laurie Garrett explained in a May 11 *Newsweek* article, as "four-legged viral mixing vessels. . . . The jackpot events in influenza evolution occur when two different types of flu viruses happen to get into an animal cell at the same time, swapping entire chromosomes to create 'reassorted' viruses."[2] Garrett astutely noted how factory farming practices—cramming billions of pigs and chickens into tiny spaces—create fertile breeding grounds for this sort of genetic reassortment.

Various web blogs, however—and one Australian scientist—also raised the possibility that the virus was a manmade concoction that might have been accidentally or deliberately released. In the aftermath of the Ivins revelations, or the Ivins allegations—depending on your skepticism about the FBI's case—such suspicions might be understandable.

Engineering dangerous pathogens, including freaky new forms of the flu, has become the germ lab version of extreme sport. Some of our germ lab daredevils have been gleefully contriving the sort of "jackpot events" (first prize, a new flu to defend against) Garrett mentions. Unfortunately, like other germ lab researchers, these adventurous souls do have accidents from time to time.

In 2004, articles in *Science* ("Tiptoeing around Pandora's Box")[3] and *The New Scientist* ("Superflu Is Being Brewed in the Lab")[4] revealed and questioned efforts at the CDC and other facilities to mix human and avian flu genes and create a bird flu chimera transmissible between humans. According to the *Science* article, the researchers in question want to know if a pandemic is possible, and if so, just how big the pandemic would be:

> Leaving nature to take its course, a pandemic could be ignited if avian and human influenza strains recombine—say, in the lungs of an Asian farmer infected with both—producing a brand-new hybrid no human is immune to. By mixing H5N1 and human flu viruses in the lab, scientists can find out how likely this is, and how dangerous a hybrid it would be.

This may seem a bit like a kid sticking a lit match into a can of gasoline to see if it will cause an explosion. The scientists apparently believed the experiments would allow them to determine the imminence of a pandemic:

> [T]he experiments would provide a badly needed way to assess the risk of a pandemic. If they indicate that a pandemic virus is just around the corner, health officials would further intensify their fight in Asia and go full-throttle in stashing vaccines and drugs; if not, they could breathe a little easier.

The article does not explain how the scientists' own creation of a dangerous virus will allow them to know how soon Nature might get around to the job itself. Presumably they'll compare the genomes of their new chimera toys to some of what's already circulating. In sober moments, however, even scientists recognize that they lack both the omniscience to know everything that's circulating and the omnipotence to make Nature follow the scientists' own blueprint. Depending on how you want to look at it, Nature either has a mind of her own, or behaves like an erratic dope fiend suffering from post-traumatic stress disorder (both the stress and the dope having been supplied by Nature's human cotenants). Wendy Barclay, of the British University of Reading, who considered and then abandoned the idea of creating a pandemic flu virus, agrees: "If you get a negative, how can you be sure that you have tested every option?"[5]

One would never have known that scientists or public health officials needed an experiment to tell them pandemic flu is a serious possibility and a serious risk. They have warned us of the dangers often enough. An experiment that suggested the contrary would be a bit embarrassing, after all. What would the flu scientists do, if they couldn't conduct experiments showing that what flu scientists do is very important, because the flu is a very dangerous disease that can kill you? In fact, the study eventually published by the CDC scientists, in May 2008, began by stating that "The emergence of an influenza virus that will cause a pandemic is inevitable and therefore preparedness is mandatory." The study also acknowledged that "we cannot predict the mechanism by which the pandemic influenza virus will emerge."[6]

The CDC study analyzed 63 possible "virus reassortants" derived from combining the H5N1 bird flu strain and the H3N2 human flu strain. It discovered that "nearly one-half of the reassortants replicated with high efficiency *in vitro*, revealing a high degree of compatibility between and avian and human virus genes." Thirteen of those reassortants were also very effective in killing mice.[7]

So, the CDC experiment served as a sort of combination E-Harmony and brothel for flu viruses. It found that two disreputable strains showed a lot of affinity for each other, put them together to see if they would procreate, found that the answer was yes and that various sorts of monstrous offspring were the result. Therefore, we as the public need to be concerned about any flu monsters we may encounter. Except, of course, the ones our adventurous scientists created, which are all locked safely away in "secure" (but not necessarily BSL-4) facilities. Or are they?

Science said of these projects in 2004:

> Such experiments can give the world a better handle on the risks, but they could also create dangerous new viruses that would have to be destroyed or locked up forever in a scientific high-security prison. An accidental release—not so far-fetched a scenario given that the severe acute respiratory syndrome (SARS) virus managed to escape from three Asian labs in the past year—could lead to global disaster.[8]

New Scientist also warned that if the scientists succeed, "they will have created a virus that could kill tens of millions if it got out of the lab."[9]

The *Science* article revealed that the CDC had begun its efforts to create a human/bird flu hybrid, under the direction of Nancy Cox, in 2000, three years before the major 2003 bird flu outbreak. Reassortment experiments were enthusiastically promoted by the World Health Organization and its principal flu scientist, Klaus Stohr. Ironically, both Cox and Stohr would serve as prominent media commentators during the 2009 swine flu episode (by then, Stohr was serving as "Global Head—Influenza Franchises" for pharmaceutical giant Novartis). Cox quickly dismissed the suggestion of Australian scientist Adrian Gibbs (reported on May 12) that the new virus might have been created by human error. The WHO followed suit two days later. This despite the fact that the source of the outbreak had not in fact been identified.

Digging Up a Vampire's Coffin

Creating human/bird flu hybrids is not the only example of risky influenza research. In the 1950s, scientists began trying to recreate the "extinct" 1918 flu virus, which had killed something like 50 million people worldwide. Initially they tried, unsuccessfully, to revive the virus from victims buried in the Alaska permafrost. Forty years later, in the mid 1990s, Dr. Jeffrey Taubenberger of the U.S. Armed Forces Institute of Pathology started to screen preserved tissue samples from

1918 influenza victims. Taubenberger's team apparently launched their efforts because it seemed like an interesting thing to do ("far and away the most interesting thing we could think of")—as opposed, say, to trying to cure lupus or Lou Gehrig's disease. By 2002, they had sequenced four of the strain's eight viral RNA segments.[10]

Teaming up with researchers from New York's Mount Sinai School of Medicine, Taubenberger's team started reconstructing the strain in various ways. A May 2009 article would call this reconstruction "rescuing."[11] (Odd form of disaster relief, this.) Initially, they combined gene fragments from current influenza strains with one or more genes from the 1918 strain.

In 2005, Taubenberger's team published the entire genome sequence for 1918 flu in *Nature*.[12] Meanwhile, a CDC team had used Taubenberger's research to actually bring the 1918 flu back to life, and they published the article describing their discoveries in *Science*. Their research included infecting mice with the "rescued" virus. Not surprisingly, the CDC discovered that the 1918 flu was highly virulent and dangerous. 39,000 more virus particles were found in mouse lung tissues four days after infection than in a current Texas strain. The mice lost 13% of body weight only two days after infection. (Weight loss from the Texas strain was brief and transient.) All mice died within six days of infection with the 1918 virus; no mice died from the Texas strain.[13]

Some scientists lauded the groundbreaking work of Taubenberger and the CDC. But quite a few others found the research projects disturbing. Even before the entire 1918 genome had been published, Peter Jahrling of *Hot Zone* fame, himself not generally the sort of person to shy away from risky experiments, said that research combining 1918 genes with those of other flu strains was like "looking for a gas leak with a lighted match."[14] After the virus was brought back to life, Richard Ebright said "the researchers have constructed, and provided procedures for others to construct, a virus that represents perhaps the most effective bioweapons agent now known." Ebright said the risk of accidental release, or of theft by a "disgruntled, disturbed, or extremist" laboratory employee—that baseline risk of the "human element" again—bordered on inevitability. Jonathan Tucker, a policy analyst at the Center for Nonproliferation Studies in Washington,

D.C., pointed out that the publication of the entire sequence made it easier for others to artificially construct the virus.[15]

Concern arose from the very beginning that experiments involving 1918 flu genes, or hybrids of avian and human flu, should be conducted at the highest containment level, BSL-4. But they were often conducted at BSL-3, or something called "BSL-3 enhanced." University of Wisconsin researcher Yoshihiro Kawaoka was criticized in 2004 for moving 1918 flu research from a BSL-4 lab in Winnipeg, Canada to a BSL-3 lab in Wisconsin.[16] The WHO's Stohr said then, "What we mustn't forget is that what they're working on is not the 1918 virus."[17] That was because they hadn't yet reconstructed the whole genome. But a few months later, what "they" were working on *was* the 1918 virus.

In October 2005, the editors of *Nature* and *Science* defended their respective decisions to publish the Taubenberger and CDC studies.[18] Shortly after the publications, the CDC added the reconstituted virus to the "select agent" list. The CDC said, however, that BSL-3 "enhanced" was sufficient containment for working with the virus. It apparently wanted the dangerous pathogen distributed as widely as possible.[19]

Nature's subsequent November 10 issue seemed at the least nervous, if not downright schizophrenic, about the CDC's approach. On the one hand, an editorial said that the decision not to recommend BSL-4 containment "seems justifiable, in the interests of rapid research." Yet the same editorial seemed apprehensive about just how widely the virus would be distributed: "Some sharing," it said, "is needed to accelerate progress in understanding its virulence—but it will also increase the risks of an accidental release." The editorial noted that the 1918 virus was hard to contain and spread rapidly between people, and that the threat of accidental release was very real. *Nature* seemed particularly concerned that other countries were not subject to the CDC's guidance, not acknowledging that its own publication of the genomes had made the virus's duplication by other countries all the more likely.[20]

An article by Andreas von Bubnoff in the same issue of *Nature* had a more consistently negative tone. The article reported that the CDC planned to mail the virus to any BSL-3 labs that requested it. (So much for locking it away in a "highly secure prison.") The article said this was contrary to the CDC's prior assurances about containment and security. The CDC had initially responded to concerns that the virus might escape by noting that only one researcher, with extensive background checks, had access to the virus, and by stating that it would not be sharing the virus with other labs.[21] Now, it was not only planning to share the virus with any BSL-3 facility that wanted it, but was planning to send it through the mail.

Harvard virologist Jens Kuhn said the virus should never have been recreated in the first place: "We have enough bugs to deal with on this planet already."[22] Kuhn's point about the proliferation of the 1918 flu virus is the same point made by the GAO about the proliferation of biodefense research generally: the greater the number of labs with access, the more likely accidental release or theft. He said access should be restricted by international agreement to a few labs worldwide.[23]

In an illustration of how likely international proliferation now was, the National Microbiology Lab in Winnipeg, Canada, indicated in November 2005 that it planned "to follow the lead of American researchers and resurrect the virus that caused the 1918 Spanish flu epidemic." Kuhn said he feared revitalization and use of the virus by American and Canadian researchers would rekindle the bioweapons race by prompting other countries to say: "Well, if you can do it, we have a right to do it, too."[24]

The CDC seemed not in the least perturbed by all this, announcing plans to test various mutations of the reconstructed virus for virulence. The University of Washington in Seattle announced its interest in testing mutations, and then the full 1918 virus, on macaques. The Winnipeg lab said it also planned to test the virus on macaques.

Ironically, an article appearing in the May 2009 *Journal of Virology*, in the midst of the swine flu scare, brought the whole 1918 virus controversy into the swine flu excitement.[25] It also indicated how far research with the 1918 virus had proceeded. Scientists had now demonstrated to their satisfaction that the 1918 virus was "highly pathogenic" in mice, ferrets, and nonhuman primates. Perhaps, having brought the 1918 virus back to life, they needed to determine what animals could serve as a new breeding ground to reintroduce it to the human race. Or perhaps 50 million bodies weren't enough to convince them that the 1918 virus was indeed deadly to humankind.

The typical germ lab double standard is at work here. In the spring of 2009, after it seemed that the swine flu scare might have been a bit excessive, officials warned that the virus might mutate by the fall and become a real killer. But this thought didn't occur to the CDC when it resurrected a germ that already had killed 50 million people. Arguing that current drugs and vaccines would be some help, they gave no consideration to the possibility of their own lab production mutating into an even deadlier form.

The new study revealed that a group of Canadian, American, and Thai researchers had successfully infected swine with the 1918 pandemic flu virus; "however, this species does not appear to be as sensitive to the virus as mice, ferrets, and nonhuman primates"[26]—or as sensitive as human primates, one presumes. What the study did not mention, having been submitted for publication in November 2008, is the possible role that accidental reintroduction of the original 1918 virus into the world's swine population—with pigs' inherent excellence as viral mixing vessels—might make for further dangerous flu mutations.

The mice of the world can rest easier, however. Just as the swine flu episode was getting started, researchers from the CDC and an American pharmaceutical company, Novavax, reported they had successfully vaccinated mice against the 1918 flu virus.[27] Presumably this gives the researchers some hope that, having "rescued" the 1918 virus from oblivion, they can now rescue humanity from the 1918 virus, should their containment protocols not prove as godlike as their gene-splicing.

In Europe, A Mysterious Accident

On February 17, 2009, reports began appearing that flu vaccines supplied by the Austrian facility of U.S. pharmaceutical company Baxter International had been contaminated with the H5N1 avian flu virus. Baxter acknowledged that it had supplied contaminated material, but wouldn't say how the contamination had occurred, except that "It was a combination of just the process itself, (and) technical and human error." The company said elaborating would give away proprietary information about its production procedures.[28]

Baxter said it had sent "experimental virus material," not vaccines intended for human use. The material was supposed to be H3N2 (human flu) virus product, for animal testing, but had been contaminated with live H5N1 virus. Baxter distributed the contaminated product to an Austrian company, Avir Green Hills Biotechnology, which disseminated the product to subcontractors in the Czech Republic, Slovenia, and Germany. The problem was discovered after ferrets inoculated with the material died. Ferrets don't normally die from human flu strains, but they *do* die from H5N1.[29]

Thirty-six or thirty-seven people were exposed to the contaminated product, but no one was actually infected. The danger was basically this: anyone who had been co-infected with the two viruses might have incubated a hybrid virus easily transmissible between people. European public health authorities called the incident a "serious error."[30] The CDC had already produced viable hybrids of the two viruses.

Baxter was on the verge of obtaining a European license for an H5N1 vaccine, which might explain why it had H5N1 virus on hand, except that the vaccine was being produced at a Czech facility, not the Austrian facility where the contamination occurred.[31] A May 22 story on Bloomberg.com revealed that the United Kingdom, France, Belgium, and Finland had agreed to purchase 158 million doses of a vaccine for the new H1N1 swine flu strain from Baxter and London-based GlaxoSmithKline.[32]

Baxter's refusal to discuss details of the accident, along with the fact that it stood to reap enormous financial gains from a flu pandemic, led some bloggers (such as Prison Planet.com) to suggest Baxter had deliberately tried to provoke a pandemic.[33] (Eerily, one blogger had even predicted, just after the Baxter episode—and just prior to the spread of H1N1—that there would be a major flu incident within the next month or so.) Prison Planet said several Czech newspapers were making similar suggestions.[34]

Ironically, Baxter's own propaganda about the safety of its work may have contributed to the suspicions. Baxter was performing the research at BSL-3. Baxter issued the standard safety assurance: "The material was stored and handled throughout under high biosafety conditions." Prison Planet relied on these standard assurances—and a related discussion at Natural News by "health expert" Mike Adams—in arriving at the headline: "'Accidental' Contamination of Vaccine with Live Avian Flu Virus Virtually Impossible." Prison Planet quoted Adams as saying: "The shocking answer is that *this couldn't have been an accident.* Why? Because Baxter International adheres to something called BSL3 (Biosafety Level 3)—a set of laboratory safety protocols that prevent the cross-contamination of materials."

The author of the Prison Planet blog then goes on to quote the Wikipedia explanation of these protocols, which he calls the BSL-3 Code of Conduct:

> Laboratory personnel have specific training in handling pathogenic and potentially lethal agents, and are supervised by competent scientists who are experienced in working with these agents. This is considered a neutral or warm zone. All procedures involving the manipulation of infectious materials are conducted within biological safety cabinets or other physical containment devices, or by personnel wearing appropriate personal protective clothing and equipment. The laboratory has special engineering and design features.[35]

Unfortunately, the Wikipedia description comes straight out of the germ lab propagandist's playbook, the standard description of state-of-the-art-containment features that is recited ad nauseam to worried germ lab neighbors in Kansas or Boston or Frederick, Maryland. Unfortunately, neither the germs, nor the scientists, nor their high-tech equipment adhere religiously to the BSL-3 Code of Conduct. Accidents are not "virtually impossible," but rather more or less inevitable.

That didn't keep Baxter from trying to wrap the incident in the warm, cozy language of biosafety protocol. On March 3, after its shares had dropped more than 8 percent in two days, Baxter granted an exclusive interview to LifeGen.de, a European online life sciences journal. The "interview" was more like an abstract of an interview, with only six questions, and a one-word answer of "No" to the question "Will the contamination in Europe lead to management changes in your company?" Baxter tended to recite, in robot-like fashion, the same answers to different questions.[36]

One of the robot answers, repeated twice in response to different questions, was this one: "Exposure was highly unlikely—the individuals working with the experimental material were operating in laboratory containment conditions specifically designed to prevent exposure."[37] The Baxter robot presumably meant that *infection* was unlikely, since journalists reported that thirty-six people or so had in fact been "exposed." And the same laboratory containment conditions "specifically designed to prevent exposure" were also *designed* to prevent contamination accidents in the first place, so one might feel less than reassured by what the laboratory containment conditions were *designed* to do.

Another robot answer—this one repeated three times—was the soothing assurance: "The root cause of the incident has been identified. It was due to a unique combination of process, technical and human error in a procedure used for this specific research project in our facility in Austria (Orth). The chances of such a confluence of events repeating itself are virtually impossible."[38]

Still another robot answer, this one repeated only twice, was the confusing declaration: "Cross-contamination of commercial product, material, or other clinical experimental material has been absolutely excluded."[39] Was the Baxter robot playing semantic games, declaring that the undeniable contamination was not "cross-contamination"? Or is it declaring that once it sent out contaminated product, there was not further contamination by the recipients?

Without knowing what precisely went wrong, one has no reason to trust any of Baxter's reassurances. Cross-contamination had already occurred within the experimental material (unless the mistake involved Baxter producing a hybrid of the two viruses and inadvertently sending the hybrid out—which would make one wonder what Baxter was up to, producing a hybrid in the first place); so how could Baxter confidently assert that no cross-contamination had occurred once the material was sent out, especially since the lab workers didn't know they were working with bird flu or a bird flu hybrid?

The robot wasn't telling: "Further details about the process will not be discussed as it is proprietary information." Furthermore, said the robot, "We see no need to speculate on how the size of a lab had bearing on this incident."[40]

The Baxter episode reminded many observers of a similar event in 2004 and 2005, when Meridian Bioscience of Cincinnati included H2N2 influenza, the strain which caused the 1957 pandemic, in a panel of virus samples sent to over 6000 labs in eighteen countries—including Saudi Arabia and Lebanon—for use in routine training and certification. The virus was mislabeled in the kits as a relatively benign H3N2 strain.[41]

Jared Schwartz, a spokesperson for the College of American Pathologists, which had contracted with Meridian to supply the kits, said Meridian's paperwork indicated that the strain was benign: "For reasons I don't understand and Meridian doesn't understand, the documentation they had was incorrect." The source of the labeling was unclear; possibly Meridian had obtained the strain from another company that had misidentified it. The spokesman said current federal regulations would have allowed Meridian to ship the pathogen even had it known it was H2N2.[42]

Five days later, however, the *Washington Post*, which had carried Schwartz's comments, stated in an editorial, "Renegade Flu Virus," that "no one at CDC is yet able to explain whether Meridian put this particular strain of virus in its testing samples knowingly or by accident."[43]

Because the virus was easily transmitted from person to person and anyone born after 1968 would have no natural immunity to it, it could have caused another deadly pandemic if a lab worker became infected.[44]

The mislabeled H2N2 flu sent out by Meridian—and whatever had been sent out by any companies who might have supplied it to Meridian—had been out there posing as H3N2 for months before a second lab screw-up inadvertently revealed the first one. The proficiency kits were designed to determine if the receiving lab could correctly distinguish influenza A from influenza B. Since H2N2 and H3N2 were both subtypes of influenza A, the proficiency exercise itself would not and did not reveal the labeling error.[45]

In Canada, routine tests on a patient sample from Vancouver showed the presence of H2N2—the first such presence in a human for 40 years. Eventually the contamination was traced to the Meridian virus panel.[46]

The World Health Organization began urging labs in possession of the virus to destroy it. Determining which labs had the samples sent out by Meridian was apparently relatively easy; the CDC reported that 83% had been destroyed by April 17. Some samples were found in FedEx warehouses in Mexico, Chile, and Lebanon. They "had been kept in locked facilities but had not been delivered to labs, for reasons that were not clear."[47]

A thornier problem—which may or may not have been resolved—was tracking down what had been sent out by companies other than Meridian. Robert G. Webster, a flu expert at St. Jude's in Memphis, told the *Washington Post*: "I have been telling WHO for a number of years that this is a dangerous virus that is still out there in more labs than they know. This may alert WHO and Homeland Security and whoever wants to know that each and every H2N2 sample from 1957 needs to be rounded up and locked down."[48]

Meridian, like Baxter, was more or less tight-lipped, simply stating that it "believes it has been and is in compliance with all applicable regulations."[49] This included placing the samples in vials, then in plastic bags, then wrapping them with absorbent and puncture-resistant material, then double-boxing the packages.[50] This was all more or less irrelevant to the main problem, that the deadly virus was shipped at all—to 6000 labs—and mislabeled to boot.

If Meridian had indeed complied with "existing regulations," this simply pointed out a problem with the regulations. As of 2005, the CDC had not yet added any version of influenza to the Select Agent list, and therefore had "no mandate to monitor [the] lab safety" of influenza research. Furthermore, the U.S.'s National Institutes of Health only recommended BSL-2 for influenza research. Canada followed the WHO recommendation of BSL-3.

The CDC said that it would speed up plans to require tighter security for dangerous flu strains, that it had not done so previously because influenza was not considered a possible bioterror agent.[51] The agency's statements illustrate the basic fallacies of biodefense proliferation. The initial act of bioterror involved the mailing of a dangerous pathogen by a member or members of the U.S.'s *own* biodefense complex. The U.S. government's response to that act involved putting anthrax into the hands of many, many more people, oblivious not only to the enhanced likelihood of another act of deliberate bioterror, but oblivious also to the possibility of accidents from this sprawling complex. Meanwhile, four years after 2001, a pathogen which had already killed more people than anthrax ever dreamed of was unregulated, roaming around freely, being mailed to six thousand labs who didn't even know what they were getting.

Even after the wild horses were already out of the barn, various assurances were offered that the danger wasn't really so great. Schwartz of the College of American Pathologists said he was not aware of any evidence that suggested a risk of infection from the strain, because companies that make the test kits typically use only diluted virus samples.[52] The CDC said, "I think what people need to understand is the very labs that receive these strains of influenza all have people trained to work safely and effectively with these viruses."[53] And the WHO said, "[T]he likelihood of laboratory-acquired influenza infection is considered low when proper biosafety precautions are followed."[54]

Yet people trained to work "safely and effectively" had without warning distributed a pandemic virus to over 6000 labs; and the mistake had only been discovered because other people trained to work safely and effectively had let loose a virus that contaminated a patient sample. All the safety talk thus seemed mere bravado.

There's Gold in Them Thar Pandemics

One reason Web blogs floated conspiracy theories about the swine flu incident was that flu hybrids were being produced willy-nilly in the labs. Another reason was the huge profits to be made by certain companies from a pandemic or pseudo-pandemic. As the business sections of various mainstream publications seemed to imply, anything that turns a profit can't be all bad. Thus, the April 28 *San Francisco Chronicle* carried a story titled "Flu's Bad News but Shot in the Arm for Biotech."[55] A few days earlier, Reuters had carried a related story, "Venture Capital Firm Set to Reap Rewards on Swine Flu."[56] The *New*

York Times followed up with a story on the same phenomenon, "Flu Fears Fuel Rally in Kleiner-Backed Firms."[57]

The *Chronicle* story noted a recent rise in the shares of Gilead Sciences Inc., and speculated that it might "have something to do with the royalties they're collecting on Tamiflu, the influenza drug the biotech company invented and sold to Swiss pharma Roche Holding AG in 1996."

The Web blogs noted something that the *Chronicle*'s business-section story didn't note, that someone named Donald Rumsfeld had once been chairman of Gilead Sciences, and that he'd continued to be the company's largest stockholder after becoming Secretary of Defense. In 2005 Rumsfeld announced he had budgeted more than $1 billion to stockpile the Tamiflu Gilead had created and was now receiving royalties on. President Bush called on Congress to appropriate another $2 billion. When news of the rather obvious conflict of interest leaked out, the Pentagon said Rumsfeld had decided to retain rather than sell his Gilead stock because to sell would have indicated he had something to hide. "That agonizing decision," noted one Web blog, "won him reported added millions as the Gilead share price soared more than 700% in weeks."[58]

All three stories noted the rise in the share prices of Novavax Inc. (79 percent as of April 28), which was developing a bird flu vaccine, and of BioCryst Pharmaceuticals (75 percent), which was developing a flu antiviral. The venture capital firm Kleiner Perkins Caufield & Byers had invested $20 million in Novavax, and $30 million in BioCryst, two of eight companies in the firm's Pandemic and Bio Defense portfolio. The share prices of both companies had previously experienced long drops from peaks reached in 2006, when avian flu made headlines.

Novavax, which uses "genetic information and 'recombinant, virus-like particle technology' to rapidly engineer a vaccine," contacted the CDC and Mexico's Ministry of Health to offer its help with the swine flu situation.[59] Novavax researchers had previously joined with the CDC in developing a vaccine that successfully protected mice against the 1918 flu virus. (One might have preferred a vaccine to depress the serotonin uptake of manic scientists and keep them from resuscitating dead viruses in the first place.) Should the 1918 virus mysteriously reappear among the human population, Novavax presumably stood ready to produce the appropriate vaccine, in the best traditions of government-subsidized American free enterprise.

Baxter, fresh out of the European contamination incident, "once again [found] itself in the [flu] action," as it requested a swine flu virus sample from WHO "to do laboratory testing for potentially developing an experimental vaccine."[60] Within a month, Baxter and GlaxoSmithKline had signed contracts to produce 158 million doses of the yet-to-be-produced vaccine for the United Kingdom, France, Belgium, and Finland.[61] The U.S., meanwhile, had ordered $811 million of components for a swine flu vaccine from GlaxoSmithKline, Novartis AG, CSL Ltd., MedImmune Inc., and Sanofi Pasteur.[62] For some of these companies, the more flus the merrier. Four months earlier, Novartis had signed a $486 million, eight-year contract with the U.S. to produce a stockpile of bird flu vaccine.[63] Sanofi Pasteur also had ongoing contracts with the U.S. to produce bird flu vaccine and other flu materials.[64]

Smallpox Chimeras and Other Dangerous Playthings

In January 2001, an Australian research team announced it had inadvertently engineered a killer mousepox virus. In an effort to create a mouse contraceptive by stimulating antibodies against mouse eggs, they had used the mousepox virus as a "carrier" for a gene that creates large quantities of the molecule interleukin 4 (IL-4). The mousepox should normally have caused only mild symptoms in the mice, but the mousepox modified with the IL-4 gene wiped out all the mice in nine days. The IL-4 had totally suppressed the immune response that combats viral infection.[65] The engineered virus also reduced the effectiveness of an existing mousepox vaccine by half.[66]

Mousepox doesn't affect humans, but it's one of several viruses closely related to smallpox (including cowpox, used to create the first smallpox vaccine), and the fear was that smallpox could be modified with an IL-4 gene to make it both more virulent and resistant to vaccines. One of the Australian researchers said, "It would be safe to assume that if some idiot did put human IL-4 into human smallpox they'd increase the lethality quite dramatically."[67]

One encouraging trait of those who rummage about in Pandora's pathogen box is their occasional willingness to criticize the potential recklessness of others. A less encouraging trait is their reluctance to let such considerations modify their own behavior.

Aside from wondering whether a mouse contraceptive vaccine is an appropriate and safe human intervention in the ecosystem, one might question the decision to publish the discoveries in the *Journal of Virology*. D. A. Henderson, who had led the fight to eliminate smallpox from the world, and served as a presidential adviser before heading the Center for Civilian Biodefense Studies at Johns Hopkins University, noted that "blueprints for making microorganisms more harmful" regularly appear in unclassified journals. The Australian researchers, after consulting their own country's Department of Defense, decided to go ahead and publish "to warn the general population that this potentially dangerous technology is available. . . . We wanted to make it clear to the scientific community that they should be careful, that it is not too difficult to create severe organisms."[68]

"Do as we say, not as we do" is perhaps not reassuring enough for the general population, even assuming they read the *Journal of Virology* (and thus, *got* the warning) in the first place. It's doubtful how many scientists were listening, the temptations of frontier-breaking research and prestigious publication being what they are. A couple of years later, a U.S. government-funded scientist at the University of St. Louis announced he had created an even deadlier form of mousepox, deliberately. This version killed all mice even if they had been vaccinated or treated with antiviral drugs. The scientist, Mark Buller, said his work was "necessary to explore what bioterrorists might do."[69]

Richard Ebright says of this type of research: "It's like the National Institutes of Health was funding a research and development arm of Al Qaeda."[70] It's easy to see his point. Scientists like Buller, "exploring what bioterrorists might do," are first discovering ways to make germs deadlier, then publishing the research so terrorists—with their alleged homemade labs in Al Quaeda caves—won't have to waste time figuring it out on their own. Presumably if we show them what to do and how to do it, they will at least be more predictable.

Buller had already applied similar techniques to the cowpox virus, which—unlike mousepox—does infect humans. These experiments were being conducted only at BSL-3. Ian Ramshaw of the Australian mousepox team said he had "great concern about doing this in a pox virus that can cross species."[71] Ramshaw's 2001 warnings had not kept him from creating his own deadlier mousepox versions, or from engineering a deadlier rabbitpox virus (like mousepox, not currently infectious to humans). Ramshaw did say the newly modified versions weren't contagious, so that they "would not cause ecological havoc by wiping out mouse or rabbit populations around the world if they escaped from a lab."[72]

Ramshaw said there was no reason to do the cowpox experiments, since his rabbitpox work had already shown that the technique which made mousepox deadlier worked for other viruses as well. In other words, Ramshaw had already blazed and mapped a promising trail for Al Qaeda; there was no need for Buller to do it as well.

Though it didn't seem to be hampering his own efforts too much, Ramshaw at least recognized some of the dangers: "While viruses containing mouse IL-4 should not be lethal to humans, recombinant viruses can have unexpected effects, he says. 'You'd hope the combination remains mouse-specific.'"

Ramshaw's concern about idiots getting it on with smallpox is unfortunately not hypothetical. The last natural outbreak of smallpox occurred in Somalia in 1977. About the same time, the WHO began trying to reduce the number of facilities holding the smallpox virus, to prevent an accidental release that could reintroduce the disease. (A 1975 survey had determined that 74 laboratories around the world held stocks of the virus.) A 1978 lab release at Britain's University of Birmingham that infected two people and killed one raised the concern level, and in 1980 the WHO adopted a resolution urging all countries that possessed the virus to either destroy their stocks or transfer them to one of four collaborating centers. In 1983 the CDC and the State Institute for Viral Preparations in Moscow became the "sole authorized repositories of the smallpox virus."[73]

In 1990, a WHO scientific advisory committee recommended that all known stocks of the smallpox virus be destroyed by December 31, 1993. Various considerations, including the possibility of a smallpox release by a rogue state or terrorist group, have caused the WHO to repeatedly postpone the date of destruction and allow continued research on the smallpox virus at the CDC and the Russian facilities at Vector. The destruction debate has continued.[74] As this book went to press, plans for destruction had been put on hold indefinitely.

Meanwhile, in late 2004, the WHO decided—in a highly controversial decision—to formally allow for the first time the genetic modification of the smallpox virus, including the creation of chimeras combining smallpox genes with other pox viruses like cowpox and monkeypox. (The CDC freezers already held crude versions of such chimeras created forty years ago. In 2002, the WHO asked the CDC to destroy those; the CDC did not comply.)[75] The Department of Homeland Security argued that genetic modification of smallpox would speed the development of drugs and vaccines. Susan Wright, a research scientist and bioweapons expert from the University of Michigan, told the Council for Responsible Genetics: "Genetic engineering could generate thousand of variations of natural pathogens. Designing drugs and vaccines against specific variations cannot strengthen defenses. On the contrary, such research serves only to increase the risk that novel pathogens will be released, either by accident or intention."[76]

Some of the CDC's smallpox research has involved efforts to artificially infect monkeys with the smallpox virus; if this proved feasible, monkeys could then be used to test vaccines and drugs. (Given the general tendencies of biodefense research, that would probably include testing vaccines and drugs against new smallpox chimeras that had not existed before the researchers created them.) As the 2004 journal article reporting on this research indicated, "There are no animal reservoirs for variola virus in nature, and most animal species cannot even be infected in the laboratory. Attempts to develop non-human primate models for variola infection and disease in the 1960s met with only limited success."[77]

The researchers found that they could infect the monkeys with extremely high doses of smallpox. They seem not to have worried that "successful" experiments of this sort might ultimately or inadvertently turn smallpox into a zoonotic disease transmissible between monkeys and humans and establish monkeys as a reservoir for the disease in the same way that chickens and pigs now serve as reservoirs for influenza.

A 2006 article in *New Scientist*, "Bioterror special, friend or foe?" noted several other genetic modification experiments in progress or being planned in U.S. academic labs: tweaking the anthrax toxin to render experimental drugs ineffective; enhancing the potency of botulinum toxin; transforming a harmless enzootic form of Venezuelan equine encephalitis into one that can kill horses and people; transferring genes that suppress the human immune response from one pathogen to another. Some of the scientists involved had already published details of their experiments, thus bringing their discoveries and techniques to the attention of potential terrorists. John Steinbruner, a security specialist at the University of Maryland, said of such efforts to keep one step ahead of the terrorists: "You're creating the threat you're concerned about with no reason to believe that anyone is doing that."[78]

The 1988 book *Gene Wars* laid out the various ways in which genetic engineering was even then producing new biological weapons (all in the name of biodefense, of course): (1) increasing pathogenicity; (2) finding ways to defeat immunity and circumvent diagnostic testing; (3) increasing controllability; (4) increasing environmental survivability; (5) creating ethnically selective weapons; (6) inventing biochemical weapons; (7) developing novel means of toxin development and manufacturing.[79]

Such research—creating bioweapons to see what an enemy might do—has been a staple of "biodefense" research ever since the U.S.'s offensive program was supposedly terminated. The authors of *Gene Wars* reviewed 329 biotechnology projects funded by the Department of Defense between 1980 and 1986. Fifty-one involved the creation of novel BW (bioweapons) agents, twenty-three creating agents to defeat vaccines, fourteen creating agents to inhibit diagnosis, seventeen creating supertoxins, five focused on aerosol delivery of BW agents, nineteen on biological vectors for BW agents, three creating drug-resistant agents, and fifteen increasing toxin production capability.[80]

Efforts to make pathogens more deadly—such as Boyer and Cohen's 1973 experiments splicing a gene conferring penicillin resistance into *E. coli*[81]—were part of the genetic revolution's very beginnings.

NBAF 9

ONE IF BY VAN, TWO IF BY HELICOPTER

SALVAGING A PROTEST

I WAS HOME ON APRIL 13, PACKING FOR A WEEKEND TRIP. IT WAS Friday the 13th, to be exact. There hadn't been a meeting of Citizens Against a Kentucky Biolab for six weeks. The call caught me off guard.

The call was from Jim Bruggers, the environmental reporter for the *Louisville Courier-Journal*. Had I heard about the planned site visit to Pulaski County by the NBAF selection team? No, I hadn't.

My name had been given him as a spokesperson for opponents of the lab. The call went on for fifteen, twenty, thirty minutes. I didn't have my arguments on the tip of my tongue, but I wasn't completely inarticulate either. I cited the problems at Fort Detrick and Plum Island and the concerns expressed in scientific magazines about the SARS escapes and some of the more apocalyptic genetic modification research.

None of this made it into the article that the *Courier-Journal* eventually ran, just that I had a concern about leaks.[1] It was a familiar newspaper stereotype by now. Nothing indicating I was a college professor with a Ph.D. and a law degree. Nothing pointing out that I had read, or even claimed to have read, hundreds of newspaper and journal articles and plenty of books on the subject. All they seemed to need from me in the newspapers was a Clem Cadoodlehopper quote that I was wurrit that that there thang might leak and leave a stain on my moonshine still. Been there, done that. Meanwhile, the congressman who said a Biosafety Level Four lab was like going to Wal-Mart, got to read his whole damned press release like Moses just back from Mount Sinai.

I wasn't happy with the reporting; I wasn't happy that the local group had done nothing now for six weeks; I wasn't happy to have this all coming at me out of the blue. I had changed my summer around in expectation of a bunch of rallies that never happened; I was damned if I would change my weekend now.

It would be three or four days before I would piece together what was happening. A DHS inspection team was visiting each of the 17 semifinalist sites; the schedule had been posted on DHS's NBAF Web site since at least early March. I discovered this after I went to the Web site looking for contact information for the inspection team. The consortium had no doubt been aware of the visit, but since they could certainly expect a protest, they obviously weren't going to give us a heads-up. They weren't sharing the itinerary with the *Courier-Journal* reporter either. And DHS was keeping the visits hush-hush too, in a weird sort of way; they were letting the press know "We're going to be making calls in your neighborhood"—but they weren't giving any details about when or where.

We would never get that itinerary, beyond knowing what day the DHS folks planned to be in town. But we would eventually—with the major assistance of Deb Bledsoe of Appalachian Science in the Public Interest—pull together an in-their-faces protest on the property adjoining the proposed Kentucky NBAF site. We invited the media; several showed up; and Channel 36 was filming away as the law-enforcement-escorted convoy drove past, and as the military helicopter dropped out of the sky and disgorged its collection of mostly Men in Black.

The black suits left by way of the vans; and this time they had to travel the gauntlet of placards just at the edge of the road.

Channel 36 would make our protest its two-minute lead story on the 6:00 news. The *Herald-Leader* and *Commonwealth-Journal* would both carry stories too. We couldn't have asked for a better outcome. The inspectors knew there was opposition; and the state knew it, too. We would even make it onto the national news scene, by way of the Associated Press.

Inspectors were also greeted by opponents in Wisconsin and Missouri. About twenty cars displaying X-ed out NBAF signs parked along Schneider and Dyerson roads near the proposed Wisconsin site, a barren cornfield. People also marched with signs "in orderly processions along Schneider and Highway 51, which connects the site to Interstate 90." The event had been organized by George Corrigan, who said, "DHS doesn't care so much what people say as what they'll do."[2]

And several members of the Osage Group of the Sierra Club put together a welcoming committee when the DHS visited Columbia, Missouri.[3] Missouri opponents had been late in getting organized, but they too would soon make their impact felt in a major way during these pivotal days in the NBAF process.

CHAPTER EIGHT

Follow The Money
Biodefense as Academic-Military-Industrial Complex

As he prepared to leave office in January 1961, President—
and former General—Dwight D. Eisenhower spoke to the American
people and warned against "the acquisition [in the councils of gov-
ernment] of unwarranted influence, whether sought or unsought, by
the military-industrial complex." He also warned against the domina-
tion of university scholarship and research "by Federal employment,
project allocations, and the power of money"—and against "the equal
and opposite danger that public policy could itself become the cap-
tive of a scientific-technological elite."[1]

Forty-six years later, a somewhat chastened Colin Powell, in a
little-noticed interview published in the September 2007 issue of GQ,
updated Eisenhower's warning by advising that America was "taking
too much counsel of our fears . . . The only thing that can really
destroy us is *us*. We shouldn't do it to ourselves, and we shouldn't use
fear for political purposes—scaring people to death so they will vote
for you, or scaring people to death so that we create a terror-industrial complex."[2]

On September 11, 2007, a college newspaper reporter caught up with former Secretary Powell as he prepared to speak at Oklahoma University, and asked him to elaborate:

> WILL PRESCOTT: What would a Terror-Industrial Complex look like, and are we headed in that direction?
>
> POWELL: Well, I think we have to beware of it. We're spending an enormous amount of money on homeland security—and I think we should spend whatever it takes—but I think **we have to be careful that we don't get *so* caught up in trying to *throw money* at the terrorist and counterterrorist problem that we're essentially creating an industry that will only exist as long as you keep the terrorist threat pumped up.** And so *that* would be the context of that comment, and I feel strongly about it—just as, many years ago, General Eisenhower warned about a *Military*-Industrial Complex. I just wanted to make a point: *defend* ourselves, screen ourselves, do everything we can to go after terrorists and defeat terrorism, because it *is* a threat, it *is* an enemy—but let's keep it in perspective. Let's keep it in context, because the United States has *many* needs. We have needs to deal with the poverty of some of our people, education, the environment. **There are lots of things America needs to do, and we have to make sure we only spend that which is absolutely essential in our military, on our police forces, and on our [anti-] terrorist activities** [emphasis added].[3]

Powell's warning may have come too late. The terror-industrial complex is already with us, and the bioterror industrial complex is right in the thick of things: a horde of pork-seeking politicians, profit-seeking pharmaceutical and security companies, university administrators and grant-hungry scientists, give-me-some-of-that-germ-lab-construction-money chambers of commerce and economic development consortiums, oh-isn't-it-nice-to-have-all-this-bio-defense-money government agencies, and bioterrorism-is-the-single-greatest-problem-of-the-modern-world think tanks. All these feeders at the public trough, after bloating themselves over the last several years with a $60 billion biodefense buffet, are still screaming that it's not enough. Because after all, we remain "vulnerable" to biologic attack. As though, if we just spent enough, we would suddenly be invulnerable.

The biodefense complex deserves a book unto itself. One can only pull out a few highlights in the present pages, point to a few tentacles of this far-flung, amorphous entity.

We might begin with some of the people who found a silver—well, a gold—lining in the 2001 attacks. Anthony Fauci of the National Institute of Allergy and Infectious Diseases, for one, whose biodefense budget grew 3000 percent between 2001 and 2004: "We've been given a great positive boon. . . . Even if we never, ever get attacked, the benefits to society will be enormous."[4]

Sure, I certainly feel that our country's new network of 1300-plus high-containment germ labs will do all kinds of things for our society. Give us something to worry about other than bioterrorists, for one thing. What the germ lab giveth, the germ lab taketh away.

Or David Franz, former USAMRIID commander, later associated with Kansas State (and its NBAF effort) and the Midwest Research Institute. Out hyping the bioterror threat, Franz had difficulty putting the five deaths and 22 illnesses caused by the anthrax letter attacks into perspective. No such problem when he got to thinking about all those nice biodefense appropriations. As Franz explained to ABC News on April 4, 2002: "I think a lot of good has come from it [the anthrax attacks]. From a biological or medical standpoint, we've now five people who have died, but we've put about $6 billion in our budget into defending against bioterrorism."

Praise the Lord and pass the ammunition money.

Then there are the biodefense corporations. The Boston-based BioDefense Corporation, for instance, maker of the Mail Defender, intended to kill anthrax and other pathogens sent through the mail by the biodefense complex. Looking like a cross between a safe and a washing machine, the Mail Defender subjects up to three pounds of mail at a time to a 90 minute cycle of heat and ultraviolet irradiation.[5] A revamped version holds up to 100 pieces of mail and "rotates for 45 minutes while evenly exposing letters to . . . microwave, ultraviolet light, and broad-beam lights."[6] It does so without destroying the letter itself, so the DHS forensics lab will find it easier to trace the next letter to its precise origin in the U.S. biodefense complex.

The Mail Defender reportedly carries a $90,000 price tag. Bio-Defense claims there are 15,000 in use. Several have been purchased by various federal agencies, the United Nations, and the royal Saudi embassy. But "BioDefense is now targeting [sic] 9,000 federal buildings that it believes [in completely disinterested fashion, of course] should scrutinize their mail stream. A potential facility for production of the Mail Defender is also being planned for northern England as a jumping-off point for its planned penetration of the European market."[7]

BioDefense is fond of issuing a press release after any new white powder hoax, which it prefers to describe as "mail-based terrorism"[8] or "bio-chemical assaults."[9] Sample headlines: "MEDIA ADVISORY: BioDefense Corporation Sees Rash of 'White Powder' Incidents at Reuters, New York Times, and Major Banks as Real Bio-Chemical Assaults, Not 'Hoaxes'";[10] "While 'White Powder' Hoaxes Hit Governors of Six States, BioDefense Corporation's MailDefender® Only System That Kills All Known Pathogens, Including Anthrax, Ricin, Smallpox and More."[11] In another press release, issued just after the 2008 presidential election, BioDefense claimed that "[a] disturbing increase in reports that gun sales are dramatically increasing now that the presidential election is over indicates that other forms of intentional mayhem such as bio-chemical assaults might also increase."[12]

Thus, the need for increased purchases of the Mail Defender.

For its marketing in the United Kingdom, BioDefense enticed General Lord Guthrie, former chief of the British Defense Staff, to partner with it. Guthrie's initial role, apparently, was to try to hype the British bioterror threat to the same level of anxiety as the U.S.'s: "The United Kingdom needs to wake up to the threat of bioterrorism, which is far more serious than defense experts realize."[13]

Take a page from the U.S., Lord Guthrie: send those boys in the Parliament a few anthrax-laced letters. Then they'll want the Mail Defender.

Other companies in the anthrax business include Universal Detection Technologies, producer of "early warning monitoring technologies," including an airborne anthrax sensor. UDT reported a 5000% revenue increase in 2008.[14]

Then of course there is the quest for anthrax vaccine money. First on the scene was Emergent Biosolutions, in the business of supplying the current anthrax vaccine—the one forced on two million members of the U.S. military—with all its history of adverse effects and lingering questions about safety. Emergent has delivered about 20 million doses of the existing vaccine since 1998; in 2007 it won a three-year $448 million contract to provide 18.75 million doses for the national stockpile.[15]

But there is also the ongoing competition for a Health and Human Services contract to deliver a new, improved anthrax vaccine. California-based VaxGen initially won that contract, then said to be worth up to $877.5 million. VaxGen committed to delivering 75 million doses by 2006. But then VaxGen had problems both with the product and with the FDA, which questioned the vaccine's reliability.[16]

A critical December 2006 *Harper's* article called the VaxGen situation Exhibit A in what it called the Project BioShield boondoggle, and cited the suspicions of VaxGen's competitors that the company had received help in winning the contract from an HHS contracting official close to Lance Gordon, VaxGen's CEO.[17]

HHS canceled the deal with VaxGen, and the company collapsed soon thereafter, after having spent some $150 million developing the vaccine. Emergent Biosolutions then bought VaxGen's vaccine-in-progress for $2 million, plus $8 million in what are called "milestone payments," plus a percentage of sales.[18]

The *Washington Independent* attributed VaxGen's collapse to the government's bungled management of the BioShield program. It charged that Emergent had used an army of lobbyists to undercut VaxGen's relationship with HHS, and pointed out Emergent's CEO and wife had donated more than $220,000 to lobbying and political campaigns. Its lobbying campaign against VaxGen had even included hiring two former aides to Vice President Cheney. Meanwhile, "Congressmen like Michael McCaul (R-Texas) and Mike Rodgers (R-Mich.), both recipients of Emergent executives' campaign donations, attacked the VaxGen contract in committee hearings, while Emergent's lawyers wrote newspaper op-eds attacking the company."[19]

The *Independent* said the government changed the rules after VaxGen's collapse, indicating it would allow several years instead of two for delivery, and providing milestone payments and development costs along the way.[20]

Several firms were now competing for the contract to produce the new vaccine, but the two companies getting the most attention were Emergent and PharmAthene. Whereas Emergent had bought out VaxGen's vaccine-in-progress, PharmAthene had bought the bio-defense unit of the British firm Avecia, with its anthrax vaccine-in-progress Spar Vax.[21] In September 2008, both PharmAthene and Emergent were awarded "development contracts."[22]

PharmAthene's CEO acknowledged that like Emergent, Pharm-Athene also "has strong relationships with homeland security officials and lobbyists who help move biodefense legislation forward, an important competitive advantage."[23]

Ah, yes, those strong relationships with homeland security officials and lobbyists that make the political and business worlds go round. Exhibit A: "Who's Hot": from the August 7, 2009 issue of *Business Week*: 35-year-old Courtney Banks, inspired by those Herman Wouk novels of heroic warfaring, *The Winds of War* and *War and Remembrance*, to devote her life—after a brief tour of duty with the civilian Pentagon—to heroic war profiteering, first as a vice-president at defense contractor Raytheon, then as head of her own consulting firm to "help clients attract federal stimulus cash."[24]

To get clients "to the right people," Banks employs "about 20 former defense and security honchos from federal agencies, as well as veterans of consulting firms and defense companies. Her team includes former FBI investigators, CIA operatives and analysts, Air Force and naval intelligence officers, and diplomats. Many still have their security clearances."[25] And those security clearances are so helpful, because homeland security is such a growth industry: "Homeland security spending is growing 10% a year and will hit $178 billion by 2015, according to Homeland Security Research, a Washington research firm. 'Because threats manifest in different ways, there's always going to be growth,' Banks says. 'If it's not hijackers, it's swine flu.'"[26]

And there are always going to be defense and security honchos, and former government officials, who want to continue their good work keeping America safe for homeland security profits. Former DHS Secretary Tom Ridge, for instance, leading Ridge Global LLC, providing "strategic and operational consulting services that advance the security and economic interests of businesses and governments worldwide."[27] Or former FBI director Louis Freeh, struggling to make ends meet on his combined salary of $335,000 as a director of Fannie Mae and Bristol Myers, moonlighting with his own hush-hush consulting firm, Freeh Group International, providing "homeland and global security and 'strategic management of complex and sensitive queries'" for "corporations operating in the global marketplace."[28] Freeh has an especially elite group of employees: a former deputy director of the FBI, a former enforcement director of the Securities and Exchange Commission, a former federal judge, and British and Italian high court justices.[29]

In 2005, PharmAthene, Emergent Biosolutions, and several other biodefense firms formed a lobbying group, the Alliance for Biosecurity, alongside the Center for Biosecurity of the University of Pittsburgh Medical Center. The UPMC Center represents another component of the biodefense complex, and a particularly insidious one at that: the pro-biodefense think tank, devoted to hyping the fear of bioterrorism. The UPMC Center was formed in 2003, by former staff members of the Johns Hopkins Institute for Civilian Biodefense Strategies, itself originally formed by Dr. Donald Henderson, who went on to serve as an adviser to the Bush administration.

In 2009, in an action deeply distressing to critics of the post-2001 biodefense expansion, President Obama chose the UPMC Center's Director, Dr. Tara O'Toole, as the Department of Homeland Security's Undersecretary of Science and Technology, a position from which she will oversee the Department's wide-flung chemical and biological division.

Richard Ebright said:

> This is a disastrous nomination. O'Toole supported every flawed decision and counterproductive policy on biodefense, biosafety, and biosecurity during the Bush Administration. . . . O'Toole is as out of touch with reality, and as paranoiac, as

former Vice President Cheney. It would be hard to think of a person less well suited for the position.[30]

George Smith, a senior fellow at GlobalSecurity.org, called O'Toole:

The top academic/salesperson for the coming of apocalyptic bioterrorism which has never quite arrived. [She's] most prominent for always lobbying for more money for bio-defense, conducting tabletop exercises on bioterrorism for easily overawed public officials, exercises tweaked to be hor-rifying. [And she] has never obviously appeared to examine what current terrorist capabilities have been . . . in favor of extrapolating how easy it would be to launch bioterror attacks if one had potentially unlimited resources and scien-tific know-how. . . . [It's a] superb appointment if you're in the biodefense industry and interested in further opportunity and growth. . . . Alternatively, a disaster if threat assessment and prevention ought to have some basis in reality.[31]

Part of critics' dismay arose out of two simulation exercises O'Toole produced, the 2001 *Dark Winter* and the 2005 *Atlantic Storm*. Both involved fictional smallpox outbreaks, and both helped to raise the fear of bioterrorism among public officials—which is exactly what they were designed to do.

J. Michael Lane, former director of the CDC's smallpox eradica-tion program, criticized *Dark Winter* for exaggerating the smallpox threat by assuming extraordinarily high rates of transmission, ten new infections for each patient. In fact, Lane said, people generally get "desperately sick"—"knocked off their feet with high fevers and severe muscle pains"—before they're contagious, so that each patient generally transmits the disease to just one or two others.[32] Ebright said *Dark Winter* produced "shock and awe among partici-pating government officials"—leading to "disastrous policies in biodefense, biopreparedness and counter-terrorism during the Bush Administration."[33]

In a report produced for the Army War College titled *Assessing the Biological Weapons and Bioterrorism Threat*, Milton Leitenberg criticized *Atlantic Storm*—which purported to show the effectiveness of an Al Qaeda attack using smallpox—for exaggeration and "'grossly misleading assumptions' about the ease of creating and dispersing . . . a dry powder smallpox preparation, a feat that neither the US nor Soviet BW programs ever achieved."[34]

In 2003, O'Toole told the *Los Angeles Times* that "Bioterrorism is a whole new terrain of national security that's going to have the same magnitude of impact as the creation of nuclear weapons" and suggested that biodefense spending should be increased to $10 billion per annum in 2004.[35] In 2005, arguing that the failed national smallpox immunization ought to be revived, she complained that "People are now back in dumb-and-happy mode . . . [in contrast with] when we were going into Iraq, and the possibility of a smallpox attack was seen as much more plausible."[36]

During the anthrax attacks, O'Toole felt confident that "the attack was clearly the work of terrorists. Unless the government began promptly investing billions of dollars to defense against bioterrorism, future germ attacks were inevitable and people worldwide could die by the millions."[37] She characterized a lesion or scar reported on one of the 9/11 terrorists as "highly suspicious" for cutaneous anthrax.[38]

Leitenberg agreed with Ebright in calling O'Toole's nomination "[t]he most absolutely catastrophic appointment conceivable, which promises to ensure continually misdirected resources in the following years."[39]

In September 2009, the *Washington Times* discovered and reported on O'Toole's role as the "unpaid strategic director" for the Alliance for Biosecurity, which had spent more than $500,000 lobbying Congress and federal agencies since 2005. O'Toole didn't report her involvement with the group in a required government ethics filing.[40]

An "ethics official" with DHS (looking for security from DHS is questionable enough; looking for ethics from DHS is simply bizarre) said she didn't need to disclose the ties, legally, because the Alliance was not incorporated.[41]

O'Toole had signed her name on more than a dozen letters to Congress and federal agencies for the Alliance, most of them seeking more money for research and vaccines. An October 31, 2008 letter to Nancy Pelosi, for instance, called on Congress to provide more than

$900 million for the "advanced development of medical countermeasures" to be administered by the Biomedical Advanced Research and Development Authority.[42]

By March 2009, according to its own press release, the Alliance was seeking $1.7 billion for such countermeasures; and David Wright, PharmAthene CEO and Alliance co-chair, testified before Congress on the "critical importance of developing drugs, vaccines and other medical countermeasures needed to protect Americans from bioterrorism and other catastrophic health emergencies."[43] Which, of course, happened to be the very products Alliance members were offering for sale.

Such hyping of bioterror demonstrates the cozy relationship between various parts of the biodefense complex: biotech companies; centers and institutes for biosecurity; universities; government agencies. All benefit from keeping the bioterror threat "pumped up" in the way Powell warned against.

The UPMC Center has been in the thick of promoting bioterror fear, running conferences, issuing press releases, publishing a journal, *Biosecurity and Bioterrorism*. O'Toole's role has been supplemented and succeeded by the work of other UPMC staff members, such as Gigi Kwik Gronvall, who has argued for loosening regulation of the already loosely regulated biotech sector, for a voluntary, anonymous accident reporting system, and for government-funded public relations campaigns on behalf of high-containment germ labs. So far, the UPMC Center's propaganda activities have been highly influential both in keeping the fears of bioterrorism stoked, and in distracting policymakers from real regulatory reform.

The UPMC is technically a "nonprofit corporation," but one affiliated with the University of Pittsburgh. Universities in general constitute an important part of the biodefense complex. Some of the most dishonest and shameless biodefense propaganda has come from university administrators and academic colleges and departments hoping to profit from the boom in biodefense research, in a blatant abandonment of the university's supposed role as a bastion of independent thinking.

Many of the reasons for this situation are addressed in a book published just before the NBAF soap opera launched itself: *University*

Inc.: The Corporate Corruption of Higher Education.[44] Jennifer Washburn's book points to the stealthy corruption of academic life by commercial values:

> [A] wholesale culture shift is transforming everything from the way universities educate their students to the language they use to define what they do. Academic administrators increasingly refer to students as consumers and to education and research as products. They talk about branding and marketing and now spend more on lobbying in Washington than defense contractors do.

One might notice, as a clearly relevant example of a trendy and thoughtless sort of consumer branding, the way colleges and universities have rushed to offer degrees, certificates, and other credentials in "homeland security." A 2008 *Boston Globe* article indicated that about 300 schools have homeland security programs—about a third of them certificate programs, the rest master's, bachelor's, and associate's programs.[45] Steven P. Lab, a director of the criminal justice program at Bowling Green State University, said that the majors are "a hodgepodge of topics that have already existed on college campuses for the most part. And they've strung them together in a meaningless whole called homeland security."[46] Lab suggests that students who see this major as a shortcut to a "homeland security" job would have better luck specializing in some specific area that interests them, like Middle Eastern studies or biology.[47]

DHS helped encourage such programs in 2004 by earmarking $10 million for university security programs.[48] And it continues to actively promote such curricula—and a university focus on "homeland security" in general—as indicated by the Web flyer for the Third Annual DHS University Network Summit, March 17-19, 2009. Among those whom the flyer suggests should attend are "research, academic, industry, government and international communities interested in homeland security science and technology research" and "academic institutions interested in adding homeland security curriculum."

The available programs range from a seven-course certificate program offered by Curry College,[49] to the associate degree offered by

Laramie County Community College,[50] to the online offerings of Kaplan University (distance learning for the future bureaucrat hoping to "secure" the country with a few keystrokes in Washington),[51] to the master's degree recently introduced by my former employer, Western Kentucky University.[52]

One would hope the rationale for such developments would be something less mindless than this one:

> In today's global climate the safety and security of our country is paramount to preserving our way of life. If you are like a lot of people, sitting back and letting someone else make key decisions about your safety is simply not enough—you want to become involved. Homeland Security college may be just what you're looking for.[53]

If you are like a few people, you may not feel comfortable letting either the people who produce such drivel or anyone who actually believes it look after "the safety and security of our country." I would prefer my security honchos to be intelligent, cagy people with an ability to take a skeptical look at the terror-industrial complex's own propaganda—all of which is more likely to be developed by an old-fashioned liberal arts education focused on developing the ability to think and write critically, research, forgo claiming certainty where no basis for certainty exists, with some awareness of the complexities of life and perspective, and the recognition that knee-jerk actions are likely to have unintended consequences. Hopefully coupled with the courage to act on these awarenesses. Maybe a sufficient nucleus of such intelligence could have kept us from such insecurity-producing actions as the rash invasion of Iraq, or the reckless proliferation of germ labs and risky research. Maybe it would help us realize that a nation's security involves such things as a stable, sustainable economy and maintaining some semblance of an equitable social structure. What the homeland security curricula do instead is help stoke a national-security-obsessed mindset leading us away from democracy toward a police state. To the extent we are obsessed with terrorists, we forget to ask questions about what else needs fixing in our country.

This sort of mindless market expansion by academia is only one small aspect of the corruption of higher education that Washburn documents. Much more serious are the troubling alliances universities establish with a wide range of corporations, agreements that often allow the corporations to skim the cream of publicly financed research, and create obvious conflict-of-interest situations for the universities themselves: Berkeley's deal with Novartis, "a Swiss-based pharmaceutical giant and producer of genetically engineered crops";[54] Harvard's entanglements with Monsanto and Enron; Columbia's with Bristol-Myers.[55] As *Nature* said in an editorial titled "Is the University-Industrial Complex Out of Control?": "The Novartis-Berkeley deal can all too easily be portrayed as an institution undermining both its motivation and trustworthiness to provide an independent and impartial view of one of the most contentious technologies of our time—genetically modified crops."[56]

Washburn points to several instances of corporate sponsorship or agreements slanting or suppressing the conclusions of university research: the Harvard Electricity Policy Group "churning out a whopping thirty-one reports promoting the deregulation of energy markets in California—this while receiving funding from Enron and other energy firms that stood to benefit from precisely such a policy"; "the influence that major chemical, oil, gas, and pharmaceutical companies wield over the Harvard Center for Risk Analysis (HCRA), which receives 60 percent of its funding from private sources, most of it stemming from large corporations . . . that have a direct stake in the outcome of its research on regulatory matters";[57] the Iowa State researcher prevented from publishing or speaking publicly about his discovery of antibiotic-resistant bacteria in air emissions from hog farms; or the UNC Chapel Hill professor attacked by a local pork growers group after he released a paper on the poor health of people living near that state's many industrial hog facilities.[58]

Washburn tells the story of how universities and their lobbying groups quashed or watered down efforts to impose conflict-of-interest restrictions on researchers who receive federal grants.[59] She highlights the weak oversight of research involving human subjects, such as in clinical trials involving new drugs. In a manner quite similar to current discussions about regulating high-containment germ labs, the universities and their organizations fiercely resisted a 2001 proposal by the Department of Health and Human Services to impose new guidelines on financial conflicts in human subject research. They argued, successfully, that they should be allowed to develop conflict-of-interest rules for themselves.[60] And this was the result:

> A 2003 study . . . revealed that nearly half the faculty who serve on university IRBs, charged with protecting human subjects, also serve as consultants to the drug industry. In 2002, another survey of 108 university medical schools found that universities were routinely signing corporate-sponsored research agreements that failed to protect their professors' basic academic freedoms: Just 1 percent of researchers were guaranteed unimpeded access to the complete trial data; only 5 percent of research contracts required that the results be published.[61]

The kidnapping of the university by commercial values[62] has led to a focus on the university as an "economic development entity"; to a fevered search for federal grants; to a de-emphasis of teaching and of those fields like the humanities without a clear economic outcome. All this is amply discussed by Washburn, along with some of the reasons. Ironically, many in the business community itself find these developments deplorable. In a 1991 survey published by the Government-University-Industry Roundtable, corporate research managers emphasized that the *most important role* for universities is "as educator and provider of talent. . . . Universities should not attempt to orient their research more closely to product discovery. . . . Rather, they must continue to teach, to foster creativity, and to advance the frontiers of knowledge through long-term basic research."[63]

As the NIH-funded microbiologists complained, "basic research" is one of the things slighted by the biodefense bandwagon. So is basic intelligence, which would help us to put the bioterror threat into some sort of proper perspective, to weigh biodefense against more urgent needs, including those of public health, and to recognize that time and money spent in the futile pursuit of magic bullets is money diverted both from more urgent public needs and from more sensible "biodefense" strategies. Universities that connive with dishonest, hysterical propaganda are destroying their own seed stock—their reputations for "objective" scholarship, the resources available for core academic pursuits—and undermining the country's intellectual seed stock in the process.

NBAF 10

WAITING FOR BAD NEWS

1

In Kentucky, biolab opponents had an immediate upsurge of energy after the protest. Several new people showed up at steering committee meetings, and we had tentatively planned several events in the event we were named as a finalist, including hosting Michael Carroll as a speaker.

We had a huge helping hand at this time from Tom Fitzgerald, director of the Kentucky Resources Council. Tom sent an eight-page single-spaced letter to DHS Secretary Chertoff indicating what he expected to see in the environmental impact statement for a Kentucky facility. I felt this put DHS on notice that there was an astute and experienced environmental advocacy organization that would keep a close eye on any further consideration of Kentucky as a site.

During the last two weeks of June, No Ky. Biolab seemed to have lapsed into hibernation again; a planned meeting was postponed because not enough people were going to make it. Chuck's counting and copying of petitions (we had over 5,000 anti-NBAF signatures we planned to send DHS) had been slowed by work-related emergencies that sent him out to California. On Friday, July 6, however, he indicated he was getting ready to mail the petitions. But that same day, a news report appeared saying the short list had already been selected.

2

Right after a DHS team visited the Wisconsin site, the full Dane County Board of Supervisors put the final touch on local opposition efforts by voting to oppose NBAF in Dunn.[1] It thus endorsed the earlier opposition votes of three separate Board committees, and the December anti-NBAF vote of the Dunn Town Board.

It based its own vote, however, solely on the fact that "the facility would be in conflict with town goals of preserving productive farmlands for long-term agricultural use and protecting farm operations from incompatible adjacent land uses." And it offered to help the university find an alternative site. The university indicated it would be in touch with DHS about that possibility, but wasn't sure DHS would go along.[2]

Meanwhile, residents of Columbia, Missouri, who had raised pointed questions about NBAF at the university's March 2006 forum, began to oppose the facility with a vengeance. The university, like its counterparts elsewhere, proposed to educate them out of their worries, but opponents felt NBAF sponsors were the ones who needed educating. On May 1, just prior to the DHS visit, opponents had sponsored a public discussion at New Haven Elementary School. On June 7, they held another meeting to begin organizing opposition.[3]

One of the organizers was Karen Onofrio, a retired pathologist who had once inspected laboratories. She said the facility could have all the high-tech features possible, but "it all rests on the weakest link, which is people."[4]

Barbara Hoppe, who represented the city council ward where the lab would be located, had left the May 1 meeting with heightened concerns. "I've gotten a lot of input and concerns," she said, "from not only residents, but doctors and physicians, and they have told me they will move if it comes to Columbia."[5]

Mayor Darwin Hindman and county commissioner Skip Elkin had signed letters of support in the spring of 2006, but now they began backing away. Elkin said, "I want to weigh all the facts and make sure it's a right fit for our community, if it's a fit at all. What I'm hearing from the public is that they don't want it. My job is to make sure the folks I represent have a voice." Hindman said he didn't know if he would ultimately support constructing the lab in Columbia, either.[6]

At the meeting, organizers distributed information packets including 21 reasons to oppose the lab and ways to contact officials with letters of opposition. They announced the launching of a new website, www.nodeathlab.org, and began collecting money to finance the opposition efforts.[7]

On June 20, the board of directors of the Missouri Cattlemen's Association voted to withdraw its prior support for the facility. Jeff Windett, executive vice president of the organization, said, "In the beginning, there was very little data. . . . At the time it looked like it was a good thing for the state until we looked close at what was going to be involved." He said a breach at the facility could jeopardize the producers' livelihoods, and they felt it should be located in a more desolate area.[8]

The end of June saw the CDC suspending all select agent research at NBAF semifinalist Texas A&M, a DHS Center of Excellence considered by most a prime NBAF candidate. When DHS missed the promised July 1 date for announcing the short list, Ed Hammond speculated as follows in a July 3 E-mail to the Sunshine Project's biodefense listserv:

As I've said before, the inscrutable schedule of DHS with respect to NBAF happenings has always seemed to have more to do with politics than science or security. I was unaware of the July 1 short list date apparently promised this Missouri paper. Honestly, I stopped paying attention when DHS says it is going to do anything with NBAF, due to its deviation from every announced plan.

So I guess I should not read too much into the latest delay; but it is tempting to think that just maybe it's because Texas A&M was on the short list and now DHS isn't quite sure what to do. It's an interesting thought.

CONDOLENCES IN SILICON HOLLER

I HEARD THE GOOD NEWS AT NOON, ON WKU PUBLIC RADIO'S Midday Edition. NBAF would not be coming to Kentucky.

In news articles following the failed bid, Congressman Rogers said, "We didn't make it to the Final Four, but we did make it to the Sweet Sixteen." He mentioned ominously, however, that the consortium might pursue similar projects in the future.[1]

Rogers said DHS considered the consortium's research partners to be too dispersed. He also said the consortium had been trying to sell the site as being "in a fairly remote area"[2]—something that suggested he knew the facility was more dangerous than he'd acknowledged. The only advantages of such remoteness are of the "out of sight, out of mind" variety. Fewer people to be harmed by an accident; easier to construct a military fortress.

The new mayor and new county judge expressed disappointment but praised "Hal's leadership."[3]

A July 20 *CJ* article revealed that the Somerset-Pulaski County Development Foundation had already exercised an "option to purchase" 168 acres for the lab, at a cost of $436,000. Now, "plans [were] to auction the land to the highest bidder so as to be fiscally responsible with taxpayers' money, according to one high-ranking figure in the local economy."[4]

NBAF opponents in Missouri, Wisconsin, and California joined us in heaving a big sigh of relief.

In the five finalist states (Texas, Georgia, North Carolina, Mississippi, and Kansas), assorted senators, congressmen, and governors issued celebratory press releases, promising to continue their energetic efforts to bring the $500 million bio-boondoggle home to roost. It was the right thing to do for the country. It was the right thing to do for their state.

In each of the five states, mainstream media devoted their attention to handicapping the state's chances, or declaring why their state was the right place for NBAF.

The *San Antonio Express-News* opined that San Antonio was a biolab finalist for several reasons. "The city," it declared, "has experience building the rare kind of clean lab that will be required."[5] Well, NBAF will house hundreds of deliberately sickened cows, pigs, sheep, etc. As one who cleaned his share of horse, cow, and pig shit as a child, I wonder what rare kind of "clean lab" the *Express-News* had in mind. Perhaps something like the Southwest Foundation for Biomedical Research, with its 6000 sick monkeys. Experience cleaning up sick-monkey shit presumably helps prepare one for sick-cow shit.

"Perhaps most importantly," the *Express-News* said, "San Antonio is showing a unified front of support for the lab, which has encountered unfounded safety concerns in other parts of the country. Kentucky, for example, didn't make the final five because its horse industry opposed efforts to attract the lab."[6]

We hadn't noticed anyone in racing colors watching our site-visit protest from the VIP lounge. If any horse breeders did express opposition, they were certainly discreet about it. The *Herald-Leader* had wondered, back in February 2006, whether a BSL-4 animal/zoonotic disease lab was an unvarnished benefit in thoroughbred country. But there was no indication that their musings struck any sparks among the horsey crowd.

The *Express-News* itself might have wondered about putting NBAF in one of the country's big cattle states, if it had been inclined to explore the subject with anything more than a dismissive "unfounded safety concerns."

"In union there is strength," the *Manhattan Mercury* exulted, and shared a photo of state and local bigwigs toasting their success. Lest the event be mistaken for a drunken orgy, the photo caption advised readers that the toast involved "a little sparkling grape juice," not to be confused with champagne or moonshine.[7]

The *Athens Banner-Herald* rhapsodized that "Biotechnology dreams could become reality."[8] "Maybe," said Executive Editor Jason Winders, "we can mess with Texas."[9] An article in the *Atlanta Journal-Constitution*, "A very dangerous field: UGA seeks bioterror role," while it mentioned the incidents at Texas A&M and Pirbright, devoted most of its attention to the "safeguards" at the University of Georgia. For instance: "At UGA, research on dangerous agents is reviewed by a biosafety committee, which looks closely at how it will be conducted, what safety protocols will be followed and the training of the researchers, says Maria Kuhn, the college's director of biosafety."[10] The University would soon be resisting Ed Hammond's effort to verify the extent of its "review" by refusing to release copies of its IBC minutes. One bit of research the University hadn't used its IBC to review was the resuscitation of the 1918 flu virus, about as dangerous a type of research as one might contemplate.

Georgia's mouthpieces would be among the more shameless of the many NBAF prevaricators. An independent Athens publication, *Flagpole*, with a few of its journalism genes still viable, carried an article which mentioned some of the safety violations at Plum Island. UGA's Dr. Corrie Brown, who had worked at Plum Island for 10 years, said of the Plum Island lapses: "I don't recall any of them being the kind of gap that would be a biosafety gap."[11] Apparently, neither the escape of foot-and-mouth disease from containment in 1978, the internal FMD releases later, the loss of power and negative air pressure during a hurricane, workers bitten and trampled by test animals, or the EPA citation for illegal discharge of animal waste into local waters constituted a "biosafety gap" for Brown. That sort of devil-may-care attitude would itself seem a serious biosafety gap, if not a sign of mental deficiency.

UGA biosafety officer Manley Kiser would talk about how "highly regulated" biolabs are, asserting that each BSL-3 experiment is "inspected" by an inspector from the CDC.[12] Duh, no. The CDC inspects labs—about once every three years, if we're lucky—not individual experiments. Just how closely they inspect the labs had just been demonstrated by the Texas A&M revelations. A&M's biosafety breakdown had been discovered not by a CDC inspection, but by Ed Hammond's inspection of A&M lab records, with no help from the CDC, thank you.

Three months later, the GAO would tell Congress that biolabs are for all practical purposes regulating themselves. Yet UGA's NBAF point man, David Lee, would explain Plum Island's problems as follows: "The federal government is incredibly vigilant about this type of work." The history of violations at Plum Island, he suggested, might just reflect the "plethora of regulations" under which all researchers work.[13] Apparently, Lee either felt that the problems at Plum Island were just insignificant "technical violations"—or that it was so difficult keeping up with the largely nonexistent regulations that biolab researchers didn't know whether they were going or coming. That *would* be a problem.

Not to worry, according to Lee—even if a safety violation should turn out to be significant, and a pathogen should be "released"— "Most, if not all, of these pathogens are actually very sensitive organisms. . . . They don't retain viability for any length of time if simply released into the air."[14] Which begs the question of why we waste all that money on negative air pressure and HEPA filtering systems in BSL-3 and BSL-4 facilities, when a little mental germ warfare would mar the fragile self-esteem of bioweapons agents and destroy their motivation to do what they were bred to do: infect and kill people and animals.

Of course, these "sensitive pathogens" would be quite happy to hitch a ride out of containment through the bloodstream of an over-confident biodefense researcher. Intimacy is what they really want, after all, not the wide open spaces. I mean, your liver is their Grand Canyon.

Athens Mayor Heidi Davison said she hadn't "heard the reports of environmental or security problems at Plum Island." Not to worry, though, according to the mayor: "We have said all along that safety issues are of paramount importance."[15]

Well, mayor, don't believe everything you don't hear. And while the oral tradition has played an important role in American folklore ranging from Br'er Rabbit to Paul Bunyan to weapons-of-mass-destruction in Iraq, it may not be the best way for responsible public officials to keep current on biosafety lapses. The much-maligned (by NBAF propagandists) but well-documented *Lab 257* might be a place to start. Yes, Virginia, there was a foot-and-mouth outbreak there, and a hurricane, and a negative air pressure failure, and a lab employee with an arrest record who went AWOL with a laptop controlling the facility's air pressure system. None of which would constitute a biosafety lapse for Dr. Brown, mark you, but it worries some of the rest of us.

And God saying "let there be light" may have been sufficient to illuminate the original, primordial darkness, but a small-town mayor (or an NBAF lab director) declaring that "safety issues are of paramount importance" won't make safety magically appear. Nor will "confidence in the experience of Drs. Lee and Brown of UGA, both of whom she [Ms. Davison] met with in April, along with Department of Homeland Security officials and others. They are experts in their field." "It's not like"—she was reported to have said—"somebody's not giving you all the information, or leading you along." Actually, in this case, it's very much like that. Lesson Two, Mayor Davison: don't believe everything you do hear from "experts." Especially not ones with dollar signs in their eyes. (But then, you presumably had the same dollar signs in yours. . . .)

Out in North Carolina, the *Raleigh News-Observer* editorialized that "this is a bid that the Triangle, and North Carolina, should not be prepared to lose."[16] In the accompanying article, "Butner may get security lab," the *News-Observer* acknowledged safety concerns to the extent of noting that the facility would "work with some of the deadliest pathogens in the world." (The *News-Observer* hadn't talked with Georgia's David Lee, of course, and didn't understand that under their gruff barbarian exteriors many of these pathogens are sensitive, gentle creatures.) The *News-Observer* went on to note that the risks of working with such materials had "triggered organized opposition in states such as Kentucky, Wisconsin and Missouri." But of course all those lamebrains had been cut from the list. Meanwhile, in North

Carolina, the usual motley crew of bigwigs was ready to let the good times roll: "But the news that North Carolina is a strong candidate for the research lab was greeted with excitement by the consortium of business, government and academic leaders who are making the pitch." Among them were U.S. Representatives David Price and Brad Miller. Miller would eat a lot of crow a few months later; Price would opt for hem-hawing noises that might easily be confused with crow-munching.[17]

State and local officials cited the presence of a federal prison, state youth center, state psychiatric facility, and National Guard camp in the area as evidence that "local residents feel comfortable with a large government presence."[18] And of course, none of the inmates of said facilities had filed formal objections, so the bigwigs assumed the area was primed for all sorts of pathogenic pursuits.

Point man Barrett Slenning suggested it was probably due to all the good work the pro-NBAF consortium had done with local environ-mentalists.[19] *That* was one of the biggest dramatic ironies since Oedipus announced he'd catch and punish the SOB who'd killed King Laertes, or since Fort Detrick expanded to better protect the country from Fort Detrick anthrax.

CHAPTER NINE

Ain't No Sunshine When It's Gone

A Watchdog Leaves the Circus

ON FEBRUARY 1, 2008, LESS THAN A YEAR AFTER HIS DOGGED persistence forced Texas A&M to disclose two different sets of unreported accidents, and the CDC to take the unprecedented step of suspending A&M's select agent research—only four months after his testimony before a House Energy and Commerce subcommittee, at a hearing partly prompted by his discoveries—Ed Hammond posted a note on his Web site that the Sunshine Project was suspending operations. Around the country, "biodefense activists" (i.e., people anywhere who had ever at any time raised a question about their prospective "black magic" neighbors) were distressed to hear the news.

Neither biodefense nor the Sunshine Project had been on my own radar before the spring of 2006, but they'd arrived at approximately the same time, thanks to the *Lexington Herald-Leader*—one of the few newspapers in would-be NBAF states that actually acted like a newspaper, instead of the private public-relations organ of the local biolab consortium. One of the sources the *Herald-Leader* had sought out for comment was the Sunshine Project. Since then, I had come to depend more and more on the SP for information about the wild and woolly world of Germland, and I anticipated depending on it into the future as I drew together materials for my book. For a while, in the fall of 2007, it looked like the biodefense tide might turn, that Congress might take a serious look at what was being wrought and make some serious midcourse corrections. I felt the Sunshine Project had been a major contributor to such a possibility, and now the organization was going to miss the ride.

I wrote Hammond and asked why. I assumed it was either burnout, or money, or both. I guess I hoped, if it were just money, some of us might figure a way to help.

Hammond wrote back as follows:

> It is, basically, a question of money. I've worked for nothing for 8 years, barely scraping by, and the foundations that support bioweapons work, including the 'progressive' one that advertises itself on NPR, are (frankly) not interested in the Sunshine Project. They would rather, I suppose, give to groups in DC and to professors. So be it. I just can't continue. I am off to do other things. What, I'm not sure yet; but I'll figure it out.[1]

Anyone wondering just how bad Hammond's financial situation really was just needed to follow the *New Scientist*'s link to the Sunshine Project's 2005 tax return. Out of a total revenue of slightly more than $53,000, only $29,000 went to compensate Hammond as director. And apparently, since then, the funding had gotten worse. The *New Scientist* noted the irony of the SP's funding problems in light of the billions flowing into biodefense coffers. If only, the magazine said, Hammond could claim part of the $1 million fine levied against Texas A&M ("a lot of money for most of us, but small change for Texas A&M University").[2] Morally, Hammond certainly would be entitled to a "finder's fee," a commission for doing the CDC's work for it—since the A&M problems would never have come to the CDC's— or the public's—attention if not for Hammond. But since Hammond had an active complaint against the CDC itself (for denying any and all public information requests)—and since Hammond's Texas discoveries had embarrassed the CDC as well as A&M—Hammond wasn't expecting any rewards from the CDC, not even a thanks or acknowledgement of his good work.

"Who is going to hold the multi-billion US biodefence juggernaut to account now?" the *New Scientist* asked.[3] And well might it wonder.

German biologist Jan van Aken had founded the Sunshine Project in 1999; Hammond and his wife Susana Pimiento, a lawyer from Colombia, had joined with him to set up a U.S. branch in 2000.[4] The name was a reference to the fact that "many biological weapons are quickly broken down and rendered harmless by exposure to bright sunlight."[5] A San Antonio native, with Masters degrees in Latin American studies and community and regional planning, Hammond had previously worked as a Program Officer for the Rural Advancement Foundation International (RAFI, now the ETC Group).[6]

During its eight years, the SP issued frequent press releases and briefing papers about such subjects as Army and Air Force proposals to genetically engineer microbes to destroy fuel, asphalt, and plastics;[7] the genetic re-creation of the 1918 flu;[8] the University of California-Davis's concealment of a research monkey escape—allegedly disease-free—at a time it was pursuing one of NIH's BSL-4 National Biocontainment Laboratories;[9] the fact that the Southern Research Institute, which had mistakenly sent live anthrax to the Oakland Children's Hospital, was in charge of safety and security for the Department of Energy's new Argonne National Laboratory in Chicago;[10] and the CDC's refusal to release any information about the Boston tularemia incidents.[11]

In a highly prescient statement issued ten days after September 11, Hammond warned how the billions just authorized for homeland defense would swell an "opaque military-scientific-corporate biotechnology bureaucracy and the instability it creates to even larger proportions."[12] Soon afterward, following the anthrax attacks, Hammond warned that "biodefense" could not offer "durable protection from the threat of biological weapons," and that the pursuit of biotech "countermeasures" was a futile, bottomless money pit—with the chief beneficiaries being the biotech industry itself.[13]

And of course there were the laboratory mishaps to discover and reveal, all with the notion of questioning the biodefense complex's arrogant claims of flawless safety and omnipotent redundancy. Hammond demonstrated a good deal of uncanny prescience here also. An April 2005 press release focused on three lab-acquired TB infections in a Seattle, Washington lab. The State of Washington's investigation had concluded that a leaky airflow meter in the Madison Aerosol Chamber was probably responsible for the infections. Hammond pointed out that a NIAID workshop had encouraged use of the device in biodefense research; and that it was currently being used to aerosolize brucella and Q fever at Texas A&M. Hammond said the case underscored "how the 'precise, clean and neat' public image of BSL-3 and BSL-4 facilities promoted by NIAID and its labs is frequently at odds with messy and risky realities."[14]

One year later, A&M researchers using the Madison Aerosol Chamber would underscore Hammond's points even more dramatically.

During its eight-year luminescence, the Sunshine Project advised and collaborated with individuals and organizations in communities affected by the biodefense explosion. These included groups resisting biodefense expansions at Dugway, Boston, Seattle, the Department of Energy facilities at Lawrence Livermore and Los Alamos, and several anti-NBAF groups,[15] including those in California, Kentucky, Wisconsin, Georgia, and North Carolina.

The Sunshine Project's primary focus, however, was transparency and public accountability:

> Because so much research with biological weapons agents is dual-use (that is, potentially has both offensive and defensive implications), transparency at biodefense labs is critical to gaining public confidence and the trust of other countries in the peaceful intent of biodefense research. Especially when it comes to experiments using genetic engineering and related new technologies. But too often, policymakers opt for secrecy or pay only lip service to the need for openness. This, in turn, is breeding a dangerous world of biological mistrust for us and, especially, our children.
>
> To combat secrecy, the Sunshine Project files requests [under] open records laws, such as the Freedom of Information Act and equivalent laws in US states and other countries,

to obtain and disseminate information about biodefense research and the (often lacking) systems to ensure its safety and accountability. In an average year we file about 300 such open records and declassification requests, frequently in collaboration with other nonprofit partners. By exercising rights to obtain and publicize information on biodefense projects, the Project seeks to increase transparency and, thereby, safety and security.[16]

Hammond ran four or four five different listservs during the Sunshine Project's eight years, the most active being the one on biodefense. I signed up for all of them in the spring of 2007, just in time to catch the widespread news coverage of the Texas A&M situation, the October 2007 Associated Press disclosures, and the congressional hearing. Hammond's listserv was a good way—probably the best way—of keeping informed on what the Germland guru-goons were up to. After Hammond threw in the towel, I'd have to make my own way through the thicket of Google alerts and news wire feeds.

Open-records requests were the SP's bread and butter. It was from such a request—obtained only after threats of legal action, and a ruling by the Texas Attorney General's Office—that Hammond obtained the records suggesting problems at A&M.

The most ambitious and revealing of Hammond's open-records information requests was the survey of institutional biosafety committees he embarked on in January 2004.

The institutional biosafety committee "system" had been established in the 1970s as an alternative to federal regulation of the emerging field of genetic engineering. The system called for institutional biosafety committees (IBCs), operating under "guidelines" promulgated by the National Institutes of Health, to review the safety of all research involving recombinant DNA—i.e., genetic engineering.

The guidelines called for the committees to meet regularly, to keep minutes of their meetings, and to make the minutes available to the public upon request. Yet there was one significant problem with the guidelines from the git-go: research not involving recombinant DNA didn't have to be reviewed by the committees; and more importantly, institutions that didn't want or need NIH funding, like the Department of Defense, and later, Homeland Security, private biotechnology companies, etc., and various "nonprofit" institutes funded by military and DHS grants rather than the NIH, could ignore the guidelines altogether.

Hammond's survey would show that ultimately, it didn't matter whether you were receiving NIH funding; you could ignore the guidelines anyway. The NIH dealt with noncompliance by speaking softly and carrying a wet noodle. The agency seemed reluctant to even display its wet noodle, much less use it.

The NIH's Office of Biotechnology Assessment was like an automobile airbag, meticulously described in millions of owners' manuals, that never inflated in any real-world collisions. An already problematic situation became even more problematic in 2003, when the National Academies of Science recommended that the IBCs oversee all "dual-use research" with biological weapons agents. In March 2004, the Bush administration adopted the NAS recommendation and established the National Science Advisory Board on Biosecurity to flesh out the details of the IBCs' new role. The NSABB would be managed by the NIH's Office of Biotechnology Activities (OBA)—the same office that supervises the IBCs.[17]

Hammond began his survey in October 2003 by obtaining an electronic spreadsheet with the names, addresses, and contact information of the 439 institutional biosafety committees registered with the OBA. In January 2004, the Sunshine Project faxed identical signed letters to each IBC, requesting the minutes of the two most recent IBC meetings, and citing the pertinent provision of the guidelines: "Upon request, the institution shall make available to the public all Institutional Biosafety Committee minutes." The letter also asked each institution to indicate whether it was or was not registered to handle select agents.[18]

About 75 fax numbers indicated on the NIH spreadsheet proved incorrect. The Sunshine Project used alternative information sources to correct as many numbers as possible; ultimately, 390 of the 439 registered IBCs were surveyed.[19]

Hammond described what happened next:

Unexpectedly, the survey quickly generated discussions on

internet listservers of the American Biological Safety Association (ABSA) and the National Association of College and University Attorneys, whose membership together include biosafety staff and/or legal counsel for the vast majority of IBCs surveyed. Despite the clear language of the NIH Guidelines at IV-B-2-a(7), some institutions decided to refuse to reply while others delayed their answer to observe the response of other IBCs. A group of IBCs replied by saying their IBC minutes were only available to the public by inspection at their offices, often imposing requirements such as several weeks' advance notice, mandatory "supervision", fees, etc.[20]

Widespread recalcitrance resulted in Hammond sending out dozens of repeat requests in late March. On May 3, he filed a complaint against nine institutions with the NIH's Office of Biotechnology Assessment.[21] On May 15, the NIH issued *Questions and Answers Concerning Institutional Biosafety Committee (IBC) Meeting Minutes,* reiterating that IBCs needed to take minutes and make them available to the public.[22]

Eventually 276 IBCs made some sort of response to the survey. Of these, only 199 provided evaluable minutes—which amounted to only 51% of those surveyed, or 45.3% of all IBCs registered as of October 2003.[23]

For those institutions that provided evaluable minutes, Hammond devised a transparency rating system which assigned a score based on four criteria: the detail of the minutes (0-25); the scope of the minutes (0-25); the extent to which the minutes adequately conveyed the nature of the research being reviewed (0-35); and the "demeanor" of the IBC's response to the request—its promptness, helpfulness, etc. (0-15).[24]

Hammond concluded that less than 5% of the nation's IBCs were operating with an adequate level of public disclosure; secrecy in biodefense and biotechnology research was the overwhelming norm. Most secret of all were the labs operated by the government and private biotech companies. Even among the better-performing sectors— public universities and non-profit research institutes—disclosure was poor: "A third of public universities could not produce IBC meeting minutes, and fully half failed a transparency evaluation. Some IBC's not only do not [disclose] the nature of specific research projects, they try to prevent revealing the existence of high containment laboratory facilities."[25]

Hammond found that the public accountability of public IBCs was "extremely poor":

> Three out of five government IBC's, including USDA, DOE, DOD committees and even an IBC of NIH itself, either did not reply to requests for their minutes or were unable or unwilling to provide minutes. Most of the minutes that were provided were rated as failing. In addition, a notable number of federal laboratories, including the U.S. Army at Ft. Detrick, Maryland, do not maintain NIH-registered IBCs.[26]

Hammond also pointed out that the government's failure to follow the NIH Guidelines—which are administered by the federal government, but are not legally binding—undermined compliance by other sectors. The government's own "disregard broadcasts a message to other sectors that the Guidelines need not be followed."[27]

Biotechnology companies underperformed even the government in complying with the NIH IBC guidelines: "Three quarters of biotech company IBC's did not reply to the survey or could not provide evaluable minutes. None of the companies that replied with minutes provided adequate disclosure." Hammond concluded that "[t]he fact that three quarters of NIH-registered biotech company IBCs do not comply with the NIH Guidelines itself demonstrates that the private sector does not adhere to lab safety rules that lack legal force."[28]

Hundreds of biotech companies—ranging from global conglomerates to boutique operations—did not even have an IBC registered with the NIH. As an example, Hammond provided a copy of a letter from global conglomerate Hoffman La Roche to NIH stating curtly: "Hoffman La Roche no longer conducts research that is subject to the NIH Guidelines for Research Involving Recombinant DNA Molecules by virtue of having no NIH funding for our recombinant DNA activities. At this time we choose not to register our Institutional Biosafety Committee with the National Institutes of Health."[29] Nor was California-based Allergan (which manufactures botulinum toxin, a noted biological weapons agent, under the familiar tradename BoTox) registered.[30]

The problem of nonregistration, while severest in the private sector, was not unique to it: "For example, neither the Lovelace Respiratory Research Institute nor the Midwest Research Institute (MRI) has a registered IBC. Both of these institutes have DOD biodefense contracts, large aerosol chambers, and BSL-3 facilities including, in MRI's case, biological labs in at least three states—Missouri, Florida, and Maryland."[31]

Hammond suspected that the additional role of IBCs in reviewing dual-use research would lead private companies to further distance themselves from the NIH guidelines and, to confirm this, requested copies of letters from institutions who had asked to be removed from the NIH registry. NIH OBA released 33 such letters, 19 of which came from biotech companies and seven from nonprofit institutes, including the major biodefense contractor Battelle.[32]

Hammond also searched NIH's awards database for private companies receiving NIH funding, and identified 37 private companies that did not have a registered IBC but had recently received bioterrorism-related funding from NIH. He concluded that:

> 1) NIH is not following its own policy in grantmaking—it is awarding biodefense grants to institutions that do not follow the NIH Guidelines; 2) noncompliance is deeply ingrained in the private sector—many biotechnology companies plainly do not feel the need to register IBCs before seeking biodefense funding; and 3) [there is] lax enforcement of NIH policy by NIH OBA and the National Institute of Allergy and Infectious Disease (NIAID), the awarding agency in most cases.[33]

What did all this have to do with safety? Clearly, Hammond said:

> If a small proportion of institutions don't play by the rules, much less a large one, the integrity of the system cannot be ensured. Immediately there is the problem of non-compliant research and the long-term trend would be for riskier

or otherwise unconductable experiments to gravitate toward noncompliant institutions.[34]

Hammond said he was frankly surprised by what the survey revealed: "[T]he Sunshine Project expected debate over the form and substance of IBC minutes; but it did not anticipate revealing so many dysfunctional IBCs and widespread, overt refusal to comply with the NIH Guidelines."[35] And the National Academies of Science and the Bush administration were deluding themselves if they expected the IBCs to keep dual-use biofense research safe: "While there are exceptions, the notion that research laboratories generally maintain effective, accountable local committees that exercise responsibility over laboratory biosafety is herein demonstrated to be false."[36]

Four years after Hammond issued his report on the failures of the IBC system, the Government Accountability Office would confirm that neither the NIH nor the CDC were doing much to assure the safety of America's biodefense behemoth: the research facilities were essentially regulating themselves.

Hammond issued his final report on the survey in October 2004. Between June and September 2004, he published online a dozen "Biosafety Bites" highlighting his discoveries about particular institutions. His first one focused on Oak Ridge National Laboratory (ORNL), which two years later would be a member of the Kentucky/Tennessee consortium bidding for NBAF. Congressman Rogers felt that the Somerset, Kentucky area could and should aspire to become another Oak Ridge.

Because of its connection with nuclear weapons, the Oak Ridge facility is under the overall supervision of the Department of Energy. But in 1998, the Army had funded construction of a BSL-3 lab there, right next to the facility's aerosol chamber, to "facilitate research links" with Dugway Proving Ground in Utah. In its contract with ORNL, the Army had agreed to inspect the facility annually as part of its safety regimen. Soon after opening, the ORNL received its first biological agent, botulinum toxin; work also proceeded on chemical weapons agents.[37]

In 1999, however, the Department of Energy's Inspector General issued a report focusing on the facility's financial mismanagement and the fact that it had ignored the National Environmental Policy Act—which requires environmental review of new federal BSL-3 and BSL-4 labs. ORNL responded by saying it would operate the lab at BSL-2 and give up its select agent permit. The Inspector General observed that the facility "was prefabricated to contain a fully functioning Biosafety Level 3 laboratory and that the future microbiological capabilities of the laboratory would not be affected by simply deregistering the facility for live biological weapons agents."[38]

Hammond determined that by 2003, the facility was once again registered to handle select agents, and was apparently engaged in some highly secretive BSL-3 bioweapons agent research. A government report—the U.S. Confidence Building Measure A for 2004, submitted to the Biological Weapons Convention—stated that the facility held biological weapons agents, but that the precise organisms were classified as secret. Curiously, however, the government continued to describe the facility as BSL-2.[39]

How was this relevant to the IBC question? Hammond laid it out pretty clearly:

> In December 2003, the ORNL Institutional Biosafety Committee (IBC) reviewed safety at the entire lab, concluding that *"the Chem/Bio Facility continues to operate properly and* [the IBC] *remains comfortable of the review and inspections of the Chem/Bio facility conducted by the CDC and the Army."*

Oops.

> In early 2004, the Inspector General returned. It turns out that CDC hadn't visited the lab since its commissioning in 1999, and that Army safety inspectors, who were supposed to come every year, hadn't been seen for three years. In effect, the IBC had declared its satisfaction with fictitious safety inspections. And the Army had neglected—for three years running—to ensure the safety of a bio (and chemical) defense lab that it built.[40]

Well, I remain comfortable with the prostate exam I haven't had for a while, too. Presumably the Army was as comfortable with the inspections it didn't give as the ORNL was in not getting them. See no evil, hear no evil, speak no evil. As Hammond's IBC survey amply demonstrated, too many IBC reviews seemed to resemble—like the rest of the biodefense universe—waving an (allegedly) magic wand and chanting Abracadabra the stipulated number of times. And these are the same people who accuse people like Hammond of "exaggerating the risks," and people like Michael Carroll, author of *Lab 257*, the Plum Island expose, of writing "science fiction."

Speaking of Plum Island, Hammond's Biosafety Bite 3 focused on the fact that DHS, since taking control of Plum Island from USDA, was on the one hand saying how "committed to positive community relations" it was, while on the other hand proposing to grant itself the authority to keep secret the environmental assessments of government activities required by NEPA. But the Plum Island Institutional Biosafety Committee had not met at all since DHS had taken over Plum Island more than a year earlier.[41]

Lots of high-powered biodefense programs resisted providing IBC minutes—often because there were no minutes to provide. Rockefeller University had held its last regular meeting on March 26, 1998—plus one special meeting on September 23, 2003, where the committee looked briefly at one project, involving injection of DNA into cancer patients, decided it didn't involve gene transfer, and therefore wasn't subject to the NIH guidelines.[42] Genomics and nonprofit powerhouse The Institute for Genomic Research, with assets of nearly $200 million and annual revenues of $40 million, met only twice between 1992 and 2004, and examined only two projects, quickly deciding that each was exempt from full committee review under the NIH guidelines.[43] Battelle, a gigantic science contractor with an emphasis on defense, including much classified research, received $1.3 billion annually in grants from the U.S. government, plus hundreds of millions in payments for services. Yet Battelle could not produce a single page of IBC minutes.[44]

At Emory University, "a biomedical research powerhouse," with 1.3 million square feet of research laboratory space, 600 biological safety cabinets, a primate research center, and several BSL-3 labs, the IBC had met only three times since 2001, and had not reviewed any projects in at least four years.[45] Tulane University, which already operated 3 BSL-3 labs and the Tulane National Primate Research Center, and had been awarded a grant by NIAID to construct a Regional Biocontainment Laboratory, could not produce a single page of IBC minutes for the last 2½ years.[46]

All the above institutions were conducting a host of risky experiments with bioweapons agents, many involving genetic engineering. Under the NIH Guidelines, their Institutional Biosafety Committees should have had frequent meetings and extensive discussions. But at too many institutions, the so-called germ safety police cowered in their offices like timid sheriffs while rowdy cowboys shot the town up.

The 1918 flu had been brought back to life by scientists from the Departments of Defense and Agriculture and private institutions such as the Mt. Sinai School of Medicine in New York. The final steps had been taken at the USDA's Southwestern Poultry Research Lab at the University of Georgia. Neither the USDA nor UGA bothered to have an IBC review the experiments beforehand.[47]

Now, labs around the country were taking 1918 flu genes and tinkering with them in various ways. Through his records requests, Hammond discovered that University of Washington researchers were planning to create new flu hybrids by replacing up to five of the six genes in "a similar (H1N1) but less dangerous type of flu that was isolated in Texas in 1991" with 1918 flu genes. They then planned to spray the genetically engineered flu hybrid onto macaque monkeys.[48]

If they were to obtain federal funding, the Washington researchers needed their IBC's approval. This they had no trouble in obtaining, though the IBC crashed through several safety fences in granting it.

Many experts felt that if the 1918 flu were going to be researched at all, the research should be conducted in a BSL-4 facility. The USDA hadn't done this at Georgia, but it had conducted the research at something called BSL-3Ag, which has some features of BSL-4 containment.

UW was planning to build a new BSL-3 animal facility for the flu experiments, but it wouldn't be BSL-3Ag. UW's IBC waved its magic wand and said the new lab would be BSL-3Ag "in principle." The IBC then approved allowing 1918 flu research to begin or continue even before the new BSL-3 facility was constructed. And the safety protocols it selected as appropriate for 1918 flu research were those used to handle HIV, a particularly irresponsible choice according to Hammond because "the virus that causes AIDS is relatively difficult to transmit, especially by aerosol, the main cause for concern with influenza. Moreover, the risk to the community posed by a lab-acquired HIV infection is trivial in comparison to the threat posed to the world by a case of potentially pandemic influenza."[49]

And in contemplating what to do in the event of a researcher dropping a tray of 1918 flu cultures, UW's IBC indulged in yet another magic wand moment, concluding that researchers "will be trained to stop breathing . . . just as they are taught to do when working with HIV."[50] Unfortunately, most of us, when we drop something, tend to gasp rather than stop breathing. But perhaps UW has developed a special negative-air-pressure form of breathing, in which, in the event of a hazardous pathogen release, researchers only exhale. Given UW's laxness about other safety issues, those living in the Seattle area may wish to obtain the negative-pressure breathing technology for their private residential use.

In an appendix or "annex" to the Sunshine Project's IBC survey, "Hall of Shame," Hammond passed out "awards" for the most evasive and contemptuous responses. The described behavior constitutes a sordid compendium of academic and institutional prevarication. The University of Texas-Southwestern placed the sought-after information in "safety plan summaries," which it didn't provide. What it did provide was this: "The committee reviewed and approved the research protocol safety plan summaries as listed in section E of the May BCSAC agenda without additional comment."[51] The Pennsylvania State University Medical College ignored Hammond's first two requests, and responded to the third by asking Hammond to "[p]lease provide the specific NIH guidelines referenced in the attached letter" (Hammond's "referencing" had both quoted and cited the guideline number).[52]

The Southwest Foundation for Biomedical Research (SFBR) suggested that Hammond needed to send requests by registered mail, "as we have no way of verifying the authenticity of a Fax that comes to a common Fax machine within our institute." (As Hammond pointed out, the minutes were supposed to be available to **any** member of the public, not just a "biodefense activist." So there was no need for the

Institute's fax machine to verify the authenticity of Hammond's fax machine, should some member of the public actually be interested in impersonating Hammond. Historically, a number of people, and perhaps even a few fax machines, have claimed to be Napoleon, but so far only Hammond himself has claimed to be Hammond.) The SFBR then acknowledged reviewing the requested records—implying that such records existed (though not providing them)—but asserted, rather confusingly, that "No new proposals were submitted during the time frame of your request."[53] Hammond's request was not for "new proposals," but for IBC minutes. If the SFBR had IBC minutes, it should have provided them; if it didn't, it should have said so.

The University of Delaware got Hammond's "Most Likely to Wear a Pocket Protector Award" for blotting out completely three full pages and all but about eight lines of the other two pages.[54] Dishonorable Mention in the Pocket Protector category went to Indiana University, which cited a state law as grounds for redacting (blotting out) "information concerning research"; "expressions of opinion"; "advisory or deliberative material"; and information "communicated for the purpose of decision making." As Hammond complained, "The redactions defeat the core purposes of the public access provisions of the NIH Guidelines and make the minutes substantively unintelligible."[55] Princeton University received Hammond's "Most Economical Use of Microsoft Word" award for similar behavior, providing for each approved project the names of the technical and medical reviewers, and the date the project was approved, but blotting out the name of the project itself, and the investigator.[56]

The New York State Department of Health's Wadsworth Center got the "Most Remarkable Biodefense Chutzpah" award for acknowledging that its IBC was "inactive," and declining to tell the public whether it, duh, handled select agents, even as it applied for NIH funding to build a BSL-4 National Biocontainment Laboratory.[57]

The University of Wyoming's Risk Management Director ("Most Contemptuous of the Citizenry" award) provided the requested IBC minutes, but then proceeded to lecture Hammond, telling him that it was not a properly "formal" or "legal" request, "nor does your request meet any test for justification of your release of information request" [sic]. (The guidelines do not specify a particular form for the request—in triplicate on NIH Form 07-A, for instance—nor do they establish a test for "justification.") The university director proceeded to snipe a bit more: "A survey inquiring as to whether or not (yes or no) an institution would honor a justified request for the release of information would have come much closer to addressing your stated premise of 'national survey of compliance with the public access provisions of the NIH guidelines.'"[58]

I think not. There is no doubt in my mind that almost every institution, if asked the **hypothetical** question of whether it would comply with the NIH guidelines and honor a request for its IBC minutes, would answer Yes, certainly, especially if the institution can require—in a way not **justified** by the NIH Guidelines—that the request itself meet a **"justification"** test, and if the institution also gets to determine what **"justified"** means. No, the only way to tell who would comply with the guidelines was actually to make the requests, and wait to see what happened.

The director could not conceal his indignation that some upstart watchdog project should try to test the University's compliance with the NIH guidelines by doing something so unheard-of as actually requesting IBC minutes:

> If the true purpose of "the sunshine project" is to stop the development of products that could be used in bioterrorism acts you should try to educate yourselves on where that development is most likely to occur. Sponsored research at institutions of higher education is, in fact, the least likely source but is the most likely source for full compliance with NIH and other federal, state, and local rules and regulations. One should know their subject and their source; if your "project" was ever hopeful of gaining support from the academic community, you have now taken the first big step in losing what little there may have been.[59]

The director's response is bizarre, to say the least. He insists that "sponsored research at institutions of higher education . . . is the most likely source for full compliance with NIH and other . . . rules and regulations." Yet Hammond's simple request for the minutes of the last two IBC meetings, a request itself made in full compliance with the NIH Guidelines, has—according to this academic spokesperson—so pissed off the academic community that he will never be able to get that community's support.

The response also seems naïve when it suggests the improbability of academic "development of products that could be used in bio-terrorism acts." Well, the products that could be so used are the bioweapons pathogens themselves, and, even as the director blus-tered, research on such pathogens at institutions of higher education was proliferating at an explosive rate—more and more pathogens, at hundreds of institutions, in the hands of thousands of researchers. And much of that research involves tinkering with existing pathogens to make them more dangerous, so that defenses can then be built against them. What more likely source for "products that could be used in bioterrorism acts"?

Tulane University, with its collection of 5000 research primates, got "The Most Frightened Monkeys" award for indicating that it had "no documents that are responsive to your request" (i.e., no minutes of IBC meetings, and thus, probably, no meetings).[60] North Carolina State dealt with the problem of no IBC minutes a different way, sup-plying a letter from the outgoing IBC chair to the University's Asso-ciate Vice Chancellor for Environmental Health and Safety, indicating that "[t]here were no major issues or problems that required the IBC action"; the chair indicated he reviewed and approved projects per-sonally. He stated that two IBC meetings had been held, the most recent to introduce the new chair; he did not reveal the purpose of the other meeting. NC State got Hammond's award for "Most Dubious After-the-Fact Manufacture of Records."[61]

On February 20, 2006 Hammond posted the Sunshine Project's Top 10 Freedom of Information Failures. About half were military-related, involving the Defense Advanced Research Projects Agency (DARPA); the Office of the Secretary of Defense/Joint Chiefs of Staff; the U.S. Navy/U.S. Marine Corps; the U.K. Ministry of Defense/U.S. Marine Corps; and the U.S. Army/USAMRIID. Most of the mil-itary requests sought information about proposals and plans for development of "nonlethal" biological and chemical weapons, offensive "anti-material" biological weapons, authorizations to use "riot control agents" (a type of chemical weapon) in Afghanistan and Iraq, and development of unmanned aerial vehicles to deliver chemical agents. The request to USAMRIID was for smallpox safety and research protocols related to Peter Jahrling's efforts to infect monkeys with smallpox.[62]

Hammond also sought unsuccessfully to obtain information from the CDC about its smallpox safety and research protocols. And pursuing the possibility that Boston University researchers might not have had the required federal permits to possess virulent tularemia at the time they were infected, he asked CDC for the dates on which a number of BU researchers were issued their select agent permits, but "CDC quickly denied this information on plainly specious grounds and has yet to process the appeal we filed."[63]

Hammond's description of what happened with respect to his request from NIH's Office of Biotechnology Activities, for records of NSABB's "review" of the resurrection of 1918 influenza, deserves to be produced in full:

> After the deeds were done and the opportunity to meaningfully intervene largely sacrificed, in 2005 the National Science Advisory Board on Biosecurity (NSABB) was called in to assist with public relations problems surrounding the resurrection of 1918 influenza. NSABB "reviewed" the 1918 flu experiments literally only hours before they were to be printed. The "review" happened in a hastily convened phone call whose purpose was to rubber stamp a project that had not been properly overseen. It was a farce of a "review" aimed at derailing opposition to the experiments, rather than fulfilling NSABB's responsibilities.
>
> NIH promised openness and transparency at NSABB, however, so we put that rhetoric to the test by asking for full documentation of the "review." Rather than putting its records where its mouth is, OBA has responded by haggling and ignoring Federal Appeals Court rulings about FOIA fees. We suspect this petty attitude from a FOIA office that is normally reasonably efficient has to do with a reluctance to release unflattering records.[64]

Hammond also speaks best for himself on his problems getting documents from the National Academy of Sciences related to non-lethal weapons:

> NAS must obey FACA, a law that says that when NAS does a study for the government, documents that are deposited in the Public Access Records File are public. That's at least what the law says. . . .
>
> We requested dozens of public documents about "non-lethal" weapons. We started getting them. Some enthusiastically endorsed illegal chemical and biological weapons. Then, a Marine Corps Colonel sent a letter to NAS with an illegal "order" for it to stop sending the papers. NAS knows who butters its bread. Violating federal law, it stopped releasing documents. Since mid-2002, NAS has ignored all queries about the issue. Our request remains standing.[65]

In 2006, Hammond also released several new Biosafety Bites. One focused on the complaint he had filed with OBA about a sham IBC at the University of South Carolina. The U of SC had ignored Hammond's first two requests for IBC minutes for the last 2½ years, responding after Hammond cc'ed OBA on his third request. USC then provided two-thirds of a sheet of paper, representing the minutes of a July 7, 2004 meeting, primarily focused on discussion of the SP's requests and the IBC's noncompliance with the NIH Guidelines, but also containing a paragraph asserting that "a portion" of the committee met May 21, 2004 (three days before Hammond's first request) to find committee members, elect a chair, and adopt a biosafety manual.[66]

Hammond again requested minutes in March 2006. The university provided the sparse minutes of one additional meeting, in September 2005, where the community again discussed the SP's request and membership problems. At no time did the committee review any substantive business.[67]

What especially troubled Hammond was that the university's president was a member of the National Science Advisory Board on Biosecurity. Hammond noted sarcastically: "One might have expected that the President would have ensured that his own house was in order before seeking to advise the nation on dual-use research compliance, but that assumption would obviously be incorrect. Similarly, it might have been expected that NIH OBA would seek NSABB members from institutions that are compliant with existing relevant federal research recommendations, but this too seems not to be the case."[68]

The University of Georgia IBC also met only after Hammond requested minutes. The minutes that it produced were those which might have been expected from a "first meeting": committee members and safety staff got to know each other, then discussed what an IBC was and what it was responsible for. Hammond dubbed UGA the "Dukes of Hazards" because of all the risky research it had engaged in without IBC supervision: recreating the 1918 flu; and work with bird flu, anthrax, botulinum, and a variety of other genetically engineered pathogens.[69]

Biosafety Bite 18 focused on the fact that the IBC of AlphaVax, a private company in North Carolina's Research Triangle Park, which had received $42 million in NIAID grants for biotechnology-related research, much of it at BSL-3, hadn't met for 3½ years. Its core research involved genetically modifying Venezuelan equine encephalitis for use in vaccines, incorporating genetic pieces from such other weapons agents as Ebola, Marburg, smallpox, SARS, and pandemic influenza. Instead of convening a meeting of the IBC, AlphaVax had a habit of sending out safety documents by E-mail and then writing memos to the file stating that the IBC had "met and reviewed" the projects in question.[70]

BSBs 15 and 17 focused on especially bizarre responses to Hammond's IBC minutes requests. East Carolina University had forwarded pages in which almost every line had been blacked out, but the blotting became transparent after it dried. And behind the blotted minutes the university carelessly included a non-blotted copy, along with correspondence between East Carolina's and the University of North Carolina's lawyers, on the theory and practice of redacting IBC minutes before supplying them to the SP. UNC, noting either that the SP had Advisory Board members from Africa and Asia, or staff (Hammond's wife Susan Pimiento) from Colombia, meditated:

What kind of worried me about the request was that some of the people on the [Sunshine Project] board were from foreign countries where there had been terrorist cells found, or where I think I remember some assertion by the feds that some rebel group in the country was allied with Al-Quaeda [sic].

UNC told ECU proudly that it had deleted the names of a few bugs/toxins that weren't necessarily select agents "but some stuff we'd just as soon terrorists didn't know we had around." Hammond pointed out that the deleted information could almost certainly be obtained from other sources.[71]

Hammond noted with amusement ECU's redactions of all information about herpes research, "as if the US was threatened by terrorist cold sores." Other deleted information included references to the fact that no glass would be used in the lab, that a new keyboard cover needed to be installed on the computer. On the other hand, the ECU IBC happily released information evidencing its command of technical terminology: "Would a vortex or a homonognizer be better than the mortor and pedestile? [sic]"[72]

Hammond thought he knew exactly why ECU's IBC had completely deleted a paragraph describing the University's malfunctioning waste incinerator, and a reference to a Brucella exposure, however.[73]

The Public Health Research Institute of Newark, New Jersey, whose problem with disappearing plague-infested mice had become the subject of international media attention in September 2005, deleted all references to select agent research from its minutes, and defended the redactions by telling Hammond that PHRI had entered into an agreement with the FBI to "not publicly disclose which specific select agent pathogens and/or strains are stored at our facility." Hammond obviously doubted the truth of the statement, but said if it was true, the "public might now be kept in the dark about research risks and accidents under apparent FBI order." But he also noted that the only way to enforce such a secrecy order would be for PHRI to cease scientific publishing and remove information from the public domain.[74]

Hammond noted that one could easily find references to PHRI's bioterrorism research on PHRI's own Web site, in the NIH CRISP database, and on the home page of the Northeast Biodefense Center. The latter featured an article written by PHRI's president and four other scientists describing their aerosol challenges of mice with plague bacteria. The other sites indicated PHRI's work with plague, anthrax, and botulinum, including aerosolizing of anthrax.[75]

The Crocodiles Shed a Few Tears, Then Get Down to the Business at Hand

Fellow biodefense skeptic Richard Ebright said the Sunshine Project's demise would "create a vacuum," and that "We'll go back to silence."[76] Ebright had spearheaded the 2005 petition drive in which a significant majority—758 of 1143—of NIH-funded microbiologists said the agency's new obsession with biodefense was imperiling important public health research. That he sincerely regretted the Sunshine Project's disappearance could not be doubted.

The same might not be true for some prominent members of the biodefense complex. Gigi Kwik Gronvall said, "There's no one else I know of that will look over at that detail and keep things transparent."[77] That Gronvall cared about transparency might be doubted, however, since she and the UPMC in general were busy promoting a voluntary, anonymous accident reporting system that would help malfeasants like Texas A&M stay out of the public eye—certainly not what Hammond (or most people) would consider transparency.

C. J. Peters of the University of Texas Medical Branch at Galveston told *Science* that while Hammond was a "pest" who "exaggerated risks" (and Peters didn't give any examples), he was "strangely, sad to see him go." "I think the country works best with watchdogs," he insisted.[78] But UTMB-Galveston would, shortly, demonstrate that the only watchdogs it had any use for were ones with muzzles.

Blessed with a broad and explicit open-records law in Texas, Hammond had done an especially good job of keeping the feet of Texas biodefense institutions to the fire. Of course, the "fire" he was keeping them to, for the most part, was a simple compliance with the public access portion of the NIH guidelines. The Texas A&M revelations were his most famous handiwork, of course. But he had followed up the A&M revelations with a September 18, 2007 press release about workers exposed to aerosolized anthrax at the University of Texas Health Science Center at Houston; workers who entered a University of Texas at San Antonio tularemia lab to inspect malfunctioning air filters without gloves or respiratory protection; and four lab-acquired infections at the University of Texas at Austin between 2002 and 2005 that had not been "properly documented, investigated, or reported." Hammond said the new reports demonstrated the "crumbling of the biodefense lobby's safety façade," and said the fact that no Texas institution had produced a single record of a select agent mishap that occurred before April 2007 (the month of the Texas A&M revelations) indicated a "de facto policy of not recording accidents with bioweapons agents, probably for fear of the potentially embarrassing and costly consequences."[79]

In July 2004, Hammond had filed a complaint with the NIH's OBA about the IBC at one of Texas's two BSL-4 facilities, the Southwest Foundation for Biomedical Research in San Antonio, after the institution had failed to reply to records requests and then had produced only a list of 12 project titles, with no IBC minutes or other indication the IBC had ever met. In a June 2006 press release, Hammond had again focused on the SFBR, noting that it still could not produce any evidence of actual IBC meetings or reviews, even though the NIH assured Hammond that the 2004 complaint had been acted upon.[80]

Biosafety Bite 21, published in two segments in January 2007, had focused upon an accident involving a genetically engineered cross of bird flu (H5N1) and H3N2 influenza at the University of Texas at Austin. The first segment, based upon a UT incident report and other materials Hammond had obtained under the Texas Public Information Act, revealed a centrifuge accident involving the genetically engineered virus on April 13, 2006. A quantity of the virus mixture had been placed in a tube; the tube had been capped and placed in a centrifuge. But the tube was actually one belonging to a non-identical but similar second centrifuge in the lab:

> The postdoc pushed a button and the centrifuge began to spin. **Because the tube was the wrong type, its cap didn't fit correctly. It cracked.** The centrifuge lost balance. Turning the machine off and then opening it, the postdoc observed that

the level of virus fluid in the tube had gone down and that its exterior had become wet, **both indicators of a leak. This was a serious problem** because as the machine spun around, **the leaked virus had become aerosolized**, at least within the centrifuge.

By now the cracked cap problem had been compounded by human error, an ever-present factor in lab work. Rather than waiting for the aerosolized flu to settle, the centrifuge had been immediately opened. In an invisible puff of air, virus particles wafted out of the machine. **Now, the virus was floating around the whole lab, stirred by air movements,** then slow[ly] settling on exposed surfaces or being sucked out the exhaust which, hopefully, had effective HEPA filtration (the UT documents are silent on this item).

. . . .

The researcher sprayed Lysol and wiped up surfaces in the work area, exited the lab, took a shower, and put on new clothes. Within hours, the postdoc was taking Tamiflu, in the hope that it would stop the virus if the researcher had been infected. For several uncomfortable days, the University of Texas staff waited to see if the researcher developed symptoms. None are reported to have appeared.

The University of Texas at Austin had dodged a bullet. It took longer for a UT biosafety team to straighten out the lab and reopen it. Under any of a variety of plausible scenarios, the accident might have resulted in disaster. For example, if the cap leaked but didn't crack, without the postdoc noticing, thereby multiplying the danger to include everyone working in the lab over a longer period of time [emphasis added].[81]

UT first fought to keep the details of the accident secret. Then it finally provided a series of documents revealing a continuing process of "revising the facts." UT first produced a version in which the facts were called into doubt. The researcher only thought the volume of the tube had changed, but had not been 100% sure. The liquid on the outside of the tube "may have been from condensation." The "doubt" then became the basis for concluding that an accident didn't happen: "There is the possibility that there was no leak and therefore no contamination occurred." And therefore no report to the NIH.[82]

Hammond's Biosafety Bite had described the university's public relations obfuscations and provided a detailed rebuttal.[83]

In October 2007, Hammond had revisited the University of Texas at San Antonio, after discovering that the university was operating a public IBC for show, while transacting most of its business through a second, secret IBC, whose records it was fighting to keep secret. It was also displaying on its Web site "a well-organized set of forms and policies, interspersed with fancy graphics flashing messages about a 'cleaner world' and 'hazard-free work environment'"—but in fact it had not used any of its Contaminated Sharps Injury report forms (for reporting nonsterile needle sticks and such) or Employee Exposure Notification forms (for other suspected exposures) in five years, even though Hammond had eventually learned, by dint of repeated pressure, that 30 workmen's compensation claims had been filed for sharps injuries or monkey bites.[84]

NBAF 12

DOG-AND-PONY SHOWS, VS. INDEPENDENT JOURNALISM

IN AUGUST AND SEPTEMBER, DHS WOULD HOLD "SCOPING MEETINGS" as part of the required environmental impact process. The scoping meetings—supposedly to allow citizens to tell DHS what concerns they'd like to see "addressed" in the environmental impact statement—mostly turned into NBAF-kissing lovefests dominated by politicians and chamber of commerce types who told DHS what wonderful big security-goon muscles the agency had, and how excited the state and community were that DHS was thinking of sticking its germ lab into little ol' Butner, Manhattan, Flora, Athens, or that big old germ lab floozie San Antonio. They had everything DHS needed for the germ lab, including people who wanted it. Like really bad.

"San Antonio ready, able to accommodate biolab," simpered the *Express-News*.[1] "This is an absolutely beautiful place for you to be," drawled Bexar County Judge Nelson Wolff.[2]

"We've got what it takes," said Mississippi Governor Haley Barbour.[3]

"There really is broad support for your proposal," insisted Kansas Lt. Governor Mark Parkinson.[4]

Ed Hammond would refer scornfully to the Kansas NBAF efforts as "the Kansas Pork Railroad." All the "scoping sessions," of course, represented efforts by the bigwig consortiums to co-opt and stage-manage an event allegedly intended—according to DHS labs director Jamie Johnson—to identify "issues" that needed "addressing" by the environmental impact statement. That the bigwigs lusted for NBAF everyone knew already; that so much of the public's first opportunity to ask questions should be drowned out by the bigwigs' public masturbation was disgraceful.

Even the *Manhattan Mercury*, knee-deep in the muck of the Kansas stage-managing, wondered how useful the dog-and-pony shows were:

> About 300 or so people showed up, the great bulk of them present in one semi-official capacity or another to urge DHS' selection of our/their home town. This predictably included a strong delegation from the university, the state, the city and county, and the Chamber of Commerce. In each case the message could be boiled down to six words: You couldn't find a better place.[5]

And in North Carolina, local resident Elaine McNeill complained, "I think some of our public officials need to listen to the public!"[6]

Nobody stage-managed better than Kansas, though. For one thing, the Kansas scoping meeting was held at the K-State Student Union, allegedly because "DHS requires the meeting to be held as close to the proposed site as possible."[7] This in itself favored commenters associated with K-State University. What made attendance even more difficult and confusing for the general public, however, was the fact that the parking structure adjacent to the Union's most obvious entrance had been blasted into rubble—a state still present when I visited Manhattan in 2009. "Unfortunately," the *Mercury* acknowledged in its editorial supporting going-through-the-motions democracy, "construction at the Union has all but eliminated parking where it would be most convenient. The university is encouraging individuals who plan to attend to park in the large lot across 17th Street from Memorial Stadium. Wisely, the university also is providing shuttle service to the Union from that lot."[8]

Something else favored the bigwigs: familiarity with the process and the need to make prearrangements. Two days before the scoping meeting, the general public would learn from the *Mercury* of the need to preregister to speak—first come, first serve.⁹ The well-connected members of the Kansas Pork Railroad, who had been lobbying for NBAF since early 2006, presumably got their places in line early.

Members of the consortium kept up a steady barrage of propaganda in the *Manhattan Mercury* and elsewhere in the days prior to the meeting. Two days prior, the *Mercury* carried an op-ed from Angela Kreps, president of Kansas Bio and a board member for the Kansas Bioscience Authority, "Let's show Manhattan is right site for NBAF."¹⁰ The next day, the *Mercury* carried another op-ed from two of K-State's Fort Detrick imports, germ lab stars Nancy and Jerry Jaax, of *Hot Zone* fame. Their contribution was titled, ironically, "Citizens need straight talk on NBAF safety." Citizens weren't likely to get straight talk from any card-carrying members of the biodefense complex, and they certainly weren't getting it from K-State administrator Jerry Jaax.

The Jaaxes emphasized the benign nature of their work at Fort Detrick, the fact that tours were given to schoolchildren and community members. And then there was this:

Although we know of no probable scenario where the NBAF mission would include ebola, opponents have used graphic descriptions of Ebola Fever symptoms to alarm the community. We are among a relative handful of people who have worked occupationally with ebola in both the laboratory and the field. Like hundreds of other employees, we returned home daily from the lab confident that our families were not at risk.¹¹

Which, of course, explained Jerry's diatribe in *The Hot Zone* about "that *fucking* Ebola!"

The Jaaxes had become heroes precisely because of their work with that dangerous Ebola. Now here they were telling Manhattanites there was nothing the least bit dangerous about Ebola, or about BSL-4 work. It's as though, years after the shootout at the OK Corral, Wyatt Earp should insist that bullets are not dangerous.

And where did those worried locals get the notion that NBAF might handle Ebola? From over two years of newspaper articles around the country, prompted by DHS's characterization of NBAF as an "integrated" human/animal/zoonotic disease supercenter. It was only recently that DHS had posted its pared-down "diseases of interest" list, omitting such scary zoonotics as Ebola and anthrax. But North Carolina NBAF mouthpiece Barrett Slenning, himself confused by DHS's contradictory information, would still be carrying the zoonootic supercenter paradigm around in his head in September 2007.

More dramatic irony: the Jaaxes would cite the NIH's Environmental Impact Statement for the Boston BSL-4 facility as some sort of safety reassurance:

> In response to local concerns, the National Institutes of Health (NIH) has just released a study concluding that building the biocontainment facility there [at Boston] "would neither elevate nor create a public health risk" for the community. That facility, unlike the NBAF, is specifically intended for research on serious human diseases like SARS, plague and ebola. For those looking for answers about safety beyond speculation, this report should help reassure citizens of Kansas about the safety of the NBAF.[12]

Three months later, a National Research Council panel would describe the NIH Boston Environmental Impact Study (EIS), with its deliberately one-sided "speculations," as about as informative—and reassuring—as a failing undergraduate biology paper. Those looking for answers about safety beyond biolab propaganda would need to look past the Boston EIS, or the Jaaxes' op-eds.

They would need to look past the *Mercury* itself. If the consortium mouthpieces ever missed any propaganda bases, the *Mercury* would sweep away the footprints so no one was the wiser. In a weekly question-and-answer column, *Mercury* editor Ned Seaton would demonstrate why DHS had ruined a Dixieland NBAF straight (with its rich collection of Southern politicians—smiles on their faces, hands in your pocket, fingers up your butt) by dealing itself a Midwestern card.

These Kansas boys knew how to sell used cars and luxury funerals as well as any slick cracker.

Four days before the scoping meeting, Seaton answered the question "If the NBAF were built here, would we become a target for terrorists?" with some cute humor:

> In my ongoing pursuit of the truth in this column, I try to ferret out the best sources of information. But I have to say that you have now saddled me with the toughest assignment yet: I'll be danged if I can rustle up a terrorist to ask. I just can't seem to find that business card for Osama Bin Laden around here anywhere. . . .[13]

Seaton then focused on one particular terrorism scenario, an attack on the facility with an airplane or car bomb, and dismissed that possibility by quoting extensively from a "report" by the Kansas Bioscience Authority, the chief lobbyist for NBAF. Naturally, this scenario just wasn't anything to worry about, even though researchers at Detrick, the CDC, and Plum Island had evacuated their impregnable facilities on 9/11, and even though DHS Undersecretary Cohen had cited—in a memo meant to remain secret—the danger of terrorism in eliminating Maryland from the Final Five selection sites.[14]

In his search for Osama bin Laden's business card—which of course will carry some other name—Seaton might have had more luck checking the applicant folders for research positions at the Biosecurity Research Institute—or any of the hundreds of other new high-containment facilities springing up around the country. Infiltration of these facilities—which would give a terrorist access to state-of-the-art technical knowledge—and state-of-the-art bioweapons pathogens—is the real terrorism threat.

Seaton closed this particular column by focusing on safety in general, with a classic version of the biodefense two-step:

> Does all of that mean there's zero chance of a problem? No. But of course there's also a chance that you'll get hit by a car, that a tornado will tear the roof off your house, and that the

double-cheese, fries and Coke you're having for lunch will eventually kill you, too.[15]

If only this admirable sense of proportion had been applied across the board—to the bioterrorism "threat" in general, for instance. Hundreds of thousands of people were dying prematurely each year from antibiotic-resistant hospital infections, AIDS, tuberculosis, influenza, obesity, etc., but research on, and measures to combat, these diseases, was languishing in the midst of the biodefense obsession. A phalanx of secretive biodefense facilities was being constructed around the country because five people had been killed by special shipments from one or more existing biodefense facilities.

Once Kansas had "won" the NBAF competition, and Congress began to display a few signs of common-sense caution (duh, why are we putting those deadly livestock germs in the heart of the livestock belt?), the Kansas Pork Railroad would be screaming the need to build NBAF right now, not a moment to lose, because Osama bin Laden was lurking out there in a Kansas feedlot, dying to try out some of the elaborate scenarios American biodefense blowhards kept telling him would be so easy. And the *Mercury* was telling worried home folks in Manhattan that *they* should get a sense of proportion?

Seaton had been a bit more defensive in the previous week's column, after a skeptical resident sent the following pointed query:

> One of the letters to the editor awhile back was titled "Defense Facility Is Something to Worry About, Not Strive For." That happens to be exactly how I feel about the National Bio- and Agro-Defense Facility, and I suspect many other local residents feel the same. But the recent front-page story titled "Meet the Competition" (about other cities trying to attract the facility) stated "K-State and city officials are enthusiastic about their prospects for landing the facility" and "To date, all signs have been positive and receptive. The city has committed $5 million in funds to land the facility, and the Chamber of Commerce has also pledged its support." Neither one of those statements reflects any polling of opinion held by the public at large.

I would like your help in understanding three things about the process that invited the NBAF to be sited here.

1. What was the nature and extent of public notices inviting public comment about this initiative?
2. What was the extent of the city's and the news media's participating in publicizing those opportunities for public comment?
3. What are the most effective ways for the public to express their opposition now that we are one of the five finalists?

"Your question," Seaton said, "is really about how we got to this point, and whether the public was really consulted about it. . . . The answer depends on what you think about representative government."[16]

The public's representatives—the City Commission—had "reviewed the matter at a meeting Feb. 6 and decided to pledge $5 million to try to help lure the facility here. The city did not place any official legal notices about that meeting; it's not required to."[17]

But the *Mercury* had published "a front-page story on Feb. 5 describing the issue. . . . So—no offense intended—anybody paying attention could have gone to the commission meeting to comment. The folks who did comment were from the Kansas Bioscience Authority, the Chamber of Commerce, and K-State. Nobody expressed any objections, according to the Mercury report on the meeting."[18]

In other words, there was a brief "opportunity" for public comment, for any City Commission news junkie, or anyone who read the *Mercury* in detail on a daily basis. One would imagine that narrowed the "public" down considerably.

Of course, the Kansas Bioscience Authority, the Chamber of Commerce, and K-State—who were requesting the financial support, and therefore knew about it already—had representatives at the meeting. They should surely qualify as "public," right? Some of them may even have voted in City Commission races.

And here's where Seaton gets a little defensive:

Now, I will grant you a point—the Mercury story in advance of the meeting did not go into detail about the nature of the materials to be studied in the lab. That's because we didn't know everything about it, based on the information available at that time. (We still don't, by the way, but we're learning more as we get our hands on more information.)[19]

This is a mealy-mouthed excuse for failing to inform the readers. The DHS hadn't been rushing out to tell potential host communities exactly what they planned to study at NBAF, to be sure. But it was pretty basic knowledge that a BSL-4 facility studied diseases with no vaccine or cure—and there were only a few of those diseases knocking about. And plenty of other publications around the country had explored what some of those diseases were—and "anybody paying attention"—say, a newspaper with some commitment to keeping the public informed rather than sedated—could have found that out for themselves.

Around the country, almost the only local newspapers acting like newspapers on the subject of NBAF were of the alternative, weekly ilk. First out of the gate, and most persistent, was the *Independent Weekly* of Raleigh, North Carolina. Soon after the finalist announcement, I received a call from Lisa Sorg of the *Indy Weekly*. She was gathering background information for a long, balanced, but probing article that would appear in the *Indy Weekly*'s July 25 issue, "Biotech or Biohazard?"

The article noted the celebration by members of the N.C. consortium, who were quoted frequently throughout, but went on to explore why, "before the champagne corks are popped, there is reason for caution, even skepticism." It then explored the various security breaches, safety violations, and accidents at Plum Island ("DHS didn't return repeated phone calls and messages seeking comment") and other high-containment germ facilities; included comments from Michael Carroll questioning the wisdom of putting this particular bit of pork in the country's second biggest swine-producing state; and noted the fact that the five finalists all had "Republican connections and/or high-placed lawmakers on key, related committees." It pointed out that the nearest Hazmat teams were in Raleigh and Durham, a half hour or more away; and that many of the lab's potential neighbors were involuntary residents who couldn't object to the siting of a germ lab in their midst: over five thousand state and federal prisoners and juvenile offenders, and another 950 psychiatric and developmentally disabled patients.[20]

A sidebar, "Feel a chill coming on?" presented capsules of the eight "Diseases of interest" posted on the DHS Web site.[21] "Gift horses" by editor Jennifer Strom introduced the Sorg article and took the *Durham Herald-Sun* and *Raleigh News-Observer* to task for their gushing, rah-rah coverage and failure to even hint at "the potential costs of this lab—in human and animal endangerment, environmental detriments and other downsides."[22]

Out in San Antonio, Greg Harman of the *San Antonio Current*, where Lisa Sorg had once worked, wrote a similarly probing article titled "Banging the drum for Bio-defense." Harman revealed that the Texas consortium had already spent $500,000 on lawyers and public relations specialists, and lamented the fact that "at a time when news of lab lapses appears to be everywhere . . . the only significant discussion the proposal has provoked locally has to do with money and jobs." Harman balanced bio-evangelist Jean Patterson, chair of virology at the Southwest Foundation for Biomedical Research, who said of the risks, "I don't see any," with commentary from Ed Hammond, Michael Carroll, and George Annas, chair of Health Law, Bioethics and Human Rights at Boston University.[23]

In Athens, Georgia, *Flagpole* editor Pete McCommons, in a commentary titled "The Pathogen City?," revealed his disquiet about the ongoing marketing of Athens "as a center for the study of strange stuff that infects animals." He said the homefolk, including himself, hadn't been paying attention to what was going forward in their name:

> Not that our paying attention would have changed a comma of the whole enterprise. When the recent news dawned that Athens has made the short list of five sites in the finals for this research facility, some of us began to go, do what? Deadly pathogens in our sewer system, in our air, in our river? A big old prison-like facility where once there were green fields? Our comfortable little town overrun by Homeland Security storm troopers?[24]

None of the homefolks in the five finalist towns had been paying a lot of attention up to this point. In many cases, there hadn't been the sort of news attention which now began to emerge. Prior to the finalist selection, for instance, the *Manhattan Mercury* hadn't given the lab nearly the ink that the *Lawrence Journal-World*, whose publisher was on the board of the Kansas Bioscience Authority, had. And despite the fact that the *LJW*'s ink had all been of the rah, rah, lab-boosting sort, a core of vocal opposition had arisen around the proposed site at Leavenworth. And residents around one of the three Mississippi sites—the Hinds County one—had dared the "ignorant gossip" epithets hurled by the *Jackson Clarion-Ledger* and mounted enough visible opposition that Mississippi had yanked that sucker just before the Homeland Security boys got into town.

There was opposition at all five finalist sites, mark you. "Community acceptance"—one of DHS's alleged criteria for choosing the final site—didn't really mean what you'd think it'd mean. Nobody polled the citizens around any of the sites to see whether they wanted NBAF. Nobody did a scientific telephone survey or anything of that ilk. DHS didn't care about all that. To the extent they cared at all, they just wanted to make sure they wouldn't be met with the sort of public-relations ass-kicking they'd get at Butner, North Carolina. They wanted official endorsements, of course; they wanted the sort of fawning, simpering, yassuhs and nosuhs the bearers of a $500 million—make that $700 million—bioboodle project had the right to expect; and of course they wanted to hear just what you were going to give *them* out of your state and local coffers by way of land and infrastructure. But as far as the people, they just wanted to know whether you could keep them sedated, cowed, bamboozled, whatever. Whatever lies you needed to tell were your business. It's what they called "managing community acceptance."

In Kansas, Texas, and Mississippi they knew how to manage community acceptance. In North Carolina and Georgia they thought they did.

"A few dissidents were scattered among the audience," the *Mercury* reported, "but the mood was generally positive."[25]

Among the dissidents was Walter Dodds of the K-State Biology Department, who warned that there was no such thing as a fool-proof facility and expressed his dismay about the lack of comments about the possibility of environmental disaster.[26]

A question-and-answer period preceding the formal comments period allowed DHS to engage in a little propagandizing. In a manner resembling the "Denny Crane" incantation of William Shatner's "Boston Legal" character, NBAF program manager Jamie Johnson kept repeating the phrase "standard operating procedures" in responding to citizen concerns about safety. And he assured those present that NBAF would not study anthrax or those scary old hemorrhagic viruses like Ebola or Marburg, even though these were noted zoonotic diseases, and NBAF had been touted as *the* lab for zoonotics. And though 350-plus American labs stored and did research with anthrax, many of them at your neighborhood universities down the street—and kept doing freaky things like shipping it to children's hospitals—and KSU's Biosecurity Research Institute had said it planned to do some things with anthrax—they simply wouldn't do such scary research at the country's second biggest BSL-4 lab, managed by an innocent agricultural/public health organization like the Department of Homeland Security.[27]

"Biolab project's hearing has few naysayers," the *San Antonio Express-News* gushed. There was a lot of gushing going on: most everybody "gushed praise for the city, its scientists and its research institutions." Not everyone was enthralled, the *Express-News* acknowledged, though it cited as example only Linda Loomis, concerned about Plum Island's safety record.[28]

Even in Flora, Mississippi, where Mayor Scott Greaves swore he "had heard virtually no negative reaction from his constituents"—"with the exception of a handful of people in the area, and I mean a handful, five or six,"[29] that handful showed up to raise concerns. Not enough to dent the "host of local and state officials who spoke out in favor"—from Governor and former Republican National Party Chairman Haley Barbour on down. A few Flora residents did point out, rather rudely, "that many of those lobbying for the facility would not have to live next door to it."[30]

THE WHEELS DROP OFF A COUPLE OF BANDWAGONS

THE NORTH CAROLINA AND GEORGIA SCOPING SESSIONS SUGGESTED right off that those states' big-wig consortiums weren't going to have it all their own way.

That didn't keep them from trying.

The North Carolina consortium, for instance, put up a well-stocked Web site. One segment addressed "Ten common misconceptions about the National Bio and Agro Defense Facility," and probably told at least ten new lies in doing so. For instance, the Web site claimed that "[r]esearch facilities in the U.S. similar to the NBAF have outstanding safety records: No community or environmental safety problems have occurred in the more than 20 years that these facilities have been in operation."[1] This book's chapters on Plum Island—the country's chief existing animal disease facility—and Detrick—one of the country's two main BSL-4 facilities—discuss in detail the multiple "community or environmental safety problems" at those two "similar" facilities.

The consortium went on to claim that "[t]he National Institutes of Health have monitored safety at federal BSL4 labs for more than 20 years and there have been no accidents or security breaches."[2] Even the CDC, fondly cited by biolab advocates as the poster child for lab safety, had just had a power loss and breakdown of negative air pressure in its new high-tech suite of BSL-4 labs—an incident which came to light only because of a "leak" to the *Atlanta Journal-Constitution*. (The CDC, definitely not a poster child for lab transparency, generally likes keeping lab accidents under wraps, purportedly in the interest of national security. Biolab neighbors get restive when they hear about accidents. *Accidents? I thought you said there wouldn't be any.*) And, though the FBI was still a year away at this point from making its Ivins allegations—about the mother-of-all-biolab-security-breaches so far—since 2002 it had been an open secret (known by the country's major media, but not widely reported) that the FBI believed the 2001 anthrax came from a government lab. And Detrick "accidents" were the subject of numerous headlines between 2001 and 2006.

Also a lie, for similar reasons: "For federal BSL 3-4 facilities there have been no community releases or contaminations for any reasons."[3]

In the flagrantly specious category: "[T]he small quantities of biological agents in such labs are kept in freezers, refrigerators, incubators, and/or biosafety cabinets." Or further on, explaining why terrorism is simply no concern whatsoever: "Furthermore, within the biocontainment area, the small amounts of microorganisms will be kept inside sealed containers and those containers will themselves be in secure refrigerators, incubators, hoods, or other devices."[4] This is misleading because the "incubators" at NBAF will include at least a hundred cows, pigs, goats, etc. deliberately infected with disease, and functioning as portable breeder reactors of biological agents, breathing and excreting millions and millions of germs daily. Definitely not sealed containers. Those animals and their germs will not be kept in freezers, refrigerators, or biosafety cabinets. (One does hope they'll at least be successfully confined to the "biocontainment area.") In addition, the animals will almost certainly be subjected to "aerosol challenges" with the germs du jour—and that won't be in a freezer or refrigerator either.

Likewise misleading: "The NBAF will have its own self-contained and redundant treatment system for waste water from the facility."[5]

A system that receives waste (from a hundred or more sick livestock) and simply treats it before releasing it into the "local sewage treatment system" is not a "self-contained" system—and to call it self-contained is not only a lie, but an especially mindless one. What one hopes is that all the waste does in fact get adequately treated, and that there's no snafu of the sort that occurred when the Winnipeg BSL-4 lab released untreated wastewater into that city's sewage system.[6]

In the it-does-not-follow category:

> DHS has said from the start that this laboratory will NOT do bioweapons work. Researchers will come primarily from the USDA, and are animal disease experts. The research will be unclassified, so we will know what kinds of work is [sic] being performed. Finally, community oversight groups will be established to provide public monitoring of activities.[7]

That DHS has said, or still says, that the lab will not do "bioweapons work" tells us nothing about whether it actually will or won't. It may depend on what you mean by "bioweapons work." Was the anthrax mailed through the U.S. Post Office a "bioweapon," even though it wasn't delivered by a B-52 or a Stinger missile? Was the Plum Island effort—referred to in a GAO footnote—to genetically modify cowpox to see if it could be made lethal like smallpox "bioweapons work"? Is DHS going to guarantee that the facility won't develop new pathogens so it can then try to develop defenses against them? Can anyone believe any guarantees DHS gives?

After all, DHS gave multiple indications it was making up NBAF as it went along, or perhaps, making up its lies as it went along. After describing NBAF for over a year as an animal-human-zoonotic disease supercenter, DHS had posted an equivocal stripped-down "Diseases of Interest" list on its Web site, which left off almost the entire universe of zoonotic diseases, and certainly all the notorious ones that people worry about (anthrax, Ebola, lassa, Marburg, etc.).

Likewise, the consortium had slight grounds if any for declaring that NBAF's research would be unclassified, or that community oversight groups would be established. Even if the research should be "unclassified," that didn't mean "we" could know what kinds of work would be performed there, given the slippery "sensitive but "unclassified" category that the biodefense complex developed under Bush, or the repeated effort by biodefense facilities, most recently by the alleged transparent wannabe UTMB-Galveston, to keep any and all information away from that pesky Sunshine Project. The fact that a community liaison group of some sort might be established does not translate into such a group having a significant oversight or "public monitoring" role. Biodefense facilities have a habit of keeping embarrassing details away from such groups, or not consulting them about such a major thing as UTMB-Galveston's effort to gut the Texas Open Records Act.

(Biodefense Two-Step Warning!): "There are no records of any U.S. biocontainment laboratory being targeted by terrorists." "We are unaware of any such facilities being targeted by terrorist groups. This lack of terrorist interest in high biocontainment facilities is probably because an attack would be complicated to execute and unlikely to succeed."[8] Since when did *documentation* of "targeting by terrorists" become a requisite for bioterrorist paranoia? The only bioterror attack in the U.S. other than the salad-bar food-poisoning incident in Oregon came from one of our own "high biocontainment facilities." Yet the WMD Commission confidently *predicts* a WMD attack—most likely a bioterror attack, it says—"somewhere in the world" by 2013. Meanwhile, bioterrorists are allegedly crouched by the water tank at the OK Corral—no documentation of this, of course—waiting to disseminate the latest high-tech genetically modified germs (most likely stolen from one of our own facilities) in our food chain.

An airplane attack on a high-containment biolab is probably a "low probability" event. Nonetheless, the CDC, USAMRIID, and Plum Island did evacuate their personnel on September 11. And a terrorist explosion by no means exhausts the universe of possible attacks on a biodefense facility. Much more likely would be something like the Plum Island incident in which a striking-worker replacement with an arrest record went AWOL for an extended period of time (i.e., until formally dismissed) with a laptop capable of tinkering by remote control with the facility's ventilation and filtering systems. At any rate,

DHS Undersecretary Jay Cohen cited terrorism as a reason for nixing Maryland from the list of NBAF finalists, so maybe NBAF propagandists shouldn't dismiss terrorism concerns out of hand.[9]

Interviewed on August 10 by *Focus*, a state government radio show, NC Consortium spokesperson Barrett Slenning revealed his own confusion—or a case of propaganda deficit disorder—about what diseases would be studied at NBAF. Early in the interview, visions of zoonotic sugarplums still dancing in his head, Slenning explained why NBAF would be "such a feather in the cap":

> This lab, as Lisa was talking about it, is supposed to become the successor to a now 50-year-old animal disease lab, which is the primary animal disease lab in the U.S., at Plum Island. But it is also supposed to pick up extra missions. One of those is to become the federal point where the research is done on what are called "zoonotic" diseases. Those are diseases which are shared between humans and other animals. Currently USDA works on some, CDC works on others in Atlanta, and sometimes things may fall between the cracks. This would become the federal home for it.[10]

A bit later one of the interviewers noted that words like "Ebola" and "anthrax" raised a layperson's eyebrows, and Slenning started backing his propaganda four-wheel out of that quicksand:

> Well, let me first back up, a little, about the premise. Smallpox, Ebola, anthrax, Marburg, none of those are going to be worked at, at this facility. That has already been made very clear. They're being worked at safely now in Atlanta, Georgia with the CDC, and it doesn't make any sense to recreate the wheel up here.[11]

Smallpox isn't a zoonotic—and only the CDC and a Russian lab are supposed to have samples of the live virus—so nothing surprising about that. But the other three diseases **are** zoonotics, and hadn't Slenning just made very clear (following DHS's earlier statements) that NBAF was to be the zoonotic "focal point," including things currently worked on at the CDC? And so far as one can see, "recreating the wheel" is what biodefense research is all about—things like anthrax, tularemia, brucella, and plague have been researched ad infinitum, redundancy within redundancy, since World War II. And there will never be a shortage of new work, since America's biodefense complex is eagerly creating new threats so it can get to work developing defenses against them. Now that's what I'd call self-contained. . . .

Slenning's continued wheel-spinning on this subject bears further examination, since it suggests a probable rationale for the "Diseases of Interest" list:

DONNA: Okay. Well, we appreciate you clearing that up for us.

RICK: Well, what will be in there?

SLENNING: What will be in there are a series of diseases, or foreign animal diseases, Foot and Mouth is the one most people would have heard of. There will be a couple of what are called "Swine Fever" diseases, and then there will be, on this zoonotic side (the diseases that we share with other animals) [or that they share with us!—KK], will be some viruses, some similar to West Nile, other kinds of infectious viruses that way, but they won't be these big ones that everybody talks about smallpox, anthrax, Ebola.

RICK: Well, Professor, if North Carolina actually is named the state, will there be some kind of public notification so that people will know exactly what this laboratory will be doing, what types of diseases will be worked on, and exactly the security that will be in place?

SLENNING: Yes, they will, and actually some of this information is new, it is, they were not, things that I knew when I last talked with Lisa, because we finished a two-day meeting Thursday and Friday up in Washington DC.

DONNA: Oh! Okay.[12]

It would seem that Slenning had just been given the latest "talking points" by DHS. DHS had looked deep within its own grimy navel and discovered that no, it didn't want NBAF to be the zoonotic super-center. It didn't want to work with anything that would arouse the natives and raise the tar-and-feathering threat level out in the heart-land, or the boondocks, whatever. So, spread the word, biolab pre-varicators. We won't do anthrax, we won't do Ebola—or a score of other zoonotic diseases requiring high containment—though the diseases-of-interest threat level is subject to revision based on current threat levels, including, one imagines, the tar-and-feathering threat meter.

We call this "bait-and-switch" in the used-car business. Or the new biolab business.

It would appear that Slenning had gone to Washington to have his NBAF propaganda chips changed, and somebody forgot to take one of the old ones out.

Either that, or DHS really didn't know what it wanted to do with NBAF when it first talked about a zoonotic supercenter, and was only just now discovering its true inner being. And so now it's going to do an NBAF-lite, or a slight Plum Island upgrade. If that's the case, does it really need the country's second biggest biodefense facility, a BSL-4 monstrosity three times as big as Plum Island, for a few piddling little animal diseases? If they're not going to be taking the zoonotic Viagra after all, maybe they don't need as big a bed to romp around in.

Of course, maybe it's just your ordinary garden-variety government lying. In October 2007, I posted portions of the Slenning interview on a listserv. Richard Ebright posted a comment saying that the "Well, let me first back up, a little" statement was "a brazen, bald-faced lie (on multiple levels)"—that the only factual material in the statement was the inclusion of the word smallpox.

I wrote back wanting to make sure I'd caught all the multiple levels (and I had). With respect to anthrax in particular, Dr. Ebright said:

> [O]ne can state with near certainty that the facility will conduct research on anthrax (like the more than 350 US laboratories that already conduct research on anthrax). (It would be incon-ceivable for a facility focused on zoonoses of relevance to

bioterrorism and biowarfare not to conduct research on the single most important zoonosis. It would be inconceivable for a facility ostensibly replacing or updating Plum Island not to conduct research on an agent researched at Plum Island. It would be inconceivable for a facility run by DHS and having large-animal ABSL-3 facilities with aerosol capability not available at other DHS facilities not to conduct large-animal aerosol-exposure tests assessing anthrax weaponization technologies [and] strategies and anthrax distribution technologies. Any spokesperson suggesting otherwise is a fool, a liar, or both.[13]

The logic seems impeccable to me.

If the NC Consortium was a little confused about what NBAF was going to do exactly, it seemed confident that it wouldn't get flak from the "community." Warwick Arden, Dean of N.C. State's College of Veterinary Medicine, said just before North Carolina was announced as a finalist, "[T]he people of Granville County and Butner really have accepted this, and I think not just because of the economic impact, which is substantial, but also because there is a history of a strong state and federal presence there. So really, making the arguments has been quite easy here."[14] And Slenning claimed he had talked to people who had successfully fought the placement of a hazardous waste incinerator in Butner, "and so far most of the people who hear about it feel much better about it."[15]

Brian Alligood, Granville County manager, also said local residents were very supportive of the project.[16] And just before the North Carolina hearing, the *Raleigh News & Observer*, calling the NBAF selection process "a national beauty contest to host America's newest germ lab," predicted that North Carolina's hearing was "likely to be dominated by a long list of supporting politicians, university officials and executives from high-profile companies."[17] The North Carolina Biotechnology Center, concerned about the *Independent Weekly*'s "huge headline" "Biotech or Biohazard?" urged "[b]iotechnology company executives, university scientists and others . . . to attend the meeting and speak or write in favor of North Carolina's advantages."[18]

And indeed, the meeting did feature what the *Independent Weekly*'s Lisa Sorg called a "conga line of high-ranking politicos, university ubermen and bioscience bigwigs" entreating DHS to bring NBAF to Butner[19]—including Republican U.S. Senator Richard Burr and Democratic U.S. Representatives Brad Miller and David Price. Ronald Alligood, chair of the Granville County Commissioners, chortled: "I don't believe I've ever seen as many senators and representatives in one place at one time in Granville County."[20]

But the blunt treatment the bigwigs got from some of the locals was a sign of things to come. Several audience members shouted "time" after Burr went past his three-minute allotment of time.[21] And Brendan Meyer, an auto body specialist, asked the experts-of-the-day, "I was just curious how much you get paid to push all this [bull] down our throats?" His question "was greeted with whoops, cheers and applause from many in the audience."[22]

But perhaps the most devastating blow of opposition was struck by Daryl Moss, the African-American mayor of Creedmoor, who'd come to the meeting intending to lend his voice to the pro-NBAF chorus. As he sat there and looked out at his neighbors—many of whom had come to oppose the facility—his conscience began to bother him. And instead of support, he offered this: "I cannot say that I support this initiative. I'm really struggling. I made some commitments to support it, and I'm going back on my word to some extent." Moss was the first politician anywhere to change his position on NBAF, and the first from a finalist state to oppose his state's official efforts. (He was not the first to oppose NBAF, since the opposition of local officials in Wisconsin, Missouri, and California had helped to keep those states off the finals list.) Three weeks later, Moss would be asking the Creedmoor Town Council to pass a resolution opposing NBAF.[23]

Moss's principled stand seemed to attract the attention and respect even of *Athens Banner-Herald* executive editor Jason Winders— though one couldn't be sure that the NBAF-fawning *Banner-Herald* wasn't just welcoming this latest fly in a competitor's ointment.[24]

Barrett Slenning tried, afterward, to suggest that a lack of support from "Granville activists" wouldn't necessarily derail the project, concluding that DHS defined "community acceptance" as the opinion of "stakeholders," which according to DHS included "biotech, academia, local and state government as well as the local community." One wondered how many of the foregoing "community group" would show up for the next Butner hog roast or raffle. Slenning also tried to suggest that unidentified "activists outside of Granville County" were spreading unidentified "misinformation."[25] This from a consortium with multiple lies on its Web site.

The North Carolina bandwagon soon faced more flat tires. The City of Raleigh, concerned about the facility's impact on the Falls Lake watershed which supplied Raleigh's water, submitted a list of 38 written questions to DHS. It was within the range of DHS's competence to blurt out one-liners about redundancy and standard operating procedures, but asking DHS to give a direct answer to 38 questions represented two optimistic assumptions: first, that DHS knew the answers, and second, that it—which regularly ignored congressional subcommittees—would actually condescend to note the concerns of a city council. Answers to the city's concerns were not in the North Carolina NBAF future.

Second, members of the Granville Non-Violent Action Team (GNAT), which had successfully resisted plans to place a hazardous waste incinerator in Butner in 1990, decided they didn't like what they were hearing about NBAF, and asked residents opposed to the lab to submit written comments to DHS.[26]

The Georgia NBAF consortium revved up the propaganda engines getting ready for its DHS scoping visit. They apparently decided it would be a good idea to take the edge off local concerns with a pre-scoping question-and-answer session on the UGA campus.

If something went wrong, somebody asked, what's the worst that could happen?

"We have trouble imagining what you would consider a terrible event," said UGA vice-president David Lee. This even though Lee—who, like North Carolina's Slenning, had apparently not had the latest diseases-of-interest propaganda chip implanted—talked about the facility combating "bio- and agro-terrorism threats and emerging disease pandemics, especially those such as avian flu, that can transfer from animals to humans."[27]

To assist Lee's imagination just a bit—just off the top of my head—how about this scenario? A "successful" experiment creating a form of the avian flu that can easily be transmitted between people—an effort a few labs are currently engaged in. (Twisted logic, I admit: they want to know how likely it is such a mutation will occur in nature. If they succeed in doing it in the lab, of course, the question is answered. 100% probability, because we just did it. Aren't we so brilliant? Guess we better work on some countermeasures now, unless we want to try cross-breeding the new avian flu with the 1918 flu, first, to make a real killer.) Then a researcher catching said flu, through carelessness or bad luck, and unknowingly carrying it into the community and spreading it. An emerging disease pandemic emerging from NBAF. Would that qualify as a "terrible event"?

Of course, DHS doesn't currently acknowledge that it will research avian flu in NBAF (they already do that in a BSL-3 lab on the Georgia campus, the Southeastern Poultry Research Center—where they also brought the 1918 flu back to life, if Lee needs help imagining a few more terrible events). But DHS does acknowledge foot-and-mouth disease research. Just let a little FMD get out (maybe through an accident like that in UGA's Animal Health Research Center) and infect a cow or pig there in the Georgia pines. Presto, the U.S. loses its FMD-free status. Even if the outbreak is quickly contained, meat exports shut down, causing significant financial losses. If the outbreak isn't quickly contained, the losses are catastrophic. Terrible enough for you? Of course, this scenario works better in Kansas, where they have the main livestock farm right there next to NBAF (and next to the football stadium), if you want to think about a perfect storm of FMD contagion.

Of course, none of this could happen at Georgia, where all biolabs exist only in the twilight zones of biodefense heaven. According to Chris King, an assistant vice president for research at UGA, you deal with mistakes by careless workers by "prevent[ing] people like that from working in the building."[28] Well, frankly, Chris, if you have a foolproof test for screening out careless workers, you need to get a patent on that baby, and forget the germ lab stuff. Of course, I note that people who make reckless statements like you and Lee have risen to positions of responsibility at UGA, so I guess the UGA careless-worker screening device still has a few kinks.

And of course, in biodefense heaven, "[i]f someone who works at the lab should get sick, he or she would be tested and quarantined."[29]

Guess no one at UGA followed the recent infection travails of NBAF competitor Texas A&M, or the earlier ones of BSL-4-wannabe Boston University. (So much for the educational value of voluntary accident reporting schemes. If shutting down Texas A&M's select agent research didn't get UGA's attention, don't think the voluntarily-assembled accident database is going to do much for them, either, assuming that reading the entries is voluntary, too.) But in biodefense heaven, the heroic germ lab jockeys are always honest, always know what they're doing, and always make the right decisions.

"Even in a power loss like the recent ones at the Centers for Disease Control and Prevention in Atlanta, the building would have an air-tight lock-down system."[30] (The CDC had an "air-tight lock-down system," too, when it had a second power loss in 2008: duct tape.)

And Lee, as he had done on other occasions, referred to the "teaspoon-size amount of pathogens that would be stored at the facility."[31] A deception like that ought to require Lee to be consigned to the Eighth Circle of Dante's Inferno, compelled to remove the pathogen-laced daily manure of a hundred cows with a teaspoon.

"The facility will not be any more equipped to make weapons than a car plant is."[32] For that comment, Lee ought to be appointed the designated mailer for the biodefense complex's next "wake-up call," preparing five envelopes for each of the NBAF's teaspoon-sized samples of pathogen, using his ungloved hands to fill the envelopes, and then licking the envelopes to seal them.

Lee attributed the British FMD outbreak to the vaccine-manufacturing portion of Pirbright,[33] even though British investigators found it impossible to assign blame between the vaccine-manufacturing and research portions of Pirbright. UGA animal pathologist Dr. Corrie Brown, who had worked at Plum Island and baldfacedly asserted that there is a "culture of safety" at bioresearch labs, also made the false assertion about the source of the U.K. FMD outbreak.[34] Ironically, the vaccine manufacturer at Pirbright, Merial, was a member of the Georgia NBAF consortium.

The day before the Georgia forum, Lee, Brown, and two other UGA NBAF mouthpieces, declared that "studies of deadly infectious diseases have been conducted around the world for decades in a range of government, academic and private research facilities, with no documented infections of community citizens we are aware of."[35] It troubles one to know that people posing as biosafety experts missed the 1979 Sverdlovsk incident, or the community death caused by a SARS lab infection and widely reported in scientific journals in 2004, or that they're unaware that the last human death from smallpox was a British medical photographer working in the same building with a smallpox lab, and that she also transmitted smallpox to her mother. (And the distraught British smallpox researcher committed suicide in remorse.) One might also consider relevant the deaths of two CDC janitors—surely not people who willingly assumed the risks of pathogen research—from Rocky Mountain Spotted Fever in 1976.

About 200 people attended the Georgia scoping session; the 40 people who spoke were split about evenly, according to *ABH* reporter Blake Aued. But apparently the pro-NBAF people were higher-class people, according to AT&T District Manager and NBAF supporter Paul Chambers: "When you think of the kind of people who showed up in favor, this really bodes well for the community."[36]

For Athens Area Chamber of Commerce president Doc Eldridge, it was a matter of not being able to imagine a viable germ-lab-free future. In such instances of parasite/host co-dependency, it is usually the germ which needs the host to live; in this case it seems to be the city of Athens which can only survive by attaching a dangerous germ lab to itself. Or at least, according to Eldridge. When you think of the kind of germs that might show up, this does not bode well for the community.

Shortly after the scoping meeting, Ed Hammond did what he did best and tested the research facility's safety claims by asking, under Georgia open-records law, for all records concerning possible or actual exposures to risk group 2 or high biological agents since 2000. UGA equivocated about whether it had such records, then provided only a one-page statement about a protocol for reporting exposure to *T. cruzi* and filing a worker's compensation claim. It claimed a "medical records" exemption from the state's open-records law. A lawyer specializing in open-records law said incident or accident reports following a pathogen exposure should be available to the public, so long as the employee's name and personal information are blacked out.[37]

On October 9, the NBAF-supportive *ABH* editorialized that the records should be provided: "This community needs to know—and it needs to know now, with a final decision on siting NBAF probably less than a year away—that the university's labs are being operated safely. Releasing information on the operation of these labs is one way in which the university can show it is worthy of the trust it seeks from the people of this area."[38]

Of course, that was and is Hammond's basic point: information about accidents lets people know whether or not labs are being operated safely, under the assumption that they want and need to know before they can take a position on the construction and siting of new labs. And Hammond suspects—and every evidence suggests he is right—that many more accidents happen than we know about. Thus the eagerness of facilities to conceal them: they prefer that the facilities be presumed safe until such time as people around them start keeling over en masse.

One week later, UGA released twenty-five pages of E-mails regarding two safety incidents in one of its labs. Hammond noted that "[t]here are no records, and Georgia thus far denies the existence of, any further records concerning any more of its dozens of labs, up to and including BSL-3 AG."[39]

By December 2007, GNAT had set up a well-stocked and well-maintained Web site, www.nobio.org, and was passing out "No Bio Disease Lab" yard signs.[40] Meanwhile, in Athens, a husband-and-wife team, Grady Thrasher and Kathy Prescott, were establishing an opposition group there, For Athens Quality-of-Life (FAQ).[41] FAQ would soon have its own Web site up. Both organizations kept up with what was happening at other NBAF finalist locations, and with the germ lab world in general.

The North Carolina opponents were becoming regulars at local government meetings. And over the next few months, the obvious and intense opposition in North Carolina would gradually peel away much of the early official support for NBAF in North Carolina. DHS's usual stonewalling tactics did nothing to endear them to lab opponents, either.

First the Creedmore City Council passed a resolution opposing NBAF, then the Stem City Council. In December, after DHS declined to send a representative to Granville County to answer questions, opponents asked the Granville County Commissioners to withdraw support for the lab. John Pike, an Oxford attorney who had been prominent in the earlier hazardous waste incinerator fight, told the commissioners, "They have spent millions of dollars of my money to sell this project across the country, but they won't spend one dime to tell us the truth."[42]

The commissioners declined to withdraw support, but they did vote to seek a federal contact to whom written questions might be addressed. Opponents weren't satisfied with the half-measure. "We ain't got time to fart around," Pike said.[43]

Joe Melamed, a Granville County doctor and managing partner of Wake Radiology, obtained the signatures of 33 other doctors on a petition opposing NBAF.[44] DHS responded to the Granville Board of Commissioners' request for a contact person by simply referring the commissioners to the NBAF Web site. In early January, the commissioners responded by formally withdrawing their support. Commissioner Hubert Gooch cited the mounting opposition: "I'm elected to represent the people of the county. If they feel that strongly about it, I don't want to force it on them if they don't want it here."[45]

U.S. Representatives Brad Miller and G. K. Butterfield then wrote a letter to DHS's National Laboratories Director, asking him to respond to the concerns of Butner residents by having officials visit Granville County or talk with county officials on the phone to answer questions.[46]

Both GNAT and FAQ brought Ed Hammond to town in January 2008. Hammond gave a PowerPoint presentation addressing some of the major myths of NBAF. In Athens, local UGA NBAF propagandist Dr. Corrie Brown showed up to heckle a bit.

The next month DHS sent a team to both sites for damage control, consisting of Jamie Johnson; DHS architect Eugene Cole; Plum Island director Larry Barrett; and Tammy Beckham, director of USDA's Foreign Animal Disease Diagnostic Laboratory.[47]

Three scientists from UGA's College of Veterinary Medicine published a commentary, "There Is Nothing to Fear in NBAF," in the *ABH* two days before the Athens session.[48]

DHS required that the questions be submitted in advance. In

Athens, controversy arose after UGA spokesperson Terry Hastings indicated that each questioner would be allowed one initial question and one follow-up question. At the session, however, moderator Tim Bryant of News Talk 1340 WGAU AM did not allow follow-up questions. He stated later that he had not been directed by anyone to allow follow-ups. A bit disingenuously, he said the panelists had made themselves available after 10:00 if anyone still had questions.[49] This ignores the essential difference between a public answer, given before hundreds of witnesses, and a private reassurance.

The difference follow-up can make was indicated by the North Carolina session two days later. In both sessions, Jamie Johnson completely ruled out studying anthrax, smallpox, Ebola, plague, or other human diseases already studied at the CDC. And he said the "Diseases of Interest" list was unlikely to change.[50]

In North Carolina, there was "follow-up":

[W]hen pressed by meeting moderator Frank Stasio of WUNC, Homeland Security officials acknowledged the list of diseases could change or expand depending on what new pathogens or threats could emerge.

"Yes or no?" Stasio asked repeatedly.

After several roundabout answers, Johnson said. "Yes. If threats change, we would look at the list."[51]

ABH editor Don Nelson showed the newspaper's inability to remain neutral on NBAF by writing a commentary, "NBAF Forum Went Well."[52] He said the forum should be considered a success because of the large crowd that showed up, the civility of the proceedings, and the fact that "[t]he panelists answered the majority of questions with what appeared to be sincerity." And he felt that "Anyone who attended the Tuesday forum with an open mind likely came away feeling relatively comfortable about the safety and . . . security of NBAF, especially after references to the Centers for Disease Control and Prevention and its safety record."

In North Carolina, the *Durham Herald-Sun*, apparently feeling that the "free speech" of government officials was in peril, took offense because audience members shouted "Liar!" after the DHS'ers repeated their usual misleading assurances that the lab "would not be involved in creating biological weapons."[53]

Two days before the North Carolina session, the Raleigh City Council withheld support for the facility until federal officials met the following conditions:

- Respond to city officials' questions about the lab;
- Explain plans for keeping sediment runoff from the lab out of Falls Lake, the source of Raleigh's drinking water;
- Respond to questions about safety posed by Granville County commissioners;
- Provide information about oversight of the facility; and
- Consider other sites in North Carolina for the facility.[54]

CHAPTER TEN

Regulation, Not
The Wild Wild West of Germland

IN OCTOBER 2007, KEITH RHODES OF THE GAO TOLD THE HOUSE
Energy and Commerce Oversight and Investigations subcommittee
that high-containment germ labs were essentially policing themselves. Since the number of such labs had multiplied many times over
in just a few years—and continued to multiply as he spoke—the situation bore certain resemblances to the 1849 Gold Rush, with germ
lab forty-niners panning for biodefense bucks on every street corner.
The situation was (and is) about as lawless as the 1849 Gold Rush,
too—the NIH and CDC being a couple of slow-on-the-draw circuit
marshals who show up for baptisms and lab christenings, but manage
to be out of town when the germs start flying.

The country currently has two primary regulatory schemes: (1)
the Select Agent Program administered by the CDC and USDA, and
(2) the Institutional Biosafety Committee (IBC) system, and Guidelines established by the NIH. The responsibility for the latter is formally vested with the National Institutes of Health's Office of
Biotechnology Activities. Neither system applies to all high-containment germ labs.

The Select Agent Program requires registration by facilities performing research on any of the listed select agents, and the filing of a certain amount of documentation with the CDC or USDA such as biosafety plans, incident reports, etc. The program also entails an initial inspection by the relevant regulatory agency, with follow-up inspections every three years or so. The Texas A&M episode revealed how just how cursory these inspections and the general monitoring of select agent research are. In addition, the Select Agent list has some notable omissions, including SARS.

Institutional biosafety committees (IBCs) are required only for facilities that both receive NIH funding and perform recombinant DNA research. The NIH supposedly requires that those institutions which fall within the IBC system's purview follow the NIH Guidelines. Failure to do so could—and should—entail losing NIH funding. The committees are supposed to review and approve their institution's recombinant research projects, meet regularly and keep minutes, and provide for public access to the minutes. Ed Hammond's study and report on the IBC system, though acknowledging that a small portion of the IBCs function as they are meant to, clearly demonstrate that the IBC system as a nationwide regulatory scheme is bankrupt. The Sunshine Project's long investigation of IBCs revealed that many existed only on paper: they did not meet regularly (some apparently weren't even established until Hammond began asking for their minutes); they didn't review and approve proposed research projects; they didn't keep regular minutes; and they resisted requests for such minutes as existed.

Even for those facilities theoretically subject to the NIH Guidelines, the Guidelines resemble an electric fence without any "juice." Many electric fences for restraining livestock consist of a single strand of wire; the wire itself is laughingly inadequate to keep in a cow, a goat, or sheep. Only the fear of being shocked keeps the animals from simply walking through or over the wire. After repeated shocks, the animals learn to keep their distance. But because the NIH looks the other way when institutions violate the Guidelines, the Guidelines resemble an electric fence that has never been electrified, a wire put up in the forlorn hope that the animals will believe the wire will shock them even though they have no experiences to suggest that.

The Texas A&M incident and the flurry of media and congressional attention that followed embarrassed the biodefense complex.
The fact that the CDC actually suspended a Select Agent program also
scared a few facilities, apparently. All kinds of previously unreported
accidents came out of the woodwork. Then the Associated Press
breached the CDC's wall of secrecy and published details of the more
than 100 accidents that *had* been reported to the CDC in the last four
years. On the one hand, it seemed a lot of accidents weren't being
reported even to the CDC; on the other hand, the CDC (as part of a
new public health strategy for reducing the number of scary-germ-
lab-accident-provoked heart attacks, perhaps) was keeping all the
accident news it could to itself. No one really knew how many accidents were occurring, but all signs indicated many more happened
than the public ever heard about.

The flurry of new accident reports that occurred after the A&M
incident suggested that one way to get more reports is to punish the
cheaters. The biodefense complex had other ideas, however. The
Center for Biosecurity at the University of Pittsburgh Medical Center
began promoting the notion of a voluntary, anonymous reporting
system, to spare institutions the embarrassment of "negative publicity
or the scrutiny from a granting agency."[1] This, UPMC implied, would
turn shy, timid, and mistake-prone labs into public-spirited safety
advocates. All the nice germ labs could then learn from each other's
mistakes, without letting the public know about those nasty accidents. Because the public does tend to get overwrought about these
things. If they hear about the accidents, they get to thinking we're not
perfectly safe. Then they don't want our new labs. And we need our
labs, no matter what the public thinks.

UPMC had actually raised the notion of an anonymous reporting
system as early as April 2007, in an article published in its journal
Biosecurity and Bioterrorism.[2] That article described some issues discussed at a July 2006 invitational meeting convened by UPMC. The
invited group were mostly prominent members of the biodefense
complex, with two or three opponents of lab expansion (Ed Hammond, and David Ozonoff of the Boston University School of Public
Health) thrown in for good measure.

The article both summarized some of the discussion at the meeting and presented the Center's own recommendations for change. The article's introductory paragraphs include a caveat that "Our recommendations are not necessarily endorsed by the participants in the July 11 meeting."[3] Interspersing the recommendations with the summaries of the meeting subtly suggests, however, that the recommendations are some sort of formal outgrowth of the meeting.

And that is misleading. Center for Biosecurity Recommendation #3, for instance, suggests:

> A system should be established by NIH or CDC to provide an analysis of mistakes and near-misses in high-containment laboratories. Institutional anonymity may be necessary for overall safety goals to be achieved; however, procedures need to define thresholds and mechanisms for reporting if mistakes pose a danger to the community surrounding the laboratory.[4]

The participants at the meeting, however, agreed only that victims and lower-level individuals should remain anonymous. They disagreed about anonymity for the institutions. Ed Hammond, for instance, vigorously criticized the concept of institutional anonymity in a *Nature* interview, arguing that it would "permit labs to escape public accountability for accidents, keep communities in the dark about dangers in their midst, and impede learning from mistakes to prevent recurrences."[5]

In fact, institutional anonymity seems designed to keep communities in the dark not only about particular facilities but about the biodefense complex in general. Ironically, the UPMC report reveals a particular reason why there may be a lot more accidents to hush up: the huge influx of new workers, which will "strain the current national capacity for biosafety training." "The workforce," the report acknowledges, "that is needed to make the high-containment laboratories productive and safe is not yet in place."[6]

All the more reason why the public might not want to be kept in the dark about lab accidents, however much the biodefense industry would prefer to keep its dirty lab linen to itself. What the report offers the public instead of the facts they need to evaluate biodefense safety for themselves is the "opportunity to be reassured that accidents are thoroughly examined and contained." And where will they get this reassurance? From the gods and goddesses of biodefense heaven. The nasty researchers who lie and keep accidents to themselves will suddenly start reporting the accidents when they realize they won't be punished or even embarrassed. The biodefense priests and priestesses of record will then simply assign the appropriate number of Hail Marys and tell the malingerers to go and screw up no more.

It should be pointed out that anonymity does nothing to assure that accidents are "contained." Only those "redundant" safety features we hear so much about—and a certain degree of luck—have anything to do with containment. In fact, under the proposed system of anonymity, the only accidents likely to be thoroughly examined are those which are not "contained"—such as the 2003–2004 SARS escapes from Asian high-containment labs. Would the Texas A&M situation have been "thoroughly examined" under an anonymous reporting regimen? Not likely. In fact, given the CDC's refusal to release information about accidents to the public, the current reporting situation is already "anonymous" except for the reports to the CDC itself. Ironically, it was Texas A&M's reporting failure—and the fact that this was discovered by someone other than the CDC itself—which triggered the CDC investigation. But once that investigation was triggered, the CDC discovered a generally slipshod and dangerous situation at A&M. That situation would likely have gone unnoted (and uncorrected) under an anonymous reporting situation. Which is exactly what the A&Ms—and it must be added, the CDCs and UPMCs—of the country would prefer.

The UPMC report does recommend some threshold for reporting to the public at large "if mistakes pose a danger to the community surrounding the laboratory." Presumably, the institution itself (or the institution's laboratory) will have to decide whether a given incident meets that threshold, since the institution would be anonymous, apparently even to the regulatory agency. Since the CDC decided that none of the hundred-plus lab accidents reported over the last four years presented any danger to the community, one has to assume this threshold is a fairly high one.

The UPMC wants to make sure that the labs do a better job of "communicating with the public":

Public engagement should be a priority for all laboratories, and federal funds should be made available specifically for that purpose. As part of a proactive public engagement program, the need for individual high-containment laboratories in the context of the overall U.S. biodefense strategy should be clearly articulated to the public by the federal government and the laboratories themselves.[7]

And why is "public engagement" so important? Because "[p]ublic resistance was experienced during efforts to build facilities in Boston, in Davis, California, in Hamilton, Montana, and in Seattle."[8] How unfortunate! The UPMC apparently believes that opposition to these NIH lab sitings was due simply to poor public engagement strategies. Not that there might be any actual risk associated with the facilities, even though the center's own report openly wonders where the needed supply of trained lab workers is going to come from. So what the UPMC contemplates is a federally funded effort to brainwash and propagandize the public. We need these labs! And they are so, so safe! And the most important part of this public engagement strategy is the "anonymity" of the labs having safety lapses. It's a lot easier to talk about safety when you can keep your accidents to yourself.

Yet what UPMC doesn't understand is that the secrecy, the lies, and the inflated safety claims made for high-containment facilities contribute enormously to the distrust and skepticism of potential host communities—as occurred with me when I investigated the *Commonwealth-Journal*'s claims that BSL-4 labs had operated in the U.S. without accidents for fifty years, or that the BSL-4 labs in Geelong, Australia and Winnipeg, Canada had operated "without incident"— or as occurred when Boston University hid its tularemia infections amidst its application for a BSL-4 facility.

There was a certain dramatic irony in the report's noting that "generally positive support was achieved for the Galveston laboratory administered by the University of Texas Medical Branch." That support was gained through the Galveston lab's promises of openness and keeping the community informed—promises which it broke first by resisting the Sunshine Project's requests for IBC minutes, and then by trying to slip a secrecy bill through the state legislature that would have exempted all its records from the Texas Open Records Act. Once Galveston revealed its true colors, that "generally positive support" was in serious question.

In her October 2007 testimony before the House Energy and Commerce Committee, Gronvall argued directly for the voluntary, anonymous reporting system without noting the dissents from that view at the 2006 meeting. So, naturally, she was pleased when Senators Kennedy and Burr proposed (in June 2008) such a system in the Select Agent Program and Biosafety Improvement Act of 2008. "It's very exciting," she said. "It has a lot of things that I completely agree with."[9]

After the Ivins revelations, the biolab window-dressing chorus would start up full-time. First up was the military, which launched a four-month review of "lab security, biosurety and safety policies."[10] The Army announced a "biological personnel reliability program"— designed to exclude alcoholics, drug addicts, or people exhibiting "an inappropriate attitude, conduct, or behavior" (apparently anyone resembling the rumors about Bruce Ivins) from working with select agents.[11] The Navy and Air Force suspended work in their laboratories while they conducted a safety review.[12]

At the end of the four months, the services announced that they would increase camera surveillance, hold surveillance videotapes for a year, perform more detailed background checks on lab workers, conduct quarterly and annual inventories of select agent inventories, and conduct new training for lab supervisors.[13] The Army said that it would offer a weeklong "security refresher" for its lab workers.[14] At the same time, Army assistant deputy chief of staff Major General Robert Lennox boldly declared that "security measures put in place since the 2001 anthrax attacks would make it difficult for someone to conduct a similar campaign today."[15]

Right after this, however, the military's chief lab, USAMRIID, seriously undercut the general's credibility by suspending most of its select agent research. Preliminary inventories had revealed numerous unlogged, unrecorded, and unidentified germ samples.

About the same time, one of the numerous government bodies which would be making reports and recommendations—the Trans-Federal Task Force on Optimizing Biosafety and Biocontainment

Oversight—announced a meeting to take public comment on the "overall goals of improving biosafety/biocontainment oversight and protecting public health and agriculture, while simultaneously encouraging the progress of research."[16]

Just before he left office, President Bush, who had piggybacked on the 2001 anthrax attacks to build fear about Saddam Hussein's supposed weapons of mass destruction, issued an Executive Order to establish a "Working Group on Strengthening the Biosecurity of the United States."[17] Up until this point, the president had shown more concern about the security of Saddam Hussein's nonexistent WMDs than he had about America's own stockpiles of real ones.

In March 2009, UPMC would rush to make propaganda points with the new Obama administration, urging it to "make a robust biodefense a top national security priority."[18] One would have thought a twenty-to-thirty-fold increase since 2001 would have been robust enough for anyone. But UPMC cited with approval the WMD Commission's recent ratcheting up of the bioterror alarm. Therefore, the country must "build the strongest feasible response to biological attacks,"[19] with—you guessed it—increased funding for all those bio-entrepeneurs—including—you guessed it—UPMC itself, which was competing for a major new vaccine facility. The administration should make sure to tuck some biodefense funds into all that economic stimulus money, because everybody knows how economically stimulating those killer pathogens can be, especially for drug companies and biodefense think tanks.[20]

On the other hand, UPMC said, the WMD Commission's worries about lab security should be taken with a grain of salt: "Recent calls for increased laboratory security by the *World at Risk* report by the WMD Commission should be carefully considered before action is taken." Because "[t]here is . . . a real danger that excessively intrusive or expensive security measures will discourage scientists from pursuing the research needed to treat emerging infectious diseases or to respond to a bioterrorism attack."[21]

In other words, give us more of that biodefense swill, but don't be fencing us in or putting rings in our noses. We'uns need to root, hog, root.

Over the next several months, elements of the scientific community waxed positively hysterical at the notion that the government might interfere with their constitutional right to keep and bear deadly pathogens free of government intrusion. Having willingly climbed into bed with Frankenstein, they seemed surprised that Frankenstein was not the world's courtliest lover. Having smashed the bioterror fire alarm every time they passed it, they seemed surprised that anyone wanted to restrict their right to play with matches.

Biodefense scientists took offense that anyone would want to probe too deeply into their background, run credit checks, do drug tests, the way it happens at some of the nuclear research facilities. "Many scientists take 'pride in their eccentricity,'" complained UTMB-Galveston lab director Stanley Lemon.[22] What did people mean interfering with a person's "inner mad scientist," after all? Some of these measures, "some academics" worried, would interfere with the "open culture" of "life sciences research."[23] Well, aside from pointing out biodefense is as much "death sciences research" as it is "life sciences research," one feels like saying: if you're a rebellious person with anarchistic tendencies, if you feel confined by all those standard operating procedures, and security types looking over your shoulder while you play around with your killer germs, if you feel like tearing off your BSL-4 space suit three minutes after you put it on—maybe you should get into some other line of work, join an MFA program or something. Or maybe you could go back to doing TB and AIDS research. I personally could do with a few less biodefense cowboys.

Biodefense scientists are such sensitive blossoms, we are told, that some are at the point of leaving their jobs, or casting off this mortal coil altogether. (Richard Ebright takes all the whining and threats to quit with a grain of salt, pointing out that "for all the grumbling, researchers entered the biodefense field in droves after the select-agent rules came into effect.")[24]

"In many cases," we are told, "the researchers are not natives of the United States, meaning they can easily relocate—along with their knowledge and expertise—outside the country."[25] Yes, that's what certain critics of the biodefense expansion have been warning—what better place to gain crucial, and dangerous, biodefense knowledge than in one of America's hundreds of shiny new biodefense labs? Probably we would all be safer if a large percentage of would-be biodefense researchers *were* discouraged by nosy investigators.

In March 2009, about the same time they issued their advice to the Obama administration suggesting an even greater expansion of biodefense, the UPMC folks were making the rounds of Capitol Hill, telling lawmakers that "[f]ederal inspections of laboratories handling dangerous biological agents have grown so intense that they may actually harm national security by driving away top-level researchers and discouraging global collaboration." "'Requirements imposed in the name of security do not equal security,'" UPMC's Gigi Kwik Gronvall was reported as saying.[26]

To which one might retort: an epidemic of biodefense facilities constructed in the name of security does not equal security, either. That federal inspections of high-containment germ labs are "intense" would surprise the Government Accountability Office, or anyone who read the GAO's reports, or watched the October 2007 hearing. That hearing was being conducted precisely because the CDC's routine inspection of Texas A&M was so lax that it missed a biosafety crisis in the brewing. Dr. Gronvall attended that hearing, and even testified at it; she gave not the slightest hint there—under oath—that she thought federal inspections too intense.

Chalk up another falsehood for biodefense truth-twisting in the service of biodefense turf expansion.

The latest UPMC salvo upped the ante considerably, not just resisting additional regulation, but complaining about the current "self-policing" regimen.

Look out for the counterattack of the killer biolabs.

June 2009: the American Association for the Advancement of Science warned against applying nuclear lab personnel security measures to "bioresearch labs." The reason: because existing employment and biosafety training practices—the AAAS claimed— may "already contribute to vetting of personnel" and prevent "malicious actors or unstable personnel" from gaining access to pathogens. Instead, the AAAS called "for greater government funding for safety training" [and for "applied biosafety research" and maintenance] and urge[d] the facilities themselves to enact training requirements for their personnel." The report also called for a national, anonymous database of "exposures." Government, give us your money but not your regulation: we will take care of ourselves.[27]

More propaganda. For one, an outrageous statement from an unidentified "source within the bioresearch field," claiming that a large portion of private biolabs—generally not subject to government regulation—"voluntarily obeys federal guidelines." Hammond's intensive survey of IBCs suggested that few of the labs supposedly subject to the guidelines give them much more than lip service; and private entities were the worst of all. So who are these people allegedly following the guidelines voluntarily?[28]

Second—in a classic example of the biodefense two-step—the AAAS itself argued that biological researchers work with "things found in nature that aren't inherently national security risks, but global health problems."[29] As though bioweapons and the concept of bioterrorism didn't exist, as though virulent anthrax was lurking in every back yard instead of in hundreds of American biological research labs, as though the stock in trade of biodefense research was not lab-produced, genetically modified bioweapons agents, as though reckless biologic researchers had not recreated from scratch something no longer found in nature, the 1918 flu virus. That's like saying nuclear weapons are not "inherently national security risks" because uranium is found in nature.

September 22, 2009: Ronald Atlas, a past president of the American Society for Microbiology, warned a congressional subcommittee against precipitous, excessive policy changes which might upset the "delicate balance," the "careful equilibrium" of wet noodle regulation which had been reached through the collective non-wisdom of Wizard of Oz biodefense regulators.

September 30, 2009: the National Research Council of the National Academy of Sciences issued a report entitled "Responsible Research with Biological Select Agents and Toxins." The gist of the report is that there is no "silver bullet" against potential terrorists working in a U.S. laboratory. Instead, the report emphasizes the "implementation of programs and practices aimed at fostering a culture of trust and responsibility."[30] Just say no, mad-scientist bioterrorists!

Another example of the biodefense two-step. There for sure is no silver bullet against a bioterrorist attack, but that doesn't prevent the biodefense complex from eagerly pursuing silver bullets, slingshots, bows and arrows, incantations, and whatever else they can get funding for. But when it comes to guarding against unbalanced biodefense researchers, all they can think of is building more bio-defense labs with more researchers to dream up more countermeasures against whatever ways of defeating the countermeasures the researchers dreamed up last week, because they were worried some-body might dream them up sometime, and guess what, somebody did, us, all with three daily choruses of can't-we-all-just-get-along-and-trust-each-other-and-be-responsible-bioresearchers, hallelujah.

It is remarkable how low-tech (and how trusting) the high-tech wizards can be about the human element of biolabs. Instead of all those intrusive "personnel reliability measures," some suggest, "[a] better approach . . . would be to trust lab managers to keep a watchful eye on employees for troubling behavior."[31] How do you turn that into a Standard Operating Procedure? How many times an hour should this watchful eye be turned toward a given employee? Should it be the left eye or the right eye? All this time, of course, the biodefense propagandists have been telling worried biodefense neighbors about all their failsafe high-tech magic, encouraging them to *repeat after me, redundant safety measures,* and the neighbors say, but what about human error? And then somebody says, you know, if a mad scientist sent all that stuff through the mail, shouldn't some-body keep an eye on the mad scientists? (It can get worse, of course, say, if you believe the government sent the stuff through the mail, shouldn't somebody keep an eye on the government—and that's when it gets really scary.) And the scientists, both the mildly eccen-tric ones and the stark raving lunatics, say, well you know, it's the human element, what can you do? Downsize the whole biodefense enterprise, for one, let you make yourselves useful working on some stuff that kills people by the boatload already—that way, if the WMD Commission should turn out to be just as dead wrong as the Saddam-Hussein-has-weapons-of-mass-destruction-crowd, you won't have wasted all that time playing Biodefense Monopoly with taxpayer money.

On the one hand, you have the scientists—please just trust us, how can I get my groove on, dancing around in my Christian Laurens moonsuit, injecting those monkeys with my designer pathogens, watching those suckers die, cutting them open and looking at those beautiful virulent pathogens under the microscope, breathing that pure HEPA-filtered air, moving so free and easy through the wild wild west of germland, if you've got people looking over my shoulder?

Then you have Joe Lieberman, who believed that Saddam Hussein had weapons of mass destruction in 2001, and who believes the WMD Commission now when they assure us that we will have a weapons-of-mass-destruction attack—*more likely than not*—by 2013. Because they have a high-tech crystal ball, and access to the country's best "intelligence." (And that there will be such an attack indeed seems quite plausible—it will have been twelve years since a government insider or insiders last attacked us with some of our own anthrax, and it should be about time for another lesson on what can happen if we don't give the government the tools it needs to defend us from itself.) Lieberman, and his fellow [sic] Republican, Collins, realize something needs to be done like right now, and so they introduce a bunch of things that are basically already being done, except for the scariest one of all, turning over a large chunk of the "regulation" of biolabs to Homeland Security.

And why is that scary? One, because many believe DHS's Katrina performance was not just a one-time fluke, but an indication of an institutional flaw, like a genetic modification experiment gone wrong. Gerald Epstein of the Center for Strategic and International Studies said DHS's system of ensuring the reliability of its own employees was so "dysfunctional and byzantine" that it "'does not have any business'" screening researchers working at other biocontainment labs."[32] In opposing the move, *Nature* called DHS "a hotchpotch assembled from 22 existing agencies in 2003," which "combines a sprawling mandate with a paucity of biological expertise." The bill, *Nature* said, "has accordingly alarmed scientists in this field, who say that their experience to date in executing contracts for the DHS has revealed a disturbing lack of expertise in its inspectors."[33]

It is also scary because of the Big Brother/secret police ambience of DHS. Even the GAO and congressional committees have trouble getting information from DHS. And DHS is running its own secretive lab at Fort Detrick, the NBACC. Keith Rhodes of the GAO suggested in the October 2007 congressional hearing that the entity overseeing labs should be one that doesn't operate labs itself. There is an obvious reason for that—conflict of interest. An agency which operates labs itself is likely to downplay the seriousness of accidents and near-misses because it doesn't want public concern redounding on its own facilities.

The Web site *Armchair Generalist: A Progressive View of Military Affairs* pointed out that most of the Lieberman/Collins bill's "initiatives" duplicate or otherwise mess around with things already in place:

> We already have a way to designate the top threat pathogens—CDC identifies the "category A" biological agents as the most dangerous. Pretty sure that we do have a national strategy for dispensing medical countermeasures—it's called the Strategic National Stockpile, another CDC program. We have procedures for communicating with the public, that's in the National Response Framework and practiced in national-level exercises. We already have a national bioforensics center, and I am sure it is well-funded and supported.[34]

One might add that the bioforensics center we already have is operated by DHS. But perhaps the Center has been "unauthorized" until now, and Lieberman and Collins felt the need to legitimize it. Or perhaps they've been so busy fretting over the WMD Commission Report they never noticed one of our country's many shiny new BSL-4 labs.

LIGHTING UP NBAF'S HIDDEN COSTS
UTILITY PLANTS AND THE GAO

DHS's Power Play

In late February 2008, the *Lawrence Journal-World* reported on the "high hopes" of Kansas state leaders who had gone to Washington a few months after the short list announcement to lobby for the Kansas NBAF bid.[1]

A couple of weeks later, one detected a certain irritation in Kansas Governor Kathleen Sebelius, who said DHS had sent a letter to each of the five sites which "added to the mix, for the first time, the notion that you needed a separate power-generating facility," and that it "was news to everyone."[2]

No, countered DHS spokesperson Amy Kudwa, DHS wasn't requiring a power plant, just a "standard utilities plant that also includes water and steam, hooked onto an existing electric grid."[3]

Sebelius, on the other hand, said the feds had asked "very specifically" for a natural gas-fired power plant. And she added: "Part of the question, I think, that can be legitimately asked by legislators and others is, what else is going to come up?"[4]

A spokesperson for Texas Governor Rick Perry, apparently trying to score brownie points with DHS, said the requirement didn't come as a surprise at all, that "infrastructure support, including power, has been a critical part of this from the beginning."[5]

UGA NBAF spokesperson David Lee said, "I don't fully understand the reaction in Kansas. It seems to me they misinterpreted the situation to some extent." According to Lee, what DHS was after was a "central utilities plant" or "node" for all the utilities to enter the facility.[6]

Once again, DHS indicated why it was the operator par excellence for a secretive defense facility. It had managed to keep even its enthusiastic, would-be partners in the dark about just what it was up to, and what it wanted.

DHS's statements that it was just reiterating what it had asked for all along sounded suspiciously thin. For even Lee acknowledged that DHS had *just* sent the finalist states "a detailed list of infrastructure needs for NBAF."[7]

And more than one consortium understood this to be a subtle or not-so-subtle effort to get the consortiums to sweeten their offers and outbid each other at this late stage.

The North Carolina consortium, for one, refused to up the ante: Barrett Slenning said the consortium would not offer to pay for a central utility plant. "If that puts us at a competitive disadvantage, that's the way things go."[8]

It remained for the *Indy Weekly* to shed some light on the subject. The *Weekly* noted that a utility plant requirement had "not been thoroughly discussed, if mentioned at all, in public forums."[9]

The *Weekly* noted that the February 2008 Final Scoping Report had glossed over any utility plant requirement by stating that the EIS would "describe the utility infrastructure needed for the operation of the NBAF." The report elaborated that "[t]he EIS will evaluate if existing facilities are adequate, or if upgrades, repairs or new facilities would be needed."[10]

Meanwhile, out in Kansas, the NBAF-supportive *Lawrence Journal-World* found itself troubled just a wee bit, complaining that the power plant request "seems to come out of left field; nothing such as this had been mentioned publicly prior to last week." It was as though all the painted NBAF curtains had opened momentarily to reveal the ugly witches' lair within: "This latest requirement should cause many to wonder just how many unknown features still are to be identified and how these surprise requirements might favor one site over another. Sounds like the political games are just beginning."[11]

The *Manhattan Mercury* and Tom Thornton of the Kansas Bioscience Authority were both in full damage control mode, glossing over the sudden friction and the sudden glimpse of a mendacious DHS. "Powering Up NBAF No Problem," the *Mercury* reported, citing Thornton. While Sebelius might have been caught off guard, Thornton said the Kansas Bioscience Authority "was not surprised by the federal government's requirement that an independent, self-sufficient power plant be included." He suggested that Sebelius might have been misquoted by the press. Thornton added that, whatever the cost might be, "this is important and it is the right thing to do for this plant."[12]

While Texas and other consortiums were standing off to the side and simpering how they'd do anything DHS wanted—and that they'd always expected to need to do anything DHS wanted—Kansas got down to the awkward business of doing it. Thornton said DHS wanted Kansas's "best and final offer" by the end of March. The legislature dutifully passed a bond issue of $105 million for the utility plant and other infrastructure. Paid off over 20 years, the bonds would cost the state $164 million in principal, interest, and fees.[13] At about the same time, the Kansas House, which had passed the NBAF bonding measure 37-0, also passed a bill allowing dealers, manufacturers, and private citizens to own machine guns. The legislators perhaps felt that the machine guns would go well with the state's growing collection of bioweapons agents.[14]

Whistling in the Dark: Questions about DHS and FMD

The release of a new GAO analysis and a May 22, 2008 subcommittee hearing on DHS's plans to move foot-and-mouth disease (FMD) research to the mainland would reveal the cavalier approach DHS took toward safety concerns, its penchant for freely reinventing its NBAF rationales as public scrutiny dictated, and its cheerful willingness to stonewall even a congressional subcommittee. The hearing would also display NBAF finalist-state politicians hard at work to twist the facts and squelch independent scrutiny of their $540 million pork-barrel-in-the-sky.

The House Energy and Commerce's oversight and investigations subcommittee had asked GAO to investigate the following: (1) the evidence DHS used to support its conclusion that FMD work could be done safely on the mainland; (2) whether an island location (like Plum Island) provided any additional protection; and (3) the economic consequences of an FMD outbreak on the mainland. GAO had reviewed available literature on FMD, high-containment laboratories, the consequences of the 2001 U.K. foot-and-mouth outbreak, and potential consequences of a U.S. outbreak, and interviewed DHS and USDA officials, representatives of farm organizations and the American Society for Microbiology, and officials with facilities conducting FMD work in other countries.[15]

Foot-and-mouth disease presents no danger to humans. It is devastating to cattle and pigs, however, and affects other cloven-footed animals such as sheep and goats. Young animals often die, and adults are debilitated in a manner causing severe losses in the production of meat and milk. It is the most highly infectious of animal diseases, with only ten organisms being required to infect a cow or sheep. (An infected pig exhales 400 million organisms per day.) Nearly 100 percent of exposed animals become infected. FMD can be spread by contaminated animal feed or water, contaminated shoes or clothing, contaminated vehicles or farm equipment, people's lungs, and even by the wind.[16]

A single outbreak on the U.S. mainland would bring an immediate ban on American meat exports and cause enormous economic damage. The GAO investigators and other witnesses said the extent of that damage would depend on the location and nature of the outbreak, and that various studies and scenarios had arrived at widely ranging figures, the higher range being $60 billion or more. The United Kingdom's National Audit Office figured that country's 2001 outbreak had cost the public sector over $5.71 billion and the private sector over $9.51 billion.[17]

The last outbreak on the U.S. mainland had occurred in 1929; a 1978 outbreak from the Plum Island research center had not spread from the island, a fact that kept the Office International des Epizootics (OIE) from revoking the United States' FMD-free designation.

Congressional law had prohibited live FMD research on the mainland until a 2008 farm bill lifted the FMD ban to make NBAF possible. DHS was insisting that current technologies and high-containment procedures made it safe to conduct FMD research on the mainland. The subcommittee had asked the GAO to conduct an independent analysis.

The GAO Report's title indicated its conclusion: "DHS Lacks Evidence to Conclude That Foot-and-Mouth Disease Research Can Be Done Safely on the U.S. Mainland." The GAO explained:

> We found that DHS has not conducted or commissioned any study to determine whether FMD work can be done safely on the U.S. mainland. Instead, DHS based its decision that work with FMD virus can be done safely on the mainland on a 2002 USDA study that addressed a different question: whether it is technically feasible to conduct exotic disease research and diagnostics, including foot-and-mouth disease and rinderpest, on the U.S. mainland with adequate biosafety and biosecurity to protect U.S. agriculture. This approach fails to recognize the distinction between what is technically feasible and what is possible, given the potential for human error.[18]

The GAO said the 2002 study had been selective in what it considered. It didn't assess the history of releases of FMD virus or other dangerous pathogens, in the U.S. or elsewhere; it didn't address in detail the issues of large animal work in BSL-3 Ag facilities; and it inaccurately compared other countries' FMD work to that in the U.S.[19]

The GAO said the 2002 study, conducted for USDA by defense contractor SAIC (Science Applications International Corporation), also had numerous methodological problems:

> Among other things, (1) the study used an ad hoc method to select its expert panel that was not necessarily free from bias; (2) the study report was written by a single third-party person under contract for that purpose who was not present during the panel discussions; and (3) no concern was taken to ensure that the expert panel members reviewed either the

draft or the final version of the report. At least one expert panel member expressed disappointment with the slant of the report.[20]

The 2002 study had not examined data from past releases of FMD, the history of internal releases at Plum Island, the general history of accidents within biocontainment laboratories, or the lessons that can be learned from such accidents. Such a survey, the GAO said,

> would show that technology and operating procedures alone cannot ensure against a release, since human error can never be completely eliminated and since a lack of commitment to the proper maintenance of biocontainment facilities and their associated technology—as the Pirbright facility showed—can cause releases.[21]

Though accidental releases of FMD virus occurred rarely, the GAO identified 13 incidents between 1960 and 2007 involving facilities in Europe, in addition to the 1978 Plum Island incident. In September 1978 FMD virus had escaped outside the Plum Island laboratory compound and infected clean animals being held in a separate area:

> The exact route by which the virus escaped from containment and subsequently infected the animal supply was never definitely ascertained. An internal investigation concluded that the most probable routes of escape of the virus from containment were (1) faulty air balance of the incinerator area, (2) leakage through inadequately maintained air filter and vent systems, and (3) seepage of water under or through a construction barrier near the incinerator area. Animal care workers then most likely carried the disease back to the animal supply area on the island, where it infected clean animals being held for future work.[22]

The GAO provided a table (based on analysis of USDA data) outlining in more detail the deficiencies that might have caused the release. The GAO said the deficiencies

were [all] related to human error and . . . none . . . to insufficient containment technology. Any one of these deficiencies
could happen in a modern facility, since they were not a function of the technology or its sophistication, procedures or
their completeness, or even, primarily, the age of the facility.
The deficiencies were errors in human judgment or execution
and, as such, could occur today as easily as they did in
1978.[23]

Another table listed internal releases within the Plum Island laboratories which caused unintended infections of animals. "These incidents," said the GAO, "show that technology sometimes fails,
facilities age, and humans make mistakes."[24]

All of which was highly relevant. DHS itself and NBAF promoters
around the country had repeatedly attributed Plum Island incidents
to the age of the facilities and insisted that "state of the art" technologies and procedures at NBAF would prevent releases. These past incidents indicated otherwise.

The GAO also questioned another part of the NBAF propaganda
spiel, that part which boasts of the CDC's "safe operation" in a
densely populated area. The GAO said the CDC's BSL-4 work could
not accurately be compared to FMD research involving large animals
in BSL-3 Ag laboratories:

In a BSL-4 laboratory, work is done within a biological safety
cabinet, which provides the primary level of containment.
Accordingly, there is no contact between the human operator
and the infective material. The laboratory provides the secondary containment and the laboratory staff is required to
wear special protective equipment to prevent any exposure to
the pathogens.[25]

Large animals cannot be contained within a biological safety cabinet, the GAO said, so the laboratory walls are the primary containment. There is extensive human contact between the human operator
and the infected animal. Handling large animals in confined spaces
can present special dangers to scientists and animal handlers, as can

the moving of carcasses.[26] All of which increases the odds of spreading FMD and the other pathogens.

The expanded scope and complexity of activities at NBAF would also increase the risk of FMD studies there, the GAO said. Proposed BSL-3 Ag space at NBAF would be twice that of Plum Island and accommodate many more large animals. (Currently, the animal holding areas accommodate 90 cattle, 154 swine, or 176 sheep.) Clinical trials with aerosolized FMD virus at NBAF will "challenge" groups of 30 to 45 large animals, as opposed to 16 large animals that PIADC can process today. The facility will also house a vaccine production plant with a capacity up to 30 liters—significantly increasing the volume of FMD virus currently handled at Plum Island.[27]

The GAO faulted the 2002 study for failing to consider the heavy loads air filtration systems would encounter, observing that high-efficiency particulate air (HEPA) filters are highly effective but do not represent an absolute barrier since the required efficiency is only 99.97%.[28]

The GAO said the 2002 study had inappropriately cited Australia, Canada, and the United Kingdom as foreign precedents for mainland FMD research. The GAO noted that Australia, despite containing "the world's premier laboratory for animal containment technology," outsourced live FMD work to other countries such as Thailand.[29] (At the end of 2008, groups such as the Australian Beef Association and the Cattle Council of Australia were roundly criticizing the government's tentative decision to reverse its 30-year ban on the importation of live FMD virus.)[30]

The Winnipeg, Canada laboratory that DHS cited as having worked with FMD incident-free had not even been operative in 2002, and though it was now conducting FMD research, it did so in an urban location, away from susceptible livestock, and dealt with only one or two animals at a time.[31]

Ironically, the 2002 study had also cited the British Pirbright facility to show that FMD research could be conducted safely in the vicinity of commercial livestock. The GAO noted the irony:

> The study participants could not have known in 2002, however, that an accidental release of FMD virus at the Pirbright

facility in 2007 [would lead] directly to eight separate outbreaks of FMD on farms surrounding the Pirbright laboratory.
This fact highlights the risks of release from a laboratory that
is in close proximity to susceptible animals and provides the
best evidence in favor of an island location.[32]

The SAIC study had also failed to note that both Denmark and
Germany confined FMD research to island facilities. Germany had
restricted FMD research to the Baltic island of Riems in the 1910s;
after World War II Riems was controlled by East Germany, and West
Germany had then conducted FMD work on the mainland. After
reunification, Germany had closed the mainland facility. The GAO
acknowledged that though an island, Riems was connected to the
mainland by a causeway, and a road bridge was being constructed.[33]

The GAO said an island location can help prevent disease spread if
a release occurs, and that experts it spoke with "agreed that all other
factors being equal, FMD research can be conducted more safely on
an island than in a mainland location."[34] The 1978 Plum Island outbreak had been confined to the island itself, something which per-
suaded the OIE to continue treating the U.S. as officially free from
FMD—something which would not have happened had Plum Island
not been an island. A committee that reviewed the 1978 incident had
identified three main barriers against the escape of disease agents: (1)
the design, construction, and operation of its laboratory buildings;
(2) movement restrictions; and (3) the island location. All but the
third barrier had failed in 1978, and NBAF was eliminating that third
barrier.[35]

Before submitting its report, the GAO had discussed its findings
about the deficiencies of the SAIC study with DHS and the USDA.
DHS officials had responded that the Environmental Impact Statement for NBAF would also be used to determine the safety of FMD
work on the mainland. The GAO found itself suspicious about this
assertion, since DHS had previously "stated categorically" that the
SAIC study alone demonstrated the safety of mainland FMD work.[36]
Indeed, written testimony supplied by the USDA for the hearing, as
well as a May 16 USDA letter to the National Grange (which opposed
moving FMD research to the mainland), continued to declare: "A

2002 study commissioned by USDA and completed by the Science Applications International Corporation (SAIC), found that the FMD virus and other exotic animal diseases of concern could be fully and safely contained within a BSL-3 laboratory, as was being done at the time in other countries including Canada, Germany, and Brazil."[37]

Shades of DHS's diseases-of-interest list.

The GAO said that without detailed information, it could not determine whether the EIS would "contribute significantly to addressing this issue, and said it had asked [for] "but DHS would not provide any information on what analysis they would do as part of the EIS."[38]

Energy and Commerce Chairman Dingell focused on such stonewalling in his opening statement, blasting the agency as "arrogant, incompetent, and secretive." He cited several instances in which DHS had provided the committee with key documents only after committee staff discovered their existence. (The committee had specifically requested all relevant documents.) He mentioned specifically the two SAIC studies on Plum Island and NBAF issues, and a 2007 Plum Island study performed by the Homeland Security Institute (HSI). The SAIC studies had now been provided—criticism of those studies was in fact a linchpin of the GAO's report—but the HSI study was still missing.

Other key documents were missing as well, including the Statement of Work for the Draft Environmental Impact Statement. According to Nancy Kingsbury of the GAO, DHS had refused to release the Statement of Work until it had been made public, arguing that the document was "proprietary." She also cited other delays in the providing of documents, as well as the fact it had taken six weeks to arrange a visit to Plum Island. (According to Dingell, the Plum Island visit had been approved only after a letter threatening DHS with contempt.) Investigators had experienced no problems, by contrast, in obtaining info and access to labs in foreign countries.[39]

Cohen responded to Dingell's "arrogant, incompetent, and secretive" remark in his own opening statement, and the hostility between the two was obvious. Dingell was openly angry; Cohen responded with smarmy composure, telling Dingell "you get more flies with honey than with vinegar." At times Cohen denied withholding information; at times he assured the committee that, if anything was missing, he would see that the committee got it; and at times he asserted that information (specifically, the draft EIS) was properly withheld because the NBAF process involved a "contract action." When Cohen said DHS would make the draft EIS available to the committee only when it became available to the public, Dingell said that sounded arrogant to him. Cohen responded that "it sounds arrogant that neither the GAO nor the CRS would show me, the deciding official, the courtesy to speak with me."[40]

Cohen said the GAO's difficulties in getting approved for a Plum Island visit occurred because Dr. Sharma contacted the wrong people. To which Dingell retorted: "Why is it that the wrong people couldn't refer him to the right people?"[41]

Stupak asked Cohen why a 2004 USDA estimate of the economic effects of a U.S. outbreak hadn't been provided; Cohen referred the question to Dr. Knight of the USDA; Knight said he had staff scrambling to find the materials, but believed that they were in the form of a PowerPoint presentation.[42]

The first witness panel had been comprised of the two GAO representatives and Dr. Tim Carpenter of UC-Davis; the second panel included Cohen, USDA Undersecretary Knight, and Plum Island director Dr. Larry Barrett. A third panel featured representatives from major farm organizations. Two, the National Grange and American Farmers and Ranchers (AFR), opposed moving FMD research to the mainland; one, the National Pork Producer Council, supported the move; a fourth, the National Cattlemen's Beef Association, supported the construction of NBAF generally and believed that such research could be conducted safely, but took no position on location.

Ray Wulf of AFR pointed out the rapidity with which infected livestock might be transported in the U.S.—possibly several days before the onset of clinical signs; and provided a study showing that in just five days livestock from Oklahoma City markets had been transported to 39 states.[43] Both he and National Grange representative Leroy Watson warned about the draconian measures a quarantine would mandate—how it might prevent or inhibit such activities as grain harvests, and even sporting or leisure events where large numbers of people gather. Dr. Carpenter of UC-Davis had earlier referred to the mass slaughtering of animals and the difficulties disposing of possibly millions of carcasses. 6.5 million animals had been slaughtered in the 2001 U.K. outbreak. Watson pointed out the jurisdictional problems involved in responding to an outbreak, and the damage that a mainland NBAF would do to adjacent farming and rural communities because of the perceived risk of an outbreak. Ironic, then, that so many of the NBAF finalists were rural states.[44]

Even the supportive farm representatives had several caveats and concerns: concern that needed funds wouldn't be provided for proper maintenance; concern that the needs of the agriculture community would be lost within the mission of DHS. (All four representatives said they would be more comfortable if the facility were controlled by USDA instead of DHS.) Dr. Howard Hill of the National Pork Producers Council said the "location of the NBAF must be decided based on assessed risk rather than on which entity is willing to build such a facility," and suggested re-examining locations with the effort to "recreate an island effect by siting in an area with low densities of livestock and wildlife."[45] DHS would ignore that suggestion in the months ahead.

While Dingell and Stupak were questioning the FMD move, representatives from NBAF finalist states were there to defend the facility and slip in a few hosannas about their respective states, even though Stupak had cautioned them early on that the hearing was not about where NBAF would go, but about whether it should be built at all, and, if so, whether FMD research should be moved to the mainland.[46]

As a member of the Energy and Commerce Committee at large, Charles Pickering of Mississippi was allowed an opening statement. Pickering argued that a new facility was in everyone's best interest, and—ignoring the GAO's caveat about such a comparison—said facilities like the CDC were already doing similar research "safely." Citing Mississippi's chief health officer, Pickering said the proposed NBAF diseases were also already being studied safely in the U.S. His comment was either ignorant or deliberately dishonest, since live FMD virus was only being studied at Plum Island, well offshore. DHS/USDA had also explicitly said that NBAF would study new, "emerging" pathogens in the future. Pickering claimed to speak for the cattlemen and farmers of Mississippi and Louisiana (whom he hadn't bothered to survey) in representing that the work could be done safely. "We believe strongly Mississippi is the better site," he said. "Mr. Moran [of Kansas] may differ."[47]

Pickering made multiple attempts to undercut the GAO report, trying to force GAO witnesses into yes-or-no answers to statements at odds with the agency's earlier testimony. He asked whether there had been any outbreaks at the Canadian facility, and Dr. Kingsbury responded that the facility was relatively new—which, as she had already indicated, meant it hadn't been around long enough to have much of a track record. "New like the new NBAF," Pickering then interjected[48]—trying to suggest that the reason the Canadian facility hadn't had any outbreaks was its new, state-of-the-art construction—a claim GAO had already addressed in both its written and oral testimony.

In another leading question, Pickering asked: Isn't the German island connected to the mainland by a road or causeway, so that there's really no distinction? (GAO had acknowledged the road/causeway connection in its report.) Kingsbury responded that the island was nonetheless largely surrounded by water, not the answer Pickering was seeking. "You would agree that it is a mainland-connected site?" he said. (Yes.) "So the possible outbreak scenarios are not that different, is that correct?"[49] A specious conclusion, since a single road off an island is nothing like the many routes out of, say, Manhattan, Kansas, with a livestock farm and football stadium adjoining the proposed NBAF site.

Later Pickering would assert "You're not saying the decision is riskier, just that more analysis needs to be done?" Kingsbury responded that most experts believe an island adds an additional level of protection. Later, when Plum Island Director Larry Barrett said the GAO's list of worldwide FMD releases "omitted" the fact that most releases involved vaccine-production plants, Pickering pounced on the opportunity. "So the hype of risk is really based on a misunderstanding of the causes of outbreaks, has nothing to do with the new NBAF," he insisted. "It is extremely significant that GAO misunderstood," he concluded, "based on not knowing what the research is all about."[50]

Moments later, Stupak would point out that the GAO list had in fact referred to "known and attributed releases of FMD virus from laboratories worldwide, *including those that produce vaccines*" [emphasis added]. The countries where accidental FMD laboratory releases had occurred included the United Kingdom, Denmark, Czechoslovakia, Hungary, Germany, Spain, and Russia. And of course the United States itself—at Plum Island.[51]

The British Pirbright location had been one of the "vaccine producers." The government-run Institute for Animal Health shared a drainage system with the vaccine-producing Merial plant. Barrett inaccurately said that the outbreak had been definitively attributed to Merial. Actually, British investigators had found it impossible to determine which facility released the outbreak pathogens. They suspected the quantity of wastewater released by Merial had contributed to the accident, but the basic problem had been failure to maintain the drainage system, a responsibility borne by the IAH.

Pickering also argued NBAF would be different because it would not be "producing the vaccine." He was once again inaccurate. NBAF would not be producing vaccine in the quantities of a Merial, but as the GAO noted, NBAF was in fact planning to include a vaccine facility producing significantly larger amounts than at Plum Island.[52]

Kansas representatives Nancy Boyda, a Democrat, and Jerry Moran, a Republican, though not members of the Energy and Commerce Committee, were allowed as a courtesy to question witnesses. They proceeded to present spot ads for NBAF and Kansas. Moran argued that risks would be minimized at a new, high-tech facility, and said the risks of intentional introduction were higher than the risks of accidental release (a fact which, even if true, wouldn't make an accidental release less dangerous or less probable). Boyda assured the committee that Kansas had a "procedure," if anything happened, to get it under control in a matter of days—even though the hearing had clearly established that "in a matter of days" hundreds of thousands of animals and billions of dollars could be lost. "When we hear that Plum Island is not well-maintained," she said, dripping huge crocodile tears, "it puts fear in our hearts." She offered the committee a letter from the entire Kansas delegation, documenting their desire to save the country from the dangers of Plum Island, by bringing Plum Island's scary big brother to Manhattan, Kansas.[53]

CHAPTER ELEVEN

Something in the Air
Dugway, Utah

1

A 1968 accident at Dugway Proving Ground may have contributed to President Nixon's shutdown of the U.S.'s offensive bioweapons program. How ironic, then, that Dugway would have a big part in this decade's biodefense frenzy. According to the FBI's narrative of the 2001 anthrax attacks, however, Bruce Ivins modified his spores from a strain initially cultured at Dugway.[1] Some skeptics of the FBI's version of events suspect an even more intimate involvement by Dugway.

If Dugway's military and germ-lab colleague Detrick boasts the country's largest metropolis of indoor biodefense laboratories, Dugway represents the military's chief outdoor testing facility, encompassing a nuclear, chemical, and germ-testing playground of 800,000 acres, roughly the size of Rhode Island. When worried biodefense bridegrooms suggest that their blushing germ-lab-to-be should go screw itself in a desert somewhere, they probably have in mind a place like Dugway. Except there are communities and Indian reservations less than twenty miles away, with ranches and livestock along the perimeter. Salt Lake City lies just seventy-five miles to the east, and there is daily highway traffic of more than 10,000 vehicles in the area.[2] One person's desert is another person's domicile.

During the '50s and '60s, the military had some high times indeed at Dugway, conducting over 1,000 open-air chemical weapons tests with GB, VX, and mustard agents, and over 200 open-air biological weapons tests involving multiple pathogens. These included, in one of the key battles of Western civilization, the August 1952 B-29 bombing of 3230 guinea pigs.[3] Over the next two months, 11,628 guinea pigs in all would be attacked with brucella.[4] (Any pigs so impolite as to survive the brucella were of course executed by other means.) The experiments showed everything they were supposed to: brucella killed guinea pigs. For bioweapons researchers, however, working with guinea pigs somewhat resembled having sex with rubber dolls. As one general is reported to have said, "Now we know what to do if we ever go to war with guinea pigs."

No, any bioweapons research program worth its salt needs human subjects. As the general rightly recognized, reducing the enemy's guinea pig population is not the ultimate goal of biowarfare. Bio-warriors need to know if the germs will kill, or at least sicken, people. In large numbers. The Japanese had experimented with human subjects, after all. Of course, they had acquired their human chattel during the course of an ongoing war. The U.S. would need to take a different route. It needed volunteers.

What better germ-warfare fodder than those misguided Seventh Day Adventists? They didn't believe in killing people, for god's sake. Yet they were patriotic and wanted to do their bit for their country. So they flocked to auxiliary operations like the medical corps. Well, goodness, germ warfare was a little like the medical corps, wasn't it?

Uncle Sam needs a few well-intentioned do-gooders, the Adventists were told. To help us find ways of protecting our soldiers, and our whole country, from these nasty germs. The script would find its echo in the twenty-first century, in the feel-good messages spouted to potential neighbors of biodefense labs. Except by that time, the message had changed a bit. Help us fight those nasty bioterrorists, and make good money doing it!

The Adventists went for it. And on July 12, 1955, seventy of them found themselves scattered among the rhesus monkeys and guinea pigs, awaiting their share of the Q-fever buffet. Breathe normally, they were told. Nothing to get excited about. Just let the germs wash over you. Jesus is coming. Breathe in, breathe out.[5]

It worked. The Adventists, like the monkeys and guinea pigs, got sick. Richard Miller, for instance. He was "working with the floor buffer when all at once and without much warning he lost his strength, keeled over, and collapsed onto his own highly polished floor. That was the last he remembered until he woke up in bed."[6]

Not to worry. After receiving an extended course of antibiotics, Richard Miller survived, and lived to breathe another day. And the biowarriors now had eyewitness proof "that aerosolized biological agents could infect not only caged lab animals but also live and healthy human beings, and that they could do so stealthily, silently, and from a distance of 3200 feet."[7]

But would it work from an airplane? Biowarriors needed to know. So a later experiment used a tank strapped to a jet plane to spread the germs across a wide area. Experiment designer Bill Patrick of USAM-RIID was "overjoyed" when the germs traveled fifty miles. The scientists' germ dispersion models proved accurate, and their team was producing "a liquid product that was very, very good."[8] Good work, boys. Of course, there was something resembling blowback. One of the pilots was sickened when he got out of the plane too soon; and three soldiers manning barricades on the public road at the edge of Dugway also came down with Q fever. None of which made it into the newspapers at the time.[9]

Over the next twenty years, in Operation Whitecoat, some 2,200 Adventists were infected with Q fever, tularemia, sandfly fever, typhoid fever, various forms of encephalitis, Rocky Mountain spotted fever, and Rift Valley fever.[10] None of which made it into the newspapers at the time, either. It's not the sort of thing you issue press releases about.

It wasn't so easy to keep the sheep kill out of the news.

On March 13, 1968, as part of Operation Combat Kid, the Army "dispensed" 320 gallons of "a persistent-type nerve agent"—later identified by scientists as VX—through two spray tanks attached to an F-4 Phantom jet which made multiple passes over the targeted area. Most of the spray was in larger drops that fell to the ground quickly, but Army documents later acknowledged that up to 20 pounds of finer droplets "may have escaped the proving ground" with the wind. Lethal doses of VX are 10 milligrams by touch or 2 milligrams by inhalation.[11]

Twenty-five miles to the north, in the appropriately named Skull Valley, shepherds were watching their flocks by night, or, in the case of rancher Ray Peck, cleaning out a ditch with a tractor.[12]

The wintry morning of March 14 was so bucolic, Peck couldn't resist eating a few handfuls of the new snow. Then he noticed the dead birds, and a writhing rabbit.[13]

Unknown to Peck, herds of sheep to the south of him were acting confused, stumbling around, pissing frequently, and falling to the ground and kicking the air. After an extended course of such dramatics, they would stiffen, and their eyes would glaze over.[14]

Later that week, Peck and his family thought they had the flu: ear aches; violent diarrhea; sore throats; trouble breathing.[15]

Finally, on Sunday, March 17, one of the ranchers and his veterinarian decided they were up against something they couldn't handle. According to Dugway's incident log for the day:

> At approximately 1230 hours, Dr. Bode, University of Utah, Director of Ecological and Epidemiological contract with Dugway Proving Ground (DPG), called Dr. Keith Smart, Chief, Ecology and Epidemiology Branch, DPG, at his home in Salt Lake City and informed him that Mr. Alvin Hatch, general manager for the Anschute Land and Livestock Company, had called to report that they had 3,000 sheep dead in the Skull Valley area.[16]

VX nerve agent is a chemical weapon, not a biological one, of course, but the Army's response to the incident shows why Utahns have trouble believing anything the Army says about the safety of anything it does.

The Army first told the press it wasn't conducting open-air nerve gas tests at Dugway at all. Then the office of Utah Senator Frank Moss found and released Defense Department documentation of the test in question. Ooops.[17]

So the Army switched to arguing that yes, it had conducted the tests, but something else could have killed the sheep. Test director J. Clifton Spendlove (no, that's not Strangelove) alleged that the ranch owner had sprayed an organic phosphate insecticide within a mile of the sheep the day before they started dying.[18]

The evidence was all "circumstantial," the Army said; no one could prove beyond a reasonable doubt that the Army had in fact killed the sheep.[19]

A CDC report issued that year had this to say about that:

> Circumstantial evidence tying the deaths to Dugway was overwhelming; the product tested was lethal; it had been released prior to the first sheep losses; the sheep in the bands closest to the test were affected first and most severely; and all of the affected sheep were apparently downwind.[20]

The Army ended up paying $1 million in damages, but didn't otherwise acknowledge responsibility until the Clinton administration.[21]

It also insisted that no humans in the Skull Valley had been affected. No humans died, and blood tests of Skull Valley residents "proved" lack of exposure because they did not indicate low levels of cholinesterase.[22]

Normal levels of cholinesterase vary widely, however, and it is difficult to detect the sort of small drops that might be caused by low levels of nerve agent exposure without a baseline from prior tests.[23]

The Army team that looked at health reports of Skull Valley residents asserted that "It is clear that there were no symptoms which could conceivably be related to organic phosphate (nerve agent) poisoning either acute or chronic." In fact, however, the Pecks' symptoms were quite consistent with low-level nerve agent exposure.[24]

And the Army ignored one-half of its own doctors' recommendation that "continued surveillance of the human and animal population in the valley is indicated." It followed up on sheep and cattle, but let the humans fend for themselves. More blood tests over the succeeding months could have established a retrospective cholinesterase baseline that would have documented low-level exposure. But an Army that said it hadn't been conducting nerve gas tests in the first place wasn't seeking documentation that the tests had sickened people.[25]

The Peck family suffered a variety of ailments afterward: severe headaches; bouts of paranoia (perhaps about things like government open-air testing); and a surprising number of miscarriages and abortive births. Similar problems were mentioned in a three-volume series of reports from a National Academy of Sciences panel studying possible long-term effects of low-level exposure to nerve agents.[26]

It wasn't as though Peck was this hysterical person eager to go public; after hearing about the Pecks, the *Deseret News* had contacted them repeatedly over a period of months and asked if they could work together to pursue relevant Army documents. Peck, not wanting to appear as a disgruntled or unpatriotic worker, didn't respond to the letters for months.[27]

According to an anonymous informant, Dugway's public statements were simply lies:

> A retired Chemical Corps officer who worked at Dugway in 1968 spoke on condition of anonymity. The sheep died from VX, a nerve agent the Army released over the base, he told me in 1995. "We killed them, and we know we killed them."[28]

Seymour Hersh described Dugway in his 1968 book *Chemical and Biological Warfare*, in a chapter titled "The Secret Bases." Hersh highlighted the facility's secrecy and isolation, and portrayed the biological-weapons program as especially hush-hush. A 1960 press tour proudly demonstrated the effects of nerve gas on goats and pigeons, but the tour avoided Baker Laboratory, the biological compound, situated eight miles away from everything else, and closed to visitors and all GIs except those regularly assigned to biological operations.[29]

A former GI who had worked at Dugway as a chemist for thirteen months told Hersh:

> [The] civilians, whether Ph.D.'s or not, were a strange breed of people. . . . They all had credentials, degrees, etc., but from the time they arrived at Dugway they just turned off. It was another world, ostensibly a scientific testing operation but in reality a home for derelicts of all kinds: people who

could not possibly cope with the demands of anything
closely approximating a real life situation.[30]

The informant also spoke of falsified research data, and tests being
conducted in the presence of strong winds, in violation of research
and safety protocols.[31]

2

In 1984 the Army decided it wanted to upgrade and expand the biological facilities at Dugway and add a BSL-4 component. The way it
tried to accomplish this did little to improve the Army's, or Dugway's,
standing with Utahns. But secretive facilities require covert appropriations requests, apparently. Just before Congress adjourned in August
1984, an assistant secretary of the army requested a routine reallocation of $66 million from the House and Senate appropriations com-
mittees, for such mundane purposes as new military housing, a
parking garage, and a physical fitness center. Buried among the routine items was a request for an aerosol test facility in Utah.[32]

In accordance with congressional custom, the request was
reviewed by the chairmen and ranking minority members of the subcommittees on military construction, and upon their assent,
approved.[33]

Two months later, Senator James Sasser, the ranking minority
member on the Senate subcommittee, discovered the deception and
withdrew his support for the reallocation, concluding the facility
could be used to test offensive biological and toxin weapons in violation of the BWC.[34]

Clarification provided by a subsequent Army information paper did
not reassure Sasser. According to the paper, the heart of the facility
would be a steel chamber for conducting aerosol studies "of extremely
hazardous viruses and other biomaterials." And this would be just the
first stage of a five-year, $300 million modernization program.[35]

The other four members of the subcommittee approved the reallocation request over Sasser's objection. But behind the scenes, Senator Mark Hatfield, chairman of the Senate Appropriations
Committee, was also putting his objections on record, discerning a
number of issues and problems that he felt should be addressed

before Congress approved funding. He took special issue with the Army's method for trying to bring the expansion about, saying the request was "sufficiently controversial as to warrant more rigorous scrutiny than is associated with a routine 'reprogramming.'"[36]

A lawsuit by the Foundation on Economic Trends, led by genetic engineering critic Jeremy Rifkin, brought a temporary halt to the project. The Defense Department delayed the start of construction and prepared an "environmental assessment" asserting that all toxic materials would be safely contained within the facility and therefore would not affect the environment. Rifkin then filed for an injunction to halt the project until the Army issued a formal environmental impact statement—a process requiring public hearings and consideration of alternatives.[37]

The Army's efforts attracted negative attention in the December 1984 issue of *Science*. Utah governor Scott Matheson, who was leaving office at the end of 1984, asked his attorney general to consider joining the suit on behalf of the state, and later described for Leonard Cole the frustrations of his own efforts to get information from the Army about radiation hazards in Utah. "The only information we have now about this new Dugway facility is what the military has voluntarily released," he said, "and this is never satisfactory."[38]

Federal district judge Joyce Green granted an injunction on May 31, 1985, admonishing the Army that "[a]n environmental assessment must offer something more than a 'checklist' of assurances and alternatives. It must indicate, in some fashion, that the agency has taken a searching, realistic look at the potential hazards and, with reasoned thought and analysis, candidly and methodically addressed these concerns."[39]

The Army indicated that it would complete an environmental impact statement sometime in 1988.[40]

Almost three years later, when a public hearing was held in Salt Lake City as part of the EIS process, the *Deseret News* joked that the Army

> suffered probably its fiercest attack in Utah since the Indian
> Wars ended. This time the arrows came from Utah's governor,
> a congressman, scientists, doctors and an overflow crowd of

about 350 in the State Office Building auditorium who almost unanimously and vigorously opposed plans for a laboratory at Dugway Proving Ground to test defenses against germ warfare.[41]

The DEIS called for building a BSL-4 facility, though the Army swore—biodefense Scout's honor—that it was only going to handle BSL-3 pathogens. Governor Norm Bangerter was having none of it. "If the Army is only going to test BL3," he said, "they should build for BL3, not BL4."[42]

Seems there was a little matter of trust, alluded to by Rep. Wayne Owens, D-Utah, whose written statement said: "Given the dangerous nature of biological weapons testing in a state that has seen dead sheep and rampant cancers from open-air testing, both nuclear and chemical, it is unbelievable that the Army can say, 'Trust us.'"[43]

That didn't keep the Army from trying. But when Col. Wyatt Colclasure II, Dugway's director of material testing, said Dugway operations were not carried out in secrecy, "many in the crowd laughed derisively. When he said Dugway operations are safe because of its dedicated scientists, more people laughed. . . . And when he mentioned that many microbes like the ones tested at Dugway are useful in making cheese and yogurt, virtually the whole crowd snickered."[44]

Medical opponents said the Army was not planning for accident contingencies because it believed, or said it did, that the likelihood of such accidents was low. But Kenneth Buchi, representing the Utah Medical Association—formally on record as opposing the facility— said low-probability events were familiar in the field of medicine.[45]

Over 140 University of Utah scientists had joined the UMA in opposing the facility.[46]

A week or so after the hearing—just missing April Fool's Day—the Army managed to get an article in the *Deseret News* with the headline "Army Embraces Openness Policy." Lee Davidson, the writer, had written some probing reports about Dugway in the past, and seemed to be taking a tongue-in-cheek approach to this less probing one: "Pssst. Don't tell anyone, because most people don't realize this. But the Army says it has virtually no secrets when it comes to the proposed lab at Dugway Proving Ground to test defenses against germ warfare."[47]

The Army had said at the March 23 hearing that records of all Dugway tests conducted since 1980 were available for public review.[48] Only one test, it said, was classified.[49] Downwinders, an opposition group, decided to test the Army's new openness claims, and filed an FOIA request for a list of all chemical and biological arms tests since 1951 and access to all unclassified documents generated by them.[50]

The *Deseret News* filed requests of its own, and got a partial response from the Army on May 3, when the Army released a three-page listing of the eight major biologic experiments conducted in the past ten years. The eight experiments, the Army said, had resulted in 173 open-air trials using simulants.[51]

The April 4 *Deseret News* article noted a couple of additional disclosures on the Army's part: "Dugway recently admitted that it doesn't know what old arms and wastes may be buried there, and it provided recent, somewhat embarrassing Air Force maps that say some Dugway areas are 'permanently contaminated' by biologic agents despite conflicting Army statements that they are now clean."[52]

A few months later, the *Deseret News* concluded that Army PR about openness was not the same thing as openness itself. In an article "'Open Records' on Tests Are Mostly Unobtainable," the *DN* revealed that, "in what is becoming a complicated paper chase despite apparent efforts by Army officials to help . . . many of those records are not available to non-government employees." In fact, the *DN* said, Dugway "cannot provide final test reports on any of the eight major biologic warfare defense field tests it conducted in the past decade—which involved 170 separate open-air trials."[53] A few months later, Downwinders reported that the military was ignoring its requests for information altogether.[54]

One possible reason for this "openness gap" was revealed by another article in the *Deseret News*. The information gradually emerging from *Deseret News* FOIA requests showed that many more open-air germ tests had been conducted than Utahns had realized— or than the Army had owned up to. The *DN* said the most recent tally showed at least 279 open-air germ tests, conducted within at least 114 experiments, since the 1940s. Headlines just a few months earlier had put the number at 60 or less.

Each of the 114 experiments might have represented dozens of open-air tests, so it was entirely possible that the total number of open-air tests actually ran into the thousands.

Dugway had a hard time playing the public relations game in 1988. Too many of its skeletons were coming out of the closet at once. The EPA joined other critics in expressing concern about the new facility.[55] California's attorney general went on record opposing the mailing of dangerous toxins through the U.S. Postal Service.[56] The State of Utah said the Army had violated state law in 1987 by conducting open-air tests of new chemical weapons at Dugway without first obtaining air pollution permits.[57] About a month later, the Utah Solid and Hazardous Waste Committee announced that it and Dugway were considering an agreement under which the state would not fine the Army for 7½ years' violation of state hazardous-waste rules, if the Army would just begin to comply with those regulations.[58] The Army signed the agreement in August, but a headline like "Army Finally Plans to Bring Dugway into Compliance with Hazardous-Waste Rules" can't have helped its PR campaign.

Deseret News stories that fall focused on various contaminations at Dugway. One story discussed the many old hazardous-waste sites at Dugway whose contents were now unknown even to the Army, and how the Army had declared an area it bombed with anthrax safe merely because some sheep had grazed there for three months and survived. (All the test really demonstrated was that the sheep had not contacted—or been infected by—any anthrax spores. It couldn't demonstrate that there were no spores still lurking in the soil.)[59]

Another story described the Army's likely contamination of 66 square miles of public land popular with hikers and rockhounds, and the Army's efforts to purchase the area.[60]

In May, just as the Army was releasing its draft EIS declaring that the worst disasters it could contemplate at the new Dugway facility wouldn't threaten the public, the Senate Subcommittee on Oversight of Government Management issued a scathing report concluding that the military's germ and chemical-warfare research programs weren't adequately protecting the public from accidental release of disease or nerve agents.[61] The report said only 10% of the Pentagon's germ warfare research contracts involving genetic engineering imposed specific federal safety requirements. For the rest, "contractors appeared 'to be under no legal obligation to possess the proper facilities' for containing infectious viruses, to monitor laboratory areas and workers for viral contamination, or to 'decontaminate facilities when research is complete.'"[62]

Two 1988 books and a prominent magazine article also focused negative attention on Dugway. Leonard Cole's *Clouds of Secrecy: The Army's Germ Warfare Tests over Populated Areas* presented convincing evidence that the bacterial simulants used by the Army in open-air tests at Dugway and elsewhere were not in fact harmless, and that the Army had willfully ignored evidence showing this. Charles Piller and Keith Yamamoto's *Gene Wars: Military Control over the New Genetic Technologies*[63] described a variety of reckless behavior at Dugway, including the deliberate release of infected animals and insects into the Dugway environment. Both books also described the 1984 effort to expand Dugway by stealth.

Piller's October 3, 1988 article in *The Nation*, "Lethal Lies about Fatal Diseases," focused entirely on Dugway. Piller was explicit in attacking the Army's misrepresentations before a 1977 Senate subcommittee:

> The Army said elaborate safety procedures kept the BW agents within the borders of Dugway Proving Grounds, a desert test range larger than the state of Rhode Island, located near Salt Lake City. Indeed, the report concludes, "no impact on the environment was ever detected nor were there any other untoward effects." No one was harmed.
>
> Documents I have recently obtained under the Freedom of Information Act prove that most—perhaps all—of these reassurances, and others, are lies. **The documents reveal gross omissions and misrepresentations in the Army's prior descriptions of its field testing of BW agents.** The Army has consistently understated the range and extent of its pre-1969 BW testing. **The heavily censored reports also display a stunning disregard for public safety in tests involving massive amounts of the organisms that cause Q fever, anthrax and other lethal diseases** [emphasis added].[64]

Piller noted that at least sixty-nine field tests had been omitted from the Army's official accounting to the Senate. Even the incomplete documents Piller had obtained showed that hundreds of gallons of agents had been "disseminated." For perspective, Piller noted that a single organism of Q fever agent can cause infection, that billions of organisms may be present in a single drop of slurry, and that in one test, 40 gallons of slurry were sprayed from an F-100A jet traveling at the speed of sound.[65]

Army assurances about the creation of blue-ribbon safety committees to establish testing guidelines were meaningless because the first committee was not even convened until scores of tests had already been conducted. And claims that the same committees prohibited spraying diseases not already endemic in local wildlife were also worthless. At least eight BW agents had been released before any surveillance program was initiated. In fact, the Army believed (and wasn't telling the public) it had introduced three diseases new to the area—anthrax, Q fever and valley fever.[66]

Piller described the recklessness of spraying anthrax in 20-to-60-mile-per-hour winds, or dropping undulant fever bombs set to detonate at 10,000 feet above the ground. The Army had told the Senate it designed the tests to ensure that no BW organisms could reach U.S. 40; in fact, the Army had known anthrax spores were reaching the highway, and recommended the tests continue so long as the human exposures were estimated at below ten spores—even as it acknowledged it could not reliably estimate the actual number of spores reaching the highway.[67]

On September 19, 1988, before releasing a final environmental impact statement, the Army announced it was dropping plans for a BSL-4 component in the new facility.[68] It proceeded to renovate Baker Laboratory without preparing an environmental impact statement.[69]

The Army had already begun a second environmental impact statement on its entire Biological Defense Research Program, supposedly for the purpose of determining whether to continue the programs or terminate them altogether. (The Army didn't pose—and therefore didn't consider—a "continue but modify" alternative. No one ever thought it would terminate the programs altogether.) The

Army published the Final EIS on its overall program in April 1989. Predictably, it concluded that biodefense research posed no public threat.

The *Deseret News* found that a number of the EIS's claims were contradicted by the documents the *DN* had obtained during the preceding year.[70] The *News* politely avoided calling the EIS claims lies— but that's what they were.

The EIS said no existing facilities had been contaminated by germ research—when other documents said a vast area of Utah's West Desert was likely contaminated "with buried, unexploded germ warfare arms used in early testing." The EIS said open-air testing had always been small-scale, when in fact the Army had sprayed hundreds of gallons of highly concentrated germs at Dugway. The Army said it used only safe simulants for open-air tests at Dugway, and had never used *Serratia marcescens* in germ research. It *had* used *Serratia marcescens*, however, and the safety of the other simulants was in dispute. The EIS said there had been "no measurable effects" of researching non-endemic germs in Utah, when in fact there was considerable evidence Dugway had introduced Q fever and Venezuelan Equine Encephalitis into the region.[71]

Finally, the EIS said the BDRP was an open, unclassified program. The *Deseret News* said, however, that it had

> requested final reports of all open-air testing at Dugway for the past 10 years. Many such reports were not written or cannot be found, according to Army spokesmen. Those that were found were not released because they were for distribution only to government employees. When the Deseret News requested final reports on every tenth Dugway open-air test conducted between 1940 and 1980, those files also could not be located, Army spokesmen said.[72]

Utah Medical Association trustees criticized the failure of the EIS to consider alternatives other than continuation or outright elimination, and protested the short period allowed for public comment. The 880-page document had been received by the Association April 7, leaving members less than a month to review it.[73]

During the debate over the BSL-4 facility, the Army had agreed to the establishment of a citizens' advisory committee, and Governor Norm Bangerter had appointed one in early 1989. As they held an organizational meeting in October 1989, committee members agreed that their focus should be the health and safety of the public, and others would have to do any necessary balancing with national security interests. They also said they would take the Army at its word that most of its activities were unclassified, and would not seek security clearances for the committee's members.[74]

3

It soon appeared that Dugway might be planning to do the same dangerous experiments—aerosolizing genetically engineered pathogens —in BSL-3 that it had planned to use BSL-4 for. In February 1990, the Army sought and obtained the unofficial sanction of the National Institutes of Health for conducting experiments with two such pathogens at BSL-3 instead of BSL-4. The approval was unofficial because only facilities receiving grants from NIH were required to follow NIH guidelines.[75] And in August 1992, the state of Utah worried openly that BSL-4 features included in the planned BSL-3 Life Sciences Facility might signal the Army's intent to conduct the same BSL-4 research it had originally planned.[76]

In April 1991, a *Deseret News* story revealed that germ testing would recommence at the revamped Baker Laboratory, after a six-year hiatus.[77] According to Dugway's response to an FOIA request by Downwinders, the tests would include anthrax, Q fever, and plague pathogens, and staphylococcal enterotoxin and botulinum toxin.[78] The Pentagon turned down the group's request for information about frozen pathogens stored at the base.[79]

On May 2, 1991, Dugway formally notified the state that it would resume testing germ and chemical warfare agents at the lab on June 3. This gave the Citizens Advisory Committee just over 30 days to familiarize themselves with the proposed experiments.[80] About three months later, the group said it wasn't getting enough background on proposed tests to determine their safety.[81] On July 2, Downwinders filed suit to stop the testing until the government had conducted an environmental impact study, citing insufficient stocks of antitoxins in case of an accident, a lack of training for outside civilian medical specialists, and a pattern of "wanton disregard of public safety."[82] The Utah Medical Association joined the suit in December.[83]

Meanwhile, the Army was preparing and reviewing a final environmental impact statement for a new Dugway facility, the Life Sciences Facility.[84] It would release the document in April 1992.[85] At that time it said it hadn't decided whether to build the new facility.

In June 1992, the *Deseret News* did a profile on Jim Keetch, Dugway's division chief for test conduct. Keetch was still questioning the extent of Dugway's responsibility for the 1968 sheep kill: "'We may have contributed. (But) the BLM was releasing a pesticide in the area at the same time to control grasshoppers,' he said, noting that the shepherds should have showed signs of nerve gas contamination, but they didn't."[86]

Keetch's remarks brought a blunt response from Alvin Hatch, who said he was general manager of the affected livestock operation at the time and participated in all aspects of the investigation. Hatch said the shepherds weren't affected because they were in the camp wagon when the agent was released. The horses (also not affected) were blanketed and fed baled hay, stored in a covered commissary wagon. They did not forage on native grasses and were tied up when not in use. The sheep, on the other hand, were in the hills eating browse and using snow for water and thus ingesting the agent from both sources.[87]

Hatch had run his own tests by fetching and painting four sheep from an unaffected herd forty miles away and bringing them to the affected area. The three that he recovered four days later showed the same symptoms as the sheep which died earlier.[88]

Keetch had also watched a veterinarian inject several sheep with atropine, an antidote to nerve agent. They had shown remarkable recovery.[89]

In February 1993, the Pentagon announced that it was considering closing Dugway.[90] Just two months later, however, the Army said it would proceed with plans to build the new Life Sciences Facility—which would suggest that the closure talk was a mere strategy to soften up potential opposition to the new facility.[91] Dugway said the new lab would replace the old Baker Lab and would not add any new missions.[92] Rep. Jim Hansen (R-Utah), who had previously sabotaged an effort by his Democratic colleague, Wayne Owens (D-Utah), to force Dugway to disclose the agents it was researching, now attacked opponents of the new facility for "putting our troops and national security at risk." He said he had confidence in Dugway's safety.[93]

Two months later, because of the Army's continued recalcitrance about releasing its "open, unclassified" information, the Citizens Advisory Committee, now renamed the Dugway Technical Review Committee, considered meeting in secret to encourage the Army to be more forthcoming. Committee members were expressing frustration about being able to assess safety risks of Dugway testing because the information they got was "too little, too late."[94] Three months later, state environmental regulators cited Dugway for 22 new waste-dumping violations.[95]

In December 1993, Senator John Glenn (D-Ohio) released an unclassified version of a report by the General Accounting Office showing that the Defense Department had secretly dropped radioactive material over six states (including at Utah's Dugway) between 1949 and 1952, in an effort to develop a bomb that could spread radiation around a limited area.[96] Documents released a year later would reveal "that six more Army weapons trials likely scattered radioactive dust to the Utah winds in the 1950s. They were so big they released 11 times more radiation than all the other 68 known similar tests combined." The Utah emissions were 9400 times the radiation released by the Three Mile Island near-meltdown.[97]

Even as Utah Governor Mike Leavitt (later to head the Department of Health and Human Services under the Bush administration) was demanding full disclosure about the radiation tests, and opposing the Army's plans to ship neutralized nerve agent to Tooele Army Depot for incineration, he was supporting Dugway's pursuit of a $150 million facility to produce germ-warfare vaccines.[98]

Meanwhile, Dugway was adding several more strains of bioweapons agents to those already being tested at Baker.[99] In January 1995, U.S. District Judge Bruce Jenkins dismissed the Downwinders' suit seeking to block testing at Dugway, stating that a scenario in which an organism could escape and cause medical problems was too speculative and hypothetical to constitute an "injury in fact."[100]

In 1998 Dugway was chosen as the military vaccine test site.[101]

The *Deseret News*, which had been publishing a number of worried articles about the possibility that the Dugway base might be closed, straddled the fence a bit on the vaccine plant selection, stating that "this page supports the 10-year project while noting the legitimate concerns of the Downwinders and others."[102]

It continued to straddle the fence with a March 2000 editorial:

> Utah's Dugway Proving Ground is a logical place for testing America's chemical and biological defenses. It should be given the opportunity to play an increased role in the testing program provided adequate safeguards are in place and the Army is open and honest about testing procedures. Unfortunately, cover-ups involving Dugway in the past have damaged the Army's reputation.[103]

The *News* had been reminded of Dugway's shady past just one year earlier, when a contractor "using a backhoe during an environmental investigation of an old Solid Waste Management Unit—or dump site—" dug up twenty-five germ warfare bomblets containing the stimulant *Bacillus subtilis*, which the Army continued to insist was perfectly harmless.[104] The *News* would be reminded (and would itself remind) again in February 2001, with an article on "just a few of the major BLM [Bureau of Land Management] areas suspected of being contaminated with old conventional, germ and chemical weapons."[105] The contaminated areas in question constituted hundreds of square miles of publicly owned lands outside the Dugway reservation. It was hard to see the *News*'s new fence-straddling as anything but a wistful wish for miraculous rehabilitation, as if Dugway were some sort of ne'er-do-well, violent drug dealer who would surely join the Boy Scouts and start selling *Grit* door-to-door if it were only given the opportunity.

4

In October 2001, following the anthrax attacks, Dugway insisted that it had "all its anthrax."[106] On December 12, however, the *Baltimore Sun* ran an article quoting unnamed sources who said Dugway's finely milled, "weapons-grade" anthrax was virtually identical to that used in the attacks. The article reported that the team of researchers led by Paul Keim of Northern Arizona University had found that Dugway anthrax and spores in the terror letters were identical at 50 genetic markers.[107]

The Army issued a statement that "all anthrax used at Dugway has been accounted for. There is a rigorous tracking and inventory program to follow the production, receipt and destruction of select agents."[108] Apparently the Army felt the tracking and inventory at Dugway was better than that at its other lab, Detrick, and better than Dugway's own tracking of all those buried germ-warfare bombs.

The FBI was surprised by how many labs and universities did research with anthrax, and also surprised by how cavalier it all was: "The way the science community works, it's very freewheeling and freesharing."[109]

The Army's 2001 assurances about the security of Dugway anthrax spores would be contradicted in 2002 by Gerhard Bienek, who'd directed biological safety for Dugway between 1989 and 1993. Bienek said Dugway had been "appallingly sloppy" in handling pathogens, that it would have been easy for someone to have stolen anthrax. Freezers containing anthrax and other organisms were kept out in the hall; one could just open a freezer and take out what one wanted. (An anthrax picnic, anybody?) He recalled taking inventory and finding less anthrax in the freezers than was supposed to be there.[110]

On another occasion he recalled the base commander being asked to sign—and signing—a document stating the base would have only a little more than laboratory amounts of anthrax. But then base personnel wanted to produce 30 gallons of wet anthrax. Bienek pointed out that with 30 gallons on hand, a few milligrams would never be missed.[111]

Asked whether Dugway actually produced 30 gallons, Bienek said it could have, because that was their way of doing business. "Dugway always said one thing and then did another."[112]

Bienek had dropped an earlier lawsuit against Dugway because of family health problems. Another whistleblower, David W. Hall, was still involved in his own lawsuit. Hall charged that the Army "not only covered up his concerns about safety and environmental issues but illegally retaliated against him."[113] In August, an administrative law judge ruled in Hall's favor and recommended actual and punitive damages of $1.47 million against Dugway, plus attorney fees.[114]

J. Clifton Spendlove, who had directed Baker Laboratory from 1975 to 1982 and who had, further back, directed the infamous 1968 nerve gas test and participated in the ensuing coverup, disputed Bienek's charges. The article reporting Spendlove's statement, "Dugway handled anthrax safely, says ex-lab chief," also reported that the FBI had asked Dugway and Detrick employees to take lie-detector tests as part of the Amerithrax investigation.[115] Unfortunately, no one was asking Spendlove to take a lie-detector test about his public statements. . . .

The anthrax attacks and the ensuing biodefense expansion meant business was soon booming for Dugway, which tests biological and chemical defenses for a variety of "customers" ranging from other branches of the Armed Forces to non-DOD organizations. One year after the attacks, Dugway was issuing a draft environmental impact statement outlining proposals to vastly expand its activities.[116]

In February 2004, the *Deseret News* reported that military officials had "quietly" authorized the construction of four new "temporary" germ laboratories—three BSL-3 and one BSL-2—using an abridged "environmental assessment" process and a "finding of no significant impact" (or FONSI) published only in the Federal Register and the legal notice section of a Salt Lake newspaper. The construction apparently came to the attention of the newspaper—and the public—only after Steve Erickson of the Citizens Education Project stumbled across the legal notice in the newspaper.[117]

Dugway refused to tell the Associated Press what BSL-3 agents would be tested in the temporary labs because it said the issue was "sensitive." The year prior, it had denied Erickson's FOIA request for similar information because it said the information "could enable unauthorized individuals to locate and acquire biological agents and could reasonably be expected to assist terrorists."[118]

Two weeks later, the *News* carried an article indicating that the trailers might not be so temporary after all.[119]

Over the next three years, Dugway would pursue and plan a number of expansions. In late 2004, Dugway began seeking an unspecified amount of new land—estimated by the *Deseret News* to be at least 55 square miles and possibly as much as 145 square miles—currently controlled by the Bureau of Land Management. The Bureau had successfully resisted such efforts by Dugway in 1988.[120] Now the *News* speculated on what Dugway was up to. It had to speculate because the Army denied the *News's* FOIA requests for documents explaining the whys and wheres of the expansion.[121]

One suggestion was that Dugway wanted to "forcibly obtain nearby land it contaminated with chemical weapons but has refused to clean." Another was that it wanted to keep UFO-hunting groups away from its secretive activities. (The UFO groups, however naïve they might seem to UFO skeptics, at least know enough to take anything the government says with a grain of salt.)[122]

In August 2005, the Army would acknowledge that it wanted to stop people such as UFO watchers from spying on its "activities" from nearby mountains. The revelation came along with an embarrassing admission that one Army database said DPG had 1192 square miles of test and training ranges, while a different database said it only had 456 square miles.[123] This 736-square-mile discrepancy from an organization that claimed, in 2001, it could account for every single spore of its weaponized anthrax.

Errors of this sort could certainly explain why Dugway had contaminated off-base land on more than one occasion. And maybe Dugway wanted the new land now because, like somebody who can't balance his checkbook, it needed a cushion in case it should get mixed up in the future about exactly how much of Utah it did own.

In 2006 the Army issued a five-year-in-the-making programmatic environmental impact statement, examining the cumulative effects of chemical and biological defense testing at its sites around the country. Predictably, it said the impacts of continued testing would be "negligible to minor and mitigable."[124]

Steve Erickson suspected the document was designed to help the Army avoid detailed future study of any new testing and missions. He said the Army could now refer to findings in the PEIS and publish simple "environmental assessments" instead of more intensive environmental impact statements. But he said the Army had been doing that anyway in recent years. (And, indeed, they had taken such a step with the modular labs constructed in 2004).[125] Reacting to the

Pentagon's finding of "negligible impact," Erickson said sarcastically, "Gee, they've never had problems at Dugway before, so why should they in the future?"[126]

Erickson's suspicions about the EIS being used to justify a pattern of reduced scrutiny for new proposals proved correct a few months later. In March 2007, Dugway issued a draft "environmental assessment" for a proposal to completely renovate the old Baker Lab with as many as 25 new biological testing areas. The renovated lab would contain as many as 11 BSL-3 labs and 14 BSL-2 labs. The renovation would triple Dugway's current BSL-3 space and double its BSL-2 capabilities. The centerpiece of the new lab would be the Whole System Live Agent Test Chamber, capable of testing detection vehicles against large quantities of aerosolized pathogens (anthrax, among others).[127]

Shortly before issuing the draft environmental assessment, Dugway had advertised for two 1500-liter fermenters and 1500 liters of *Anthrax sterne var.*[128]—hardly laboratory amounts. Dugway had some big anthrax plans, it seemed.

Dugway's current plans are being opposed by Erickson's Citizens Education Project; by the Utah Chapter of the Sierra Club; by the Healthy Environment Alliance of Utah; and by Bev White, a former Utah state representative and organizer for the Dugway League, seeking justice for the victims and survivors of Dugway's past testing. Says White: "They do what they want out there, then claim they never did it, and now they just wait until the witnesses are dead."[129]

KANSAS PROPAGANDA, A KANSAS TORNADO, AND THE DEIS

THE ASSOCIATED PRESS COVERED BOTH THE GAO REPORT AND the Energy and Commerce hearing on DHS's NBAF plans, along with the information about prior Plum Island FMD incidents released by DHS Undersecretary Jay Cohen under the threat of congressional subpoena. The stories revealed that there had been previously undisclosed internal FMD releases at Plum Island in addition to the famous 1978 incident.[1]

DHS spun the incidents to the Associated Press by declaring that "laboratory animals would not be corralled outside the new facility."[2] This assurance would prove to be ironic in the extreme. When I visited Manhattan, Kansas over a year later, I would discover that the chosen NBAF site adjoins the existing KSU livestock farm, almost guaranteeing an FMD outbreak in the event of even the smallest release. The livestock farm itself adjoins the KSU football stadium and a field used for overflow parking for KSU football games, an obvious means for quickly disseminating the highly contagious virus throughout this pivotal region for American livestock production.

The significance of all this is that in 1978, the animals corralled outside the Plum Island lab were uninfected animals being held for future experiments. Inexplicably, the animals became infected by FMD; the precise source of infection was never determined. Only Plum Island's separation from the mainland kept the outbreak from becoming a catastrophic event. Plum Island learned its lesson, and stopped keeping animals outside the lab. DHS wasn't worried about bringing the dangers inland, though, because it believed that NBAF— unlike any other human enterprise—would be failsafe.

In 2007, FMD escaped from a site shared by the main British FMD research facility and a vaccine plant and infected herds on nearby farms, causing significant economic losses because of the embargo on British meat exports and the ensuing mass slaughter of infected, and potentially infected, livestock. Investigators concluded that vehicular traffic most likely carried FMD to the infected farms, a risk clearly posed by the Kansas NBAF location. As the Associated Press story (and the GAO report) pointed out, FMD is so contagious that it can be carried on a worker's breath or clothes, or vehicles.[3] It is also believed that it can be carried by winds.

Placing NBAF in Manhattan doesn't just remove the extra layer of protection that an island location offers; it removes the additional safety steps that were taken at Plum Island after 1978. KSU has a substantial group of livestock grazing outside, quite near to the new NBAF site, and quite near to a football stadium with thousands of people and vehicles departing for various spots around the state of Kansas (and at least some of those vehicles associated with visiting teams from Big Twelve Conference states and elsewhere).

The AP story touched lightly on the ironic fact that Senator Pat Roberts, a chief political lobbyist for Kansas's NBAF efforts, had played the U.S. president in a fictitious 2002 simulated FMD outbreak, Operation Crimson Sky. That simulation had people rioting in the streets after National Guardsmen were ordered to kill tens of millions of farm animals, ran out of bullets, and had to dig a Kansas ditch 25 miles long to bury all the carcasses. "It was a mess," Roberts had said during a 2005 hearing.[4] Now Roberts—presumably believing all the biolab safety propaganda—was happily running the risk of just such a mess in his home state.

The Kansas propaganda machine got into high gear in an effort to drown out the sudden upsurge in concerned letters to the editor. A *Manhattan Mercury* story, "KSU: Concerns Unfounded," completely avoided mentioning the GAO report and attributed the concerns to "Democrats in Congress." KSU Vice President Ron Trewyn called the claims—essentially the concerns raised by the GAO's independent investigation—"sensational." He claimed that "We have decades of experience" and so far "we've gotten it right"[5]—with no indication of what experience he was referring to. Neither KSU nor USDA had any experience working with live FMD virus on the mainland; nor did KSU have extensive experience with high-containment germ labs.

Jaax and Trewyn followed up with an op-ed in the *Mercury*, falling back on the standard claims about "state-of-the-art" equipment, "proven safety protocols" and the fact that the CDC and Detrick hadn't caused any community outbreaks (except for the anthrax letters).[6] The "proven safety protocols" hadn't always worked at Plum Island, though, and KSU was eliminating one of those protocols by putting the lab next to its existing livestock farm. Jaax and Trewyn also cited the fact that the Winnipeg BSL-4 lab had been working with FMD, and it hadn't spread from that facility. As the GAO had pointed out, however, the Winnipeg facility had only been operating for a few years, was in an urban city—not a major livestock production area—and only worked with very small numbers of livestock.

Dr. Gary Conrad, KSU's Distinguished Professor of Biology, argued against the Manhattan location and said there were "normal people who are afraid to speak out because they feel that this is such a big thing—so much money involved, so many powerful political people involved—that they would lose their jobs."[7] Trewyn scoffed at the notion: "No one would lose a job for speaking their opinion on any matter."[8] Yet a comment by U.S. Senator Sam Brownback, R-Kan, for a different newspaper, suggested the extraordinary pressure to follow the NBAF party line. "You're not going to find a politician who will say he's against it," Brownback boasted. "It's all hands on deck."[9]

On June 11, a tornado passed through Manhattan, causing significant damage to parts of the city and damage of over $20 million to the KSU Campus. The tornado passed within a quarter mile of the Biosecurity Research Institute; but a KSU press release boasted that the building emerged unscathed and continued to function without a glitch.[10] A letter writer, however, citing National Weather Service reports that the EF4 tornado had weakened to an EF1 by the time it reached campus, questioned whether this building—or the future NBAF facilities—had been truly tested by what the Kansas weather could throw at them.[11]

On June 20, the Draft Environmental Impact Statement was released. The DEIS had been prepared for DHS by a hired contractor, Dial Cordy. The DEIS estimated the costs of an FMD outbreak from the facility as being greatest at the Kansas ($4.2 billion) and Texas ($4.1 billion) sites, and the least at the Plum Island location ($2.8 billion).[12]

The Kansas public comment hearing was held July 31, once again in the K-State Student Union, with the problematic public parking. The usual collection of bigwigs came out in support of NBAF, from Kansas governor and future Health and Human Services secretary Kathleen Sebelius on down. A significant number of opponents, though outnumbered by the pro-NBAF Kansas juggernaut, did show up and speak out. KSU VP Trewyn was up to his usual efforts to undercut bad news. Trewyn said the modeling used in the DEIS which made Kansas the most expensive site for an outbreak was "unrealistic"—even though the modeling came from an organization with every incentive to keep the estimated costs of an outbreak as minimal as possible.[13] Nancy Jaax suggested it was downright "criminal" to oppose NBAF just because of "an infinitesimal risk."[14]

Manhattan Mercury executive editor Bill Felber launched an extensive series focusing on KSU's competition and handicapping their various chances, highlighted by trips to the DEIS hearings at the other finalist sites, in Texas, Mississippi, Plum Island, Georgia, and North Carolina.

NBAF 16

A TALE OF TWO HEARINGS

THE NORTH CAROLINA HEARING ON THE DRAFT ENVIRONMENTAL Impact Statement was rife with irony.

It was Tuesday, July 29, 2008. That night, in Frederick, Maryland, Bruce Ivins was taking his own life, either a lone U.S. biodefense researcher gone off the deep end or the fall guy for something more sinister within the U.S. biodefense complex. Three days later, the *Los Angeles Times* would disclose Ivins's death and the FBI's version of events.

If that version was true, we were all in Butner that night because of what Ivins had done seven years earlier.

Some of us knew that the FBI had concluded early on that the anthrax letters came from within the U.S. biodefense complex. But we could not know what the FBI was getting ready to announce.

That made the public statement of Jesse Wilkinson, a retired engineer, a bit prescient: "Things will happen. Machinery will break. People will make mistakes. Murphy's Law is true." Then he added: "You only saw on television what you wanted to see. You didn't see that if you brought dangerous diseases here to study, and you let them out, you yourself are the terrorists."

Then there was the irony of DHS Undersecretary Jay Cohen's Final Selection Memorandum (from July 2007), leaked to the press a few days later, noting the "strong community acceptance" for the North Carolina efforts.

Another example of how little the experts knew of reality. For the North Carolina NBAF consortium was on the verge of giving its efforts up, in the face of an intense and heated opposition. Not a single person that night—assuming there was a person in the audience who wanted to—dared rise to say a kind word about NBAF.

The pro-NBAF consortium had made a sort of last-ditch effort by obtaining public funds—a $262,248 grant from the Golden LEAF Foundation—to fund an "education" campaign. Golden LEAF had been created in 1999 to collect money from the state's tobacco settlement and use it to make grants and investments promoting North Carolina's economic health.

The consortium claimed it needed to correct "misinformation" spread by lab opponents. In applying for the grant, it had offered some samples of the "corrections" it intended to provide. One statement said: "The center **WILL NOT** do bioweapons research." That statement was itself misinformation, if you consider research on "bioweapons agents" (pathogens capable of being used in a bioweapons or bioterror attack) to be "bioweapons research."

An article by Lisa Sorg in *Indy Weekly*, "Big Dough Goes to NBAF PR," revealed that more than $100,000 would be paid to PR firm French West Vaughan, with offices in Raleigh, New York City, and Tampa. One of its better known clients was Pfizer (also a member of the UPMC-sponsored Alliance for Biosecurity). The French West Web site boasted that Pfizer had hired it to influence "the mindset and behavior of politically influential individuals and groups around the state. FWV developed a three-pronged approach to shaping opinions."[1]

The decision to use public funds to support one side of a disputed issue was widely criticized in state media outlets such as *The Charlotte Observer*[2] and *Carolina Journal*. John Hood, president of the John Locke Foundation, wrote in the *Carolina Journal* that he hadn't made up his mind about NBAF itself, but was opposed to the Golden LEAF funding decision:

Doing a sales job on a biodefense lab for Granville County hardly qualifies as a core public service.

The more I learned about the interests behind the sales job, the worse the Golden LEAF decision looked. For one thing, the public and private institutions behind the pro-lab coalition include the likes of Merck, the EPA, major agribusiness associations, and large swaths of state government. These organizations already spend millions of dollars getting their messages out via public speeches, press stops, websites, and lobbyists. It strains credulity to suggest that without $262,248 from Golden LEAF, they were powerless to rebut the awesome public-relations power of the anti-lab coalition, with the apt acronym of GNAT, run by activists and local residents.[3]

Locke also pointed out that two members of the Golden LEAF board, who hadn't recused themselves, had a clear conflict of interest because of their connections with North Carolina State, the lead university in the state's NBAF efforts.[4]

Shortly after the Butner DEIS hearing, the North Carolina consortium declined the grant funds because Golden LEAF was requiring, in the proposed grant agreement, that it review the materials first to ensure that they represented "an independent and objective exposition"—something legally required for Golden LEAF to retain its nonprofit status.[5] But independent and objective exposition was not what the consortium had in mind.

More bad days were coming the consortium's way. On the very day of the hearing, the Raleigh Public Utilities Department recommended the City Council oppose the lab because the DEIS hadn't answered the city's 38 questions, about agents being studied, oversight, worst-case scenarios, engineering controls, solid and liquid waste and plans for treatment, sampling and monitoring, emergency response, and security.[6] Soon after the hearing, the City Council would adopt that recommendation and formally oppose placing the NBAF in Butner.[7] State Senator Doug Berger had pulled his support for the lab earlier that summer.[8] U.S. Rep. Brad Miller would withdraw his support on August 5. The Butner Town Council went on record opposing NBAF on August 7.[9] The Durham County Commissioners followed suit on August 11, and the Durham City Council on August 18.[10]

DHS held all the DEIS hearings in the space of about two weeks. The logistics of traveling to all of them—at my own expense—in such a short time seemed daunting. So I chose the two locations where newspaper coverage indicated there was strong opposition: Butner, North Carolina and Athens, Georgia. But even I, by now an experienced NBAF opponent, never expected what I found in Butner.

Butner is "classic" small-town America. Moreover, it is small-town Dixie. One of those towns you miss if you blink as you drive through late at night. A small town with a decent collection of small farms clustered around it. Not so different from Pulaski County, Kentucky, really. Bill Felber, the executive editor of the *Manhattan Mercury*, who ran a series of articles that fall sizing up the competition, said the Granville Non-Violent Action Team (GNAT) featured "a core of 50-ish and 60-ish former hippies who had originally coalesced in 1990 to successfully oppose the siting of a hazardous waste facility in Granville County."[11]

I attended Berea College in the seventies at a time it held a decent contingent of hippiedom—I know hippies—some of my best friends were hippies—and I'm telling you, Felber, I saw few signs of hippiedom, former or present. I've looked at pictures from the 1990 protest, and I saw no signs there, either. These were the sort of folks to show up for a barbeque, some of them in gun-rack-toting pickups, about as far from hippies as one can get. The "former hippies" is a wishful part of Felber's world view, I think—the notion that only "counterculture people" would oppose an innocent little germ lab. No, the people most likely to oppose it are the ordinary people who live next door, whatever their political stripe. There are plenty of them there in Manhattan, too—even if they have been mostly drowned out by the card-carrying members of the Kansas NBAF juggernaut.

I suspect it makes a difference where you hold the hearing. This one was at the Butner-Stem Middle School. Schools are important places for community interaction in small towns. Hearings held on the campuses of the NBAF-seeking institutions—especially during the summer when faculty and students who might question the enterprise are likely to be absent—favor the bigwigs. At their local schools, on the other hand, people not wrapped up in the state and local power structures find it easier to show up and voice their thoughts, concerns, opinions.

Just before the hearing, things seemed ripe for a confrontation. In a bit of street theater, NBAF was given its last rites and hauled away in a hearse. A small brown goat, with a sign *Skip Says No to NBAF* around his collar, watched closely. DHS had its information tables; GNAT had its. The T-shirts: *Whatever It Takes!* A slogan left over from the hazardous-waste fight, and the GNATs meant it. Many had gone to jail during that battle. Many were prepared to go to jail during this one. As one speaker said: *I'm willing to stand in front of your damn bulldozer and keep it off the damn property.*

This was the procedure for the evening, pretty much the same everywhere. DHS and Chuck Perdler, spokesman for Dial Cordy, the contractor who prepared the DEIS, took the first forty minutes to educate the public. Then the rest of the first hour, about twenty minutes or so, was available for questions. Then the next two hours were devoted to public comments. Statements were presented in the order that people signed up. Each speaker was limited to three minutes, and a no-nonsense young woman from DHS enforced the time limits. There was a substantial law enforcement presence. A couple I spoke with while people were still gathering said the police had shoved an elderly woman away from the mike during the morning session. During the meeting, the police stood up during moments of raucous applause to make their presence known. Perhaps they feared an on-the-spot lynching.

So, picture the Homeland Security seal up front, and picture folks with T-shirts and jeans, the orange "GNAT" lettering on the front of the black T-shirts, the orange "Whatever It Takes" lettering on the back, families there with kids, the evening's profanity not going past the occasional "damn," the signs scattered about the auditorium: *ACCIDENT? They'll help us like they helped New Orleans.* Or *Go Away DHS!* Or *If It Is So Safe Put It in Washington D.C.*

Jamie Johnson, head honcho of DHS's National Labs, invited the group to "visit our posters and ask questions of our subject matter specialists."

Dial Cordy spokesman Perdler said the DEIS showed "no sites with major adverse effects under normal operations." There was a "low-accident risk" at all sites. "This is a high-risk facility, no doubt about it," but the facility would be constructed to "rigorous requirements"—and if procedures were rigorously followed would be low-risk. (A speaker at one of the hearings near Plum Island said, "That's why they call them accidents, not appointments.")

The young DHS lion-tamer warned the questioners during the twenty-minute Q&A session to restrict their questions to the presentation just given. DHS declined to answer the question of how the more than 5,000 state and federal prisoners housed near Butner would be evacuated in an emergency, declaring that such matters were "operational considerations." Anyway, any "release" would be "extremely unlikely," and even a release of a virus affecting humans wouldn't matter because of the small quantities of virus being worked with. (It's scary that DHS should say this: small quantities might reduce the probability of a release, but wouldn't necessarily make a difference once a release has occurred. Jamey Johnson declared that he could promise NBAF would only examine foreign animal diseases, but he couldn't speak for what Congress might do.

The locals had some fun with DHS's confused answers about "deer eradication should a whiff of foot and mouth" get out. DHS suggested that it might put the deer "under surveillance" (nightscopes and wire-taps?) to see if they were commingling with and transmitting FMD to cattle. "Are you aware," one questioner asked scornfully, "that in this area deer jump in and out of pastures constantly and travel quite a way? A lot of people are trying to eradicate them on their own." (Laughter.) "Well, in this situation," the DHS mouthpiece mumbled, "we'd probably just let them go. In a worst case scenario, we'd fence in all the livestock, resort to hunting."

"Good luck," said the questioner, to more laughter.

The Q & A ended. The first statement was by Daryl Moss, African-American mayor of Creedmoor, who had shocked the conga line of bigwigs at the scoping session a few months earlier by saying he couldn't support NBAF "at this time." Tonight he referred to that moment, to thunderous applause: *Then I said I couldn't support the facility because of what I didn't know; now I can't support it because of what I do know.*

The ass whipping proceeded. My notes are a jumbled blur of criticisms, questions.

How are you going to contain mosquitoes? Insecticides? I'm sure they'll be as safe as DDT and Agent Orange. Laughter.

Our communities don't want your NBAF! Loud applause.

A teacher from Durham: *I'm sorry to say I don't trust my government.*

NBAF is another way for the current administration to award expensive no-bid contracts to private companies and avoid legislative oversight. Dingell had to threaten DHS with subpoenas for not providing info to GAO. Is there any transparency? Senators Clinton and Schumer are opposed to building a BSL-4 on Plum Island. What do they know, and where are our leaders in North Carolina? You should be called the Department of Homeland Insecurity. LOUD APPLAUSE. *It's a risk we cannot afford to take.*

Butner, NC is a poor choice. Water and sewage capabilities are insufficient to support the current population. You didn't tell us the rest of the story about the British foot and mouth outbreak. It was not a spontaneous outbreak, but was traced back to state of the art biosafety facilities, to heavy rains and flooding and broken sewer pipes. Butner, NC is not the proper place. Applause.

An older African-American woman: *I've been here from before I was eight years old, when Camp Butner was here. This has really had my mind a-going. I had cancer surgery Thursday. What do you do with the carcasses? Bury them? Burn them? Camp Butner left a lot of surplus stuff getting into the water. Lot of cancer deaths and cancer people in this area. Unique place, so many people incarcerated, in hospitals. With as many isolated places, why would you want to be in such a densely populated place as this?* Applause.

Susan Dayton from the Blue Ridge Environmental Defense League: *Twenty minutes is not enough for question and answer when it comes to a project of this magnitude.* Applause. *One of the problems is this project being run by Homeland Security.*

Someone cites the DEIS's claim that the facility would be a "significant benefit for wildlife." *I'm sure the facility will be fascinating to deer and wildlife.*

A nurse at one of the psychiatric facilities, a wife and mother of two beautiful girls. *Things keep coming up that you have failed to acknowledge. Avian flu. Insecticide spraying. I'm not letting you spray my family. I do not see this coming here. GO HOME. GET OUT OF TOWN.* Raucous applause. A policeman stood up.

A woman who has worked in the area twenty years: *In the event of a mishap, how will DHS and the government ensure that all losses on the part of residents and businesses in the area are compensated?*

This is our home. This is a county of farmers. We had no part whatsoever in the choosing of this site. There is a terrific failure of justice here.

Dr. William Lawson, a physician for 15 years, a Republican candidate for Congress. *How can you assess the impact on air quality without knowing the disposal method? Senator Clinton opposes having the lab on Plum Island. Why aren't our leaders opposing this obvious case of corporate welfare gone wild in the 13th District?*

One woman turned her back on the panel. *I'd rather stand here and speak to the people. I've spent thirty years of my life living near a landfill that leaks, a landfill that the state of North Carolina guaranteed us would not leak. Guaranteed us was state of the art. You can't guarantee us anything. You don't address disposal because you dare not tell this community that you're going to burn it. . . . This is a broad community. We have contested, we have been through the fires, and we will not accept this facility.* Loud applause.

Dr. Joseph Melamed, one of 43 other local physicians who practice in this area who oppose NBAF. Long discussion of Plum Island. The alarm clock went off. *Shouts: let him speak!* Is there something you can provide to us in writing? the DHS enforcer said. Someone offered his three minutes. The enforcer shut his mike off anyway. *You're a coward. You're afraid to let him speak.*

Comments about the failure to address storm water runoff into Falls Lake, the main water reservoir for the area. Concerns about aerial spraying of insecticides. Lack of itemization in the DEIS of infrastructure the host community is expected to provide. Maximum economic benefit of the facility estimated at $3.5 billion. Estimates of potential losses from FMD outbreak in California of $8.5 and $12.5 billion.

*What part of NO don't you understand? Would you please address it in the DEIS? Do you remember David in the Bible? He was small, all alone, but had lots of time to practice. We are your David, you are our Goliath, and we will **bring you down**. We have drawn your line in the* *sand; you need to know that. I want you to give more weight to public comment. I want you to give more weight to what the citizens think, not people in Raleigh. I am one of those people who will stand in front of the bulldozer. And there are a whole lot more people who will stand with us.*

I was here in the afternoon, I asked to come back tonight. I'm a little tired, but not too tired to stay here and fight this thing. This is not helping my pacemaker. Take it back wherever you people came from. This is not the place for it. God knows, I'm angry for a good reason. This is our home. We don't need it here. Please don't bring it here, please.

Perhaps you folks have noted over the evening that there's a certain lack of trust in Homeland Security. You did come here and tell us that you're going to do a marvelous job. Suggestion One: A little bit more truth and openness. Only way to monitor safety is with adversarial safety monitors. 100% removed from the culture of Homeland Security.

Someone from the River Foundation of Raleigh: *Our upstream water supply serves 400,000 people. You were severely admonished before the House Energy and Commerce Committee for making the CDC comparison. I resent you still coming here making that comparison. The fact that previous USDA studies are no longer available—what does that say about transparency?*

You didn't put the town meeting in your DEIS. NO is spelt N-O. Webster's: Not in any degree, or not at all. Reject; refuse approval. Three wishes: You to go away. You to go away. You to go away. NO NO NO. Loud applause.

Another example of small communities overtaken by powerful interests. A sprawling, complex, completely unaccountable corporate supply chain. We're not just trying to stop you, we will stop you. Do whatever it takes. At first, speak at hearings. If not heard, we will protest. If protests not heard, take our fight to the next level. Farmers don't have to stand in front of bulldozers. They understand how bulldozers work.

A woman who is a volunteer advocate for the mentally retarded: *Please, for the final draft, supply us with a flow chart of the dollars, including political fundraisers, between the members of the consortium, elected officials in favor of NBAF, lobbyists, and everyone else who spearheaded the push to bring NBAF to Butner. Why were some state employees in Butner given permission to endorse, and others put under a gag order? Again, what part of "no" don't you understand?*

A staff psychologist at the Butner Correctional Institution: *A pro-ponent accused opponents of scare tactics and hysteria. Butner people don't scare easily. Most have dangerous jobs. The irony is DHS empha-sizing the danger of pathogens, while opponents who raise objections are accused of knee-jerk reactions and hysteria.*

Someone consulted an air-quality expert, who said there wasn't enough hard data to draw any sort of conclusions. *Wonder how you were able to draw conclusions from that lack of data. The Triangle Area's ozone noncompliance not mentioned in the DEIS.*

DHS was chastised for arrogance by the House committee, and I just want to second that emotion.

Bill McKellar, of GNAT, said there were 4300 local signatures against NBAF.

Suzanne Smith: *I drove all the way to Washington. I told you last meeting, I've been trying to tell you: Bless your hearts, you don't have community support. But one thing I have learned is that you value redun-dancy. You do not have community support. Is that not redundant enough? I would like to take this opportunity to tell my community, how absolutely proud and thrilled I am to be part of this community. We do understand that people in Washington have good intentions—the same type that paved the road to hell.*

Someone who raises sheep said *NBAF is a b-a-a-d idea.* Audience *b-a-a-a-ed.* Jamie Johnson, bored, disgusted look on face, took a drink of water.

A few days later, someone from DHS told the North Carolina papers: message received, loud and clear.

Meanwhile, down in Athens, Georgia, the hearing was held in the plush Center of Continuing Education (which contains a hotel and a fairly high-class restaurant), on the university campus. Since the Uni-versity leads the Georgia consortium, opponents who wished to speak had to enter enemy territory. Many of the supporters, on the other hand, just needed to walk across campus.

Nonetheless, in the session I attended (there was another session in the afternoon) the speakers for and against were about equally divided: I counted twenty in favor, and twenty-three against.

Both sides applauded vigorously for their speakers.

The main differences were the socioeconomic origins of the two sides. The pro-NBAF speakers represented the "movers and shakers" from the university, town, and state. And also, for the most part, people who stood to gain financially from the facility, either directly or indirectly. The president of the Athens Area Chamber of Commerce; the senior vice president for academic affairs at UGA, speaking on behalf of President Adams, who had a prior commitment; the president of Georgia Bio, a nonprofit association of 300 pharmaceutical, biotech, and medical companies; an Athens-Clarke County Commissioner; the regional vice-president for Georgia Power; another commissioner and former city council member serving on the Citizens Advisory Board of UGA's new Animal Health Research Center; the president and CEO of the Georgia Research Alliance; Heidi Davidson, Athens mayor; a member of the Athens Economic Development Foundation; the new head of the Department of Veterinary Pathology at UGA; the president of the Georgia Agribusiness Council; the Director of Environmental Health and Safety at UGA; a UGA Ph.D. student, who once worked at Plum Island; a faculty member from the College of Veterinary Medicine; an oceanographer/microbiologist from UGA; a representative of U.S. Senators Isaakson and Chambliss; the Associate Dean for Research and Head of UGA's College of Veterinary Medicine; Paul Chambers, the district manager for AT&T, who had been so impressed with the "class of people" who turned out in favor of the facility at the scoping meeting; someone representing the University System of Georgia; the president of the Athens Economic Development Foundation.

Everyone either titled or associated with the university.

The opponents a much more diverse group—a goodly number of retired people, including a former college president; two UGA faculty members and two staff members—brave people, all four, I would suspect. A former employee of DuPont; a former securities lawyer; a homesteader and carpenter; and several mothers speaking on behalf of their children.

Doc Eldridge, of the Athens Chamber of Commerce, repeated his question from the scoping session that the *Banner-Herald* found so persuasive: if not this, then what? Since the 1970s, he said, this community has focused on developing the bio sciences and life sciences;

this is what we said we wanted. What message does it send to the outside world if we say no after all these years of seeking this kind of business? Alarm clock went off; someone shouted *Down!* Applause and boos followed.

UGA's vice president for public affairs spoke of the "minimal risks outweighed by the far greater benefits" and the "obligation of the university to apply its resources to serving the people of the state."

The president of Georgia Bio was interested in the hundreds of jobs. He cited the existing labs at CDC, Emory, and Georgia State. He kept going after the alarm went off. After fifteen seconds people were shouting: *You're done, you're done.*

Doug Lowry, a county commissioner, said, "If I believed any economic development proposal would harm the people or resources of our region, I would lead the opposition." His next comment—"I have been gratified to learn that every one of the elcted representatives, government representatives, every employee of the University of Georgia with whom I have spoken, feels exactly the same way"—was greeted with scornful laughter. (Lowry talked twenty-two seconds over.)

Mayor Heidi Davison spoke of the steps she had taken to reassure herself, how she came to appreciate the science and research and safety protocols. "If you can build a BSL-4 facility in Atlanta, you can build one in Athens." Both applause and boos at the end.

The Ph.D. student who worked at Plum Island said "the facility itself is engineered in such a way it is virtually impossible for anything to escape." Scornful laughter. . . .

Cecilia Purvis, an opponent, who taught environmental courses at UGA, was dismayed at how many opponents of NBAF in Athens had been portrayed as lacking proper information and seen as somehow irrational. She said this was typical treatment for people concerned about the environment.

Matt DeGenero, a young man who was reportedly ejected from the afternoon session for too-vigorous heckling, said there "was and continues to be no demonstrated community acceptance of this project. . . . In March 2006, Athens GA not only didn't want NBAF, we had no idea what NBAF even was. The first meeting open to the entire community was held almost a year and a half later, on August 30, 2007."

De Genero said Jeff Wilson, the publisher of the *Athens Banner-Herald*, sat on the board of the Athens-Clarke County Economic Development Commission, that the *ABH*'s headlines and editorials were all about cheerleading, and that anyone looking for facts had to look at newspaper coverage in other states.

The alarm went off; DeGenero kept talking; after eight seconds, his mike was cut off.

Grady Thrasher, one of the founders of For Athens Quality of Life, the chief opposition organization, a former securities lawyer, said he was here representing concerned citizens:

> We do not represent the pandering politicians, disingenuous academics, or financially interested gain seekers that comprise the Georgia Consortium. We represent the people of our community whose consent never has been sought and whose consent never will be given to the proposed NBAF being developed in the midst of our pleasant peaceful and progressive town.
>
> Do not mistake the politeness or ever natural congeniality of our people for apathy or absence of resolve.
>
> To bring NBAF to Athens you will have to climb a mountain of legal briefs and live with a lifetime of negative reaction.
>
> We the people will fight you and your irresponsible enablers at the University of Georgia and in the Athens/Clarke County government by every lawful means, every step of the way, should you choose to impose NBAF on us.

Thrasher's wife, Kathy Prescott, focused on DHS's "reluctance to talk about the zoonotic diseases," and quoted a series of documents in which DHS had spoken of NBAF as an integrated facility focused on human, animal, and zoonotic diseases—a description DHS backed drastically away from just before it announced the finalists and started the DEIS process. *What is the mission of NBAF?* she asked. *That's something we've never gotten an answer to. Tell me why we, including our local representatives in government, should believe anything you say now or ever.* Loud applause.

Tom Jenkins, a retired DuPont employee, said most of our food problems were related to the mistaken application of industrial methods to agriculture. NBAF didn't address this issue: is defense against microorganisms a logical way to proceed toward guaranteeing food security when at least 50% of our food is imported?

A woman married to a plant pathologist said she came to the meeting with an open mind, read posters, etc., but was not convinced Athens is an appropriate site for this facility. There was no local referendum or similar process for determining the feelings of the community (LOUD APPLAUSE); she didn't know how the mayor and other political representatives could presume to be representing the people of Clarke County. She was also disappointed that none of the local officials had addressed the issue of water supply. Where was the NBAF's water going to come from?

Jenny Culler, associated with an environmental law firm in Atlanta, cited documents showing that it's completely unknown whether all pathogens are destroyed by tissue digesters and other disposal methods, and an article stating that Athens-Clarke County intended to start selling sewage sludge, including waste from NBAF, for fertilizer.[12] DHS must conduct a more honest study of all the risks involved. Because of this new info and all the other info presented during the public comment period, it should prepare a supplemental DEIS so that the agency can properly consider all the info before a decision is made. Finally, the DEIS was a large document, and to allow the public only a 45-day period to digest and comment was absurd.

A man who used to work in economic development said one answer to Eldridge's "If not this, then what?" question would be a biotech project that wouldn't bring the unnecessary risks of NBAF. He understood that there is usually competition between Georgia and North Carolina, a competition that North Carolina usually won. In this case, it was interesting that North Carolina no longer wanted this particular biotech project, and maybe the economic development people should ask themselves why.

Dr. Rebecca Cook of UGA said, *When I hear people say they can have a perfect laboratory and never have anything escape from it, I think they should look for another job.* The DEIS didn't address who's going to pay the health care costs if anything gets out. She predicted government budget cutting a few years after the lab is built; when the lab's budget is cut, security is usually the first thing that suffers. The alarm went off; she kept talking; after ten seconds, her mike was cut off—the second such event for an opponent, none for proponents—though three went longer than the ten-second and eight-second cutoffs for DeGenero and Dr. Cook.

A former college president said if he were hired by the U.S. government or worked for the company that produced the lowest bid, he would work with the same bias. *If I worked for the university here, I sure would want to get the pay that you all offer because it's a heck of a lot better than it is at the university. The emphasis on potential economic rewards reminded me of something said by my father: Follow the greed. And if it sounds like it's too good to be true, it is. It's more than slightly obvious that many of the university and its affiliates were swayed by prestige, by egocentric rewards, by peer pressure, and I assure you, pressure only warrants McDonald's coffee at senior prices.*

Ed Tant, a columnist for the *Athens Banner-Herald* (living proof, he said, *that not everybody at ABH supported NBAF*). It's all about money, he said:

> All this talk about wanting to make the University of Georgia a great research institution. . . . How about trying to make it a great teaching institution?
>
> I don't trust anything that comes from the PR quacks of this university, I don't trust anything that comes from the President of this university that makes more money than his pal the President of the US, I don't trust a facility that's supposed to cost so many millions of dollars not to go into cost overruns for us taxpayers.

(He said something about UGA Athletics Director Vince Dooley's salary going up from $1 million to $2 million.)

> I don't trust the University of Georgia, I don't trust the local politicians that are pushing this malarkey.
>
> Since the Bush administration has in the past altered and edited scientific reports that conflict with its positions, led us into an Iraq war with dubious, ever-changing rationales and pushed no-bid contracts for its corporate friends, how can the

citizens of Athens and America be expected to trust a Homeland Security Department that is the creation of an untrustworthy administration?

A woman thanked Thrasher and Prescott, said we're here for all the people that couldn't be here. An amazing number of people at the university were afraid to come forward and speak. She believed there were a few less dangerous and more life-affirming sources of revenue out there.

Jo Ellen Childers, a staff member at the university, wanted to thank people who felt their backs against the wall and created rules and regulations requiring a DEIS and the need for public comment. *It's annoying to me when people talk again and again about transparency . . . We're coming to a point where there are no requirements for transparency . . .entering a zone where we may not know a hell of a lot about what's going on . . . The human factor in this that really is hard to calculate . . . The incidents at Texas A&M, the discovery way after the fact . . .Very much human nature to cover things up . . .We may be living in an age of public relations . . . A very prominent feature of institutions everywhere . . . Leaves those of us concerned about the public good very much in the lurch.*

A UGA undergrad noted the flyers all summer about water conservation tips for the drought; yet this facility would use 100,000 gallons of water a day. *How dare our leaders then turn around and offer accommodations for NBAF? After the last meeting, I was approached by David Lee, who was surprised by what I said earlier, that UGA students were in the dark about NBAF. How can students know anything when the information we're given is highly filtered by the university and politicians through the local newspaper? There's still a huge population of Athens that has no clue what's going on. . . . It seems to me this project is financially driven with no care or concern for the people of Athens.*

Another speaker asked: *Why are all the sites in the Southeast where there's a tradition of impoverishment and lack of vigorous community involvement? Why not Cornell University, which has an outstanding agro-science area? . . . I'm very disappointed in the county government here and politicians that I voted for. As for the University of Georgia, they've proved themselves to be a very poor citizen in the past year.*

Some of the other opponents included a building contractor who had spent twenty-five years building a homestead for his family in Oconee County; a representative of the Upper Oconee Watershed Network; and an elderly woman with pulmonary disease who lived on South Milledge Avenue near the proposed site.

Unlike the folks in Butner, the Georgia NBAF opponents had not succeeded in scaring the NBAF cheerleaders away, nor did they ever succeed in forcing pro-NBAF politicians to withdraw their support. Even for the DHS spin doctors, however, they had certainly raised serious questions about the degree of "community acceptance" for NBAF in Athens. It still remained to be seen whether community acceptance really mattered to DHS.

CHAPTER TWELVE

The Gangs Are All Here
The Fort Detrick Expansion

IN 1943, THE BIOWEAPONS GODS QUIETLY SHEPHERDED THEIR U.S. headquarters into the green pastures of Frederick County, Maryland. By the end of World War II, that idyllic rural outpost had become a germ-warfare metropolis of 250 buildings and living quarters for five thousand researchers and their families, dedicated to the pursuit of new and effective ways of killing people with germs. Fences, towers, floodlights, and machine-gun-toting guards would keep out intruders, and secrecy oaths and drop-of-the-hat lying would keep the details of what went on there from the public.

Tens of thousands, hundreds of thousands, of lab animals would be infected with anthrax, plague, tularemia, Q fever, brucella, and the details of their painful journeys recorded with proper scientific detachment before they reached their final destinations of dissecting room and incinerator.

From Detrick would come the interrogators anxious to spare Japanese counterparts embarrassment for the trifling matter of having used humans as the lab animals du jour. From Detrick would come the wistful wish for human guinea pigs of our own, and the Q-fever experiments with Seventh Day Adventists. From Detrick would come the reckless spraying of Americans everywhere with low-grade pathogens and heavy-metal tracers.

And from Detrick, the FBI eventually told us, would come the anthrax attacks that helped frighten a nation into the Patriot Act, a stupid war in Iraq, and the bloated expansion of the very biodefense complex which had just attacked its own citizens. And just as Detrick had been the main "distributor" (to other facilities) of the Ames strain of anthrax, so too did Detrick supply the seed stock for other aspects of the biodefense abscess as well. Personnel, for one thing. C. J. Peters would head the new BSL-4 lab in Galveston. Robert Shope, who participated in the Q-fever experiments on Seventh Day Adventists, would also go to Galveston, one year after a researcher in the Yale Arbovirus unit became infected with a Brazilian hemorrhagic fever virus and exposed eighty members of the public. David Huxsoll would head Plum Island.[1] The Jaaxes would go to Kansas State and promote NBAF; David Franz would spend some time at K-State as well before heading to the Midwest Research Institute. He would also serve on multiple federal panels defending labs from oversight. Peter Jahrling would stay in Frederick to become the head of the NIH's new BSL-4 lab on Detrick's new biodefense campus. Tom Geisbert would become associate director of the embattled Boston BSL-4 facility

Each alumnus of Detrick would take with him or her a "culture," a way of thinking about the dangers of deadly pathogen research, and the "standard operating protocols" meant to diminish those dangers.

In the stated opinion of some Detrick alumni, that culture was a "culture of safety." They'd had accidents at Detrick, but they hadn't killed anyone lately, not even themselves. (At least not until Ivins, or someone, turned their anthrax loose on congress and the U.S. Postal Service.) A few Detrick researchers had died from lab accidents in the old offensive bioweapons days, but official records showed only a few LAI's (laboratory-acquired infections) since then, and all the researchers had recovered. Still, it was perhaps a bit much for C. J. Peters, a former deputy commander of USAMRIID, to declare that USAMRIID had an excellent safety record. And it was perhaps way too much for him to attribute that "excellent" safety record "partially to a culture of safety carried over from USAMRIID's previous work on offensive biological weapons before 1969."[2]

And it was ironic since the comments appeared in a 2006 newspaper series on an April 2002 "breach of containment" at USAMRIID.

After the Ivins revelations, the Defense Department launched a "four month review of lab security, biosurety and safety policies" at the Armed Forces' biodefense labs. At the end of the review, in December 2008, the Department announced that it would be tightening up, that it had to "change the culture of the scientists and the workers in the labs to be more vigilant after each other."[3]

The Army's 361-page investigative report on the 2002 release of anthrax outside containment areas at Detrick had already admitted that "USAMRIID had a comprehensive system of procedures in place that should prevent exposures if followed, but adherence to and enforcement of those procedures was lax."[4] The "culture of safety," it seemed, was actually a "culture of complacency," as even USAMRIID's own safety officer seemed to acknowledge: "People can get complacent. Familiarity can breed complacency."[5]

That familiarity did often breed complacency was suggested by the 161 mishap reports filed by USAMRIID employees between April 1, 2002 and Dec. 1, 2005.[6]

Even C. J. Peters was troubled by an incident in which a worker handling anthrax spores failed to turn on the hood blower of the biological safety cabinet, making it possible for spores to escape the cabinet. The anthrax strain in question was non-pathogenic, but what, Peters said, "if the worker did it again with a highly pathogenic anthrax strain"?

And the commanding general at Fort Detrick didn't exactly provide reassurance when he tried to explain why mishaps happened from time to time, why the whole safety thing was so difficult. Enforcing safety among laboratory personnel, he told the *Frederick News-Post*, was "similar to police officers attempting to enforce seatbelt and speeding laws."[7] Well, we all know about the culture of safety out on the highway.

It seemed Detrick was in a never-ending cycle of finding the flashing blue lights in its rearview mirror, then pulling itself over and writing itself a ticket. And mumbling to itself all the while like Gollum in *The Lord of the Rings*. Bad Detrick, no, no, must not put nice hobbitses at risk with nasty lab practices.

In 2000, after a researcher contracted glanders (a disease usually fatal to humans if left untreated), and continued to work for six weeks while failing to report his symptoms, Detrick launched a review of safety procedures.[8] In December 2001, after Detrick had been identified as a possible source of the anthrax mailed to government and media outlets, inventory control was "re-emphasized."[9] In April 2002, after reports emerged of "shoddy record-keeping and unprofessional conduct" during the 1990s, and after an accidental release of anthrax spores within the Detrick facilities, Detrick announced a new biosurety program. 2005 brought a review of the facility's environmental policies and practices following notices of environmental violations.[10] And following the FBI's Ivins announcement, the Army created a twelve-person task force to once again review the "biosurety" program.[11] Bad Detrick.

Keeping track of its germs and what had happened to them proved to be an especially vexing problem for Detrick. You'd think that would discourage Detrick from confident assertions about the whereabouts of those germs, but no.

In 1991, test wells detected carcinogenic industrial solvents in the groundwater beneath a Detrick chemical dump. The chemicals were later found in the well water of nearby homes. Detrick red-facedly connected the homes to city water.[12] (You'd think incidents like this would discourage biodefense promoters from declaring that biodefense facilities had *never* "contaminated" communities, but no.) Nine years later, the Army and environmental regulators announced a $4.9 million plan for excavating and burning nearly 815 tons of material from the dump.[13]

One year later, the Army said preliminary excavation indicated cleaning up the dump would take two years instead of the nine months they had planned for, and $10 million to $15 million.[14] (It would end up taking three years and $26 million.)[15] Officials assured locals that no germ warfare agents had been dumped in the waste pits "since all biological research waste was sterilized and incinerated."[16]

In January 2002, workers encountered vials of bacteria in the pit, reportedly harmless. Eighteen vials found in April were not so harmless, with preliminary tests indicating meningitis, encephalitis, and pneumonia bacteria. Gerald Toomey, the civilian co-chair of a community advisory board established by the Army, said he was "highly concerned" about the finding, "the second time in which workers have unearthed live bacteria despite Army assurances that no such material had been buried there."[17]

More "hazardous surprises" were in the works. In May 2003, workers found more than 100 glass vials containing live bacteria in the dump, including "nonvirulent anthrax." "The documentation for where this came from doesn't exist," Detrick's safety director stammered.[18]

About the same time the first vials of bacteria were dug up, reports from former Detrick researchers of lax security policies and missing samples began appearing in the media. (Some of the researchers, but not all, were involved in litigation against Detrick.) The whistleblowers described situations in which "you could walk out with anything"; in which there was a "steady stream of researchers from China, the former Soviet bloc and other nations"; in which a visiting scientist without U.S. citizenship or a security clearance was "promoted to a senior position overseeing the development of liquid aerosols used in exposing lab animals to anthrax"; and in which a lot of unauthorized projects and off-the-books weekend assignments were being conducted.[19]

"7-Eleven keeps better inventory than they did," one scientist said. "It wasn't just a matter of security lapses," another researcher said—"there was no security."[20]

Then the *Hartford Courant* obtained documents from a 1992 Army inquiry about missing lab samples of anthrax, Ebola, and other lab specimens—27 sets of missing specimens in all.[21] The *Courant's* story prompted interest by the FBI and what was then the General Accounting Office.[22] Initially a Detrick spokesperson said one of the 27 sets had been located, along with portions of others, but could give no details because of incomplete records.[23] Eleven days later, however, another spokesperson said the Army had rechecked its records and found that all but three of the missing samples had been located, and that those samples were harmless.[24]

So it all seemed to be a record-keeping thing, like the live bacteria in the Detrick dump for which no documentation existed. Sometimes it was germs for which no records existed; sometimes it was records that were hard to find. There was more record-keeping stuff to come.

In February 2009, the *Science Insider* blog reported that USAMRIID would be suspending all select agent research until a complete inventory was taken of select agents and toxins at the facility.[25] (In fact, however, "ongoing" animal research was continued.[26]) A spot check in January had found 20 samples of Venezuelan equine encephalitis in a box of vials instead of the 16 listed in the institute's database. The institute's commander said there was a strong possibility of there being additional vials of BSAT not captured in the database.[27]

It seemed USAMRIID itself was satisfied with being able to find all the items listed in the database, whereas the Army and Department of Defense also wanted everything stored at USAMRIID to be logged into the database.[28] The institute attributed the problems to "unused, older samples of research materials that were in storage before the institute's records were computerized in 2005." Prior to that time, inventory had been kept on paper.[29]

Caree Vander Linden, the current public relations mouthpiece for the institute, said, "It's a record-keeping thing."[30]

Other Army officials insisted there were "no missing vials of lethal substances and no danger to the public."[31] But what sort of mental telepathy with uninventoried pathogens enabled them to say that? How does one know anything about a vial not recorded in inventory: what it is, whether it's missing, or how dangerous it is? As Richard Ebright pointed out, "If there are samples that are undeclared, unaccounted for, those samples can be stolen or diverted with no means of detection." And while such discrepancies happen all the time in normal lab environments, they're a serious concern in facilities handling bioweapons agents.[32]

The "record-keeping" problems are symptomatic of the whole biodefense enterprise, a giant sprawling behemoth that is not even properly inventoried, much less controlled. No one knows how many high-containment germ labs are out there, or why they are out there, or what germs they have on hand, or what is being done with those germs. Germs gone wild indeed.

The really scary news, according to Ebright, is that security measures at Detrick, inadequate as they are, are "substantially more stringent" than the sprawling spiderweb of university and other civilian labs which now handle lethal agents.[33] As Ebright pointed out in an E-mail to *Wired* magazine's Danger Room blog, there are more than 400 U.S. institutions and 15,000 persons authorized to possess bioweapons agents; security measures at the overwhelming majority of these institutions are inadequate; few provide comprehensive video monitoring of work areas; few have a rule requiring that at least two persons be present when agents are handled; few perform psychological screening and psychological monitoring of personnel.[34]

Several USAMRIID researchers reportedly were grumbling about the research shutdown. And in an another instance of the biodefense two-step, former USAMRIID researcher Tom Geisbert, now associate director at the controversial Boston facility, complained that it "was extremely difficult to completely account for replicating agents because, by definition, they replicate. . . . You can make a large amount from a small amount."[35] Yet one of the ways biodefense promoters at Boston and elsewhere try to reassure worried neighbors is by referring to the "small amounts" of agents the new facilities will contain.

The Army itself seemed to be doing the biodefense two-step with its assurances that no lethal pathogens were missing, and with the boast of spokesman Michael Brady (in February 2009) that "We have made it incredibly more difficult for another Bruce Ivins to happen." An April 22 story in the *Frederick News-Post*, however, said the Army's own Criminal Investigation Division was investigating "the possibility of missing virus samples" from USAMRIID.[36]

About four months after starting the inventory, USAMRIID announced it had finished, and had identified some 9200 specimens not yet recorded in the existing database of 66,000 items. This meant that something like one-eighth of the germ samples at the country's chief biodefense facility had been sitting about anonymously for years, possibly decades, with no one officially aware of their existence. The unrecorded vials included Ebola, anthrax, and botulinum toxin, which made Deputy Commander Mark Korpeter's likening the process to cleaning out an attic just a little too cute.[37]

Can we hope Detrick, after sprucing up the attic, isn't planning a surplus-germ yard sale?

After all, according to the FBI, it was vagrant Detrick germs that prompted the whole bioterror scare in the first place. And wasn't it all so good for biodefense in general, and for Detrick in particular? As a June 2002 *Science* article explained:

> Now, everything is different. In the post-9/11 world, the expertise built up at the U.S. Army Medical Research Institute of Infectious Diseases (USAMRIID), the main research institute at Fort Detrick, has proven invaluable—and suddenly everyone is grateful. The lab is working closely with the Federal Bureau of Investigation (FBI) to help unmask whoever sent the anthrax letters last fall; the so-called Brokaw, Daschle, and Leahy letters were sent here to be investigated; and the lab serves as the repository for anthrax strains subpoenaed from other labs. Tens of thousands popped Cipro last fall because USAMRIID's Arthur Friedlander showed it could protect monkeys from anthrax.[38]

"This has pulled us out of intellectual backwaters and into the mainstream," crowed senior Detrick researcher Peter Jahrling. "I'm glad we no longer have to defend it [our work] to people who don't believe there's a threat." ("This"—the Detrick-propagated anthrax attacks—would pull Jahrling into the directorship of the new NIH BSL-4 lab at Detrick.)

Of course, we now know that one of the people working closely with the FBI to "unmask the perpetrator" was the alleged perpetrator. But back in December 2001, Detrick's commander, Major General John Parker, had scorned suggestions that Detrick might be involved: "We don't have that capability here nor do we have the scientists who know how to do that."[39] And in June 2002, in response to media reports the FBI was investigating the possibility of Detrick's involvement, post spokesperson Charles Dasey had said the theory the anthrax came from Detrick was "just more speculation."[40]

There might not have been documentation at Detrick for the 100-plus vials of buried germs, or for the 9200-plus samples not recorded in inventory, but the FBI said there was documentation, of the DNA sort, which allowed them to trace the anthrax attacks to a single flask at Detrick. Assuming at least this part of the FBI narrative is correct, Detrick would turn out to be the world's first one-stop anthrax facility, supplying the anthrax and knowledge for the attacks, many of the principal investigators of said attacks, and the research that showed Cipro could protect Detrick monkeys from Detrick anthrax. (And could therefore help the human victims of Detrick anthrax as well.)

Detrick was not just "On Biowarfare's Frontline"—Detrick was on both sides of the frontline, bravely defending the nation against the savage onslaughts of Detrick anthrax.

But Detrick could not adequately protect the nation against the dangers of Detrick bioterrorism with existing facilities. (It's a tough job, protecting oneself against oneself.) So the spores were hardly lodged in the lungs of Detrick victims when plans for a massive new biodefense "campus" began to take place. In April 2002, the Associated Press reported that the Army and the National Institutes of Health would build "sophisticated new laboratories" at Detrick and collaborate on vaccines and other bioterrorism defenses. "This is a new need for us, to be moving into biodefense research," an NIH spokesperson said.[41]

By fall, the partnership plans had fleshed themselves out a bit. NIAID, or the National Institute of Allergy and Infectious Diseases, a division of NIH, would begin constructing a $105 million BSL-4 facility in 2004, the agency said. USAMRIID would then seek $1 billion to build its own new BSL-4 facilities.[42]

The partnership idea would evolve into a National Interagency Biodefense Campus, incorporating new high-containment facilities for USAMRIID, the National Institutes of Health, Department of Homeland Security, and Department of Agriculture.[43] The concept was aimed "at coordinating research and reducing duplication"[44] (a goal which obviously applied only within the Detrick microcosm, since universities were now bursting at the seams with their own bioweapons agents). In May 2005, the CDC announced that it too would join in the Detrick biodefense orgy even as it planned its own new BSL-4 facilities in Atlanta.[45]

One could see that the campus concept would take some of the load off USAMRIID itself. USAMRIID could now confine itself to producing and disseminating the germs; forensics at Homeland Security could identify the germs as Detrick's before they ever went out; and NIH, as it treated the victims, could run over to USAMRIID and borrow a little Cipro. Then roosters supplied by the Department of Agriculture could crow daily about the success of the whole biodefense enterprise.

The NIH facility was first out of the blocks and the first to declare, by way of a February 2004 Environmental Impact Statement, that it posed "little threat to lab workers and the community." It would incorporate design and operational safeguards, after all.[46] It was only later, after projects at Lawrence Livermore and Boston University had their hands slapped by the courts, that environmental impact statements for biodefense projects began to state that the facilities were, if not overly so, at least a little dangerous.

The NIH might think the facility failsafe, but Paul Gordon, a former Frederick mayor who had firsthand experience with Detrick's version of the biodefense two-step, felt facilities like this should be built in rural areas where fewer people live.[47] (Others, like Edward Hammond of the Sunshine Project and Michael Carroll of *Lab 257*, believe locating the facilities in isolated areas encourages researchers to be even more reckless. The ultimate answer, as Hammond and others have pointed out, is not to build so many in the first place.)

DHS followed with its own version of "Ain't No Danger in Our Germs" a few months later. It said its labs "will result in no significant adverse environmental, socio-economic or human health impacts." With respect to questions about congressional oversight, the agency said it would develop formal procedures for "informing" Congress about the facility's work. (In other words, DHS would oversee the facility itself and tell Congress what it wanted Congress to know.)

DHS called its facility the National Biodefense Analysis and Countermeasures Center. It said the facility would occupy about 160,000 square feet, include BSL-2, BSL-3, and BSL-4 labs, employ about 120 people, and cost about $128 million. And according to its Web site, the facility would—abracadabra—"employ state-of-the-art biosafety procedures and equipment to prevent biological material from escaping into the environment." "Rigorous safeguards"—abracadabra —"protect those who work in the laboratory and the surrounding community."[48]

Not only would the safeguards be state-of-the-art, but they would also be "proven safeguards, applying time-tested rigorous design safeguards, special safety equipment, and exacting operating procedures to govern all research,"[49] abracadabra. A skeptic might wonder how a single facility can be both the "latest and greatest" and "old reliable," but the secret is in the abracadabra. So long as one adds the magical formula, the old assurances and the new ones work equally well.

After all, the Department of Homeland Security didn't even exist before 2002. One perhaps has to grant a new agency some latitude to act like a grade-school kid with a new chemistry set, to be naïvely enthusiastic about its inherited "safeguards," as when the NBACC Web site declared: "These standards, equipment, and procedures have been demonstrated to provide the highest level of safety in existing laboratories elsewhere at Fort Detrick and throughout the United States."[50]

What the time-tested USAMRIID lab at Detrick has admitted, however, is the difficulty of enforcing the biolab equivalents of the speeding and seatbelt laws. It's not what's written in the shiny new safety manual that counts, it's what people do in practice.

The NBACC will have two parts: the Biological Threat Characterization Center (BTCC) and the National Bioforensic Analysis Center (NBFAC). The NBFAC will be "the lead Federal facility for conducting and facilitating the technical forensic analysis and interpretation of materials recovered following a biological attack."[51] Once the NBFAC has identified the perpetrators, it will then, presumably, consider going partners with them. At least, that's the way things worked out vis-a-vis the 2001 anthrax letters. Undoubtedly certain savings in fuel economy occur when the perpetrator is next door. And if it is important to research bioweapons agents at over 400 separate geographic locations, perhaps it might be advisable to build a branch bioforensic analysis center at each location.

The BTCC "is charged with defining the characteristics of biothreat agents and conducting rigorous biodefense risk assessments in order to guide national biodefense research, development, and acquisition efforts, and to provide scientific support to the intelligence community."[52]

One could wish that a "rigorous biodefense risk assessment" had been conducted prior to the post-2001 boom that brought forth the NBACC and the Detrick biodefense campus. One can hardly expect a facility controlled by an agency which owes its very existence to the war on terror—and a facility which owes its own existence to the war on bioterror—to draw any conclusion except that the risks of bioterror for the nation are huge.

What is hidden in the innocent-sounding description of the BTCC's work is the fact that it will "conduct threat assessment research, a controversial type of biological research in which new types of biological weapons are developed by researchers in order to determine their potential viability and how one might defend against them."[53] This is the sort of research that blurs the boundaries between offensive and defensive research. It is as if a supposed manufacturer of bulletproof vests kept inventing deadlier guns and deadlier ammunition that would defeat existing vests, and then tried to develop new vests to defend against the new weapons, and then tried to develop new weapons to defeat the new vests, and so on. Is the company in the vest business, or the gun-and-ammunition business? How does anyone not employed by the company really know?

Even assuming that the company's motives are truly pure, that it really wants to develop vests to defend the helpless citizenry against the onslaughts of gun-toting thugs (though we may note that people who wear bulletproof vests usually carry guns as well), do we really want that company out there developing deadlier means of killing, just because some hypothetical thug might try to do so? After all, isn't there the danger of the plans for the new weapon getting out to someone we don't want to have it? (Including one of our own rogue employees?) And do we really want our government manufacturing deadlier germs and deadlier ways of delivering them, considering the possibility of another incident like the anthrax letters, or of infiltration by one of the foreign terrorists so anxious to launch a biological Armageddon, or the possibility of a serious accident despite all those state-of-the-art things? (State-of-the-art, by the way, simply means the best we've got, and the best we've got is not usually what we get, and even if we did get it, it's still not fail-safe.)

A 2006 *Washington Post* article, "The Secretive Fight Against Bioterror," indicated that DHS planned to occupy the entire facility as a "Sensitive Compartmented Information Facility," or SCIF. As the *Post* explained:

> In common practice, a SCIF . . . is a secure room where highly sensitive information is stored and discussed. Access to SCIFs is severely limited, and all of the activity and conversation inside is presumed to be restricted from public disclosure. There are SCIFs in the U.S. Capitol, where members of Congress are briefed on military secrets. In U.S. nuclear labs, computers that store weapons data are housed inside SCIFs.[54]

Making an entire facility a SCIF was apparently unprecedented, even for nuclear labs.[55]

The article also carried a frank but rare acknowledgement from Penrose "Parney" Albright, former Homeland Security assistant secretary for science and technology, that "De facto, we are going to make biowarfare pathogens at NBACC in order to study them." A 2004 computer slide show posted on a DHS Web site suggested that some of those pathogens would be versions genetically engineered to make them more virulent, resistant to antibiotics, etc.; that the facility would engage in "aerosol challenges" of nonhuman primates; and that it would engage in "red team" exercises that simulated attacks by hostile groups.[56]

DHS abruptly removed the slide show after critics discovered it.[57] The critics were concerned not only about the proposed activities themselves but also about their secrecy. Such work, they said, is arguably a violation of the 1972 Biological Weapons Convention, and will make it harder for the U.S. to object to similar research by hostile countries.

Albright's response: "How can I go to the people of this country and say, 'I can't do this important research because some arms-control advocate told me I can't?'"[58] The dismissive attitude toward "arms control" considerations and toward questioning in general reminds one of Jay Cohen's arrogant appearance before the House Energy and Commerce Committee. DHS, it would seem, prefers to interpret the law, science, scriptures, and the nation's security needs for itself, without the hampering shackles of democratic oversight. In the words of the *Post*, "Homeland Security spokesmen insist that NBACC's work will be carefully monitored, but on the department's terms." Bernard Courtney, then the Center's scientific director, said: "We have our own processes to scrutinize our research, and it includes compliance to the bioweapons convention guidelines as well as [DHS-selected] scientific oversight."[59] In other words, the foxes will guard the henhouse themselves, and would prefer to do so without the chickens' loud squawking noises.

The *Post* article indicated that the NBACC had already begun research in Detrick lab space borrowed from the Army and in unidentified "outsourced" lab space.

A bit over two years later, in October 2008, DHS was dedicating the new facility, which, according to an admiring *Associated Press* article, resembled "a battleship being readied for war against biological weapons."[60] It is not clear whether the battleship resemblance arose from the fact that DHS Undersecretary Cohen was a former admiral, or from DHS actually believing the boasts of biodefense promoters that new state-of-the-art facilities are like "submarines in bank vaults."

The article rejoiced that the new facility would give the FBI a forensic capability it lacked after the 2001 anthrax mailings. And how clever to have located it at the very site of the mailings, and next door to a facility with over 9000 uninventoried germ samples.

Cohen himself exulted in the dedication ceremony that this was "an incredible demonstration of what we can do when we're challenged, when we're threatened." Hopefully the incredible demonstration will put the fear of God into those would-be bioterrorists over at USAMRIID.

Because there will soon be quite a few more of them: USAMRIID's own contribution to the new biodefense campus involves nearly tripling the existing space and adding 1300 people to the existing staff of 750.[61] The Army held the first scoping hearing on its plans in February 2006, expecting a routine public relations event like that enjoyed earlier by DHS and NIAID. Instead, thirty or so people—mostly opponents—showed up.[62]

At first, opponents were not optimistic about stopping the USAMRIID construction. But after the Army published its Draft Environmental Impact Statement in August 2006, stating that terrorist attacks were "not reasonably foreseeable," and that an evaluation of their impacts was not required by the National Environmental Policy Act,[63] the Ninth Circuit Court of Appeals ruled that the Department of Energy's environmental assessment for a proposed biodefense facility at Lawrence Livermore was inadequate because it failed to consider the possibility of a terrorist attack. Opponents of the Detrick expansion began to threaten their own lawsuit.

The Army responded by changing the language in the final EIS (now stating that terrorist attacks "may be credible, remotely possible threats") and declaring, for the benefit of the media, that it had properly studied the potential environmental impacts of terrorist attacks, but couldn't disclose details without tipping off "potential terrorists."[64]

In August 2007, 75 people or so attended a demonstration organized by the Frederick Progressive Action Coalition. One of the participants was Dena Briscoe, the president of Brentwood Exposed, an advocacy group for workers in Brentwood's post office during the 2001 attacks. Two of Brentwood's employees died after the facility processed an anthrax-filled letter sent to Capitol Hill, and Briscoe was one of several others hospitalized. She said she believed—and the FBI would later say she had believed correctly—that the spores were created at Detrick.[65]

In September 2007, opponents gained a sympathetic ear in County Commissioner David Gray, who issued a statement saying that federal officials had skirted the requirements of NEPA by not seeking or considering alternate sites for the labs.[66] Gray then asked Detrick officials to attend a public forum to answer questions; Detrick at first agreed, then backed out, offering a private meeting with county commissioners instead.[67]

Gray scheduled his public forum for November 19 anyway. The *Baltimore Sun* ran an article on that date, "Biodefense lab causing qualms." Base officials refused to be interviewed for the article; a base spokesperson responded instead in an E-mail: "Because we already provided numerous opportunities for the public to comment and provide input during the EIS process, we see no need to participate in another public meeting."[68] In other words, the public had had its NEPA-mandated chance to yap; now, the public be damned.

More than 100 people, mostly opponents but some supporters of the expansion, attended the forum. Among the opponents was Paul Gordon, former Frederick mayor, who cited Detrick's slow response to carcinogen-contaminated wells as an example of the facility's reluctance to inform residents about problems: "Detrick—it's need to know . . . until it's so bad they have to tell you what it is."[69]

Some of the opponents called for an independent court review of the EIS, i.e., a lawsuit. commissioner Kai Hagen supported the notion,[70] as did the *Frederick News-Post* in a November 12 editorial.[71]

The divided and uncertain commissioners ended up settling on a compromise: they wouldn't sue, but they would ask U.S. Senator Barbara Mikulski to request a review by the National Research Council, a branch of the National Academy of Sciences—a move obviously suggested by the earlier NRC review of the Boston NIH lab.[72] On April 10, 2008, the commissioners voted 4-1 to request such a review. Mikulski immediately issued a statement saying she would forward the request to the National Academy.[73]

A leader of a citizen "support group," the Fort Detrick Alliance, complained that the review would add needless costs and delays, and that the opposition was instigated by people who were "antimilitary."[74]

The next few months saw haggling over the scope of the study and its funding. In June 16, the U.S. Army Medical Command indicated that it wouldn't support funding such a study, stating that public concerns were adequately addressed by the EIS.[75] The Command's letter suggested that it might not cooperate with a study even if the funding were provided by someone else.

Meanwhile, the commissioners and the National Academy had different conceptions of what a study ought to focus on. In August, the NAS proposed reviewing the adequacy and soundness of risk assessments, scenarios, and mitigation plans for all three of the campus facilities under construction. The commissioners voted to send the NAS instead a proposal focused specifically on the risk assessments for the Army facility, but adding a review of the lab's operational procedures, including its handling of hazardous materials.[76]

In September, Senator Mikulski announced she had included an instruction in the final spending bill requiring the Secretary of Defense to contract with the National Academy of Sciences "for a review of the public health and safety risk assessments associated with the expansion of biodefense labs" at USAMRIID.[77]

While these discussions were under way, the NIH made the news with a proposal to transport workers accidentally exposed to infectious agents at its new Detrick labs to the NIH's Bethesda facility 35 miles away. On July 8, the NIH announced its intent to prepare an EIS for the proposal, and its plan to hold scoping meetings to seek public comment both that night and two nights later. The NIH did not respond to calls and E-mails from the *Frederick News-Post* seeking an explanation of why it was pursuing off-site treatment for potentially infected lab workers.[78]

USAMRIID had previously quarantined workers exposed to infectious agents in its BSL-4 patient care suite, the "slammer." But USAMRIID spokesperson Caree Vander Linden said there had been no discussions about other agencies using that suite. USAMRIID didn't plan to include a BSL-4 patient care suite in its new building, and hadn't yet decided how it would handle exposed workers in the future.[79]

The National Academy of Sciences launched the independent assessment project in May 2009, and held public hearings in September and November of 2009.[80] The Army wasn't waiting around to hear what the NAS experts had to say, however; it broke ground for the new facility in August 2009. In a peculiar example of circuitous reasoning, USAMRIID commander Colonel John Skvorak said, "It does not look like the review will hold up construction of the new building." Obviously, it wouldn't and hadn't, since the Army had rushed to start construction before the NAS panel could pose its first questions. Not to worry, though: Colonel Skvorak was happy to explain why the NAS "independent" assessment was an irrelevant exercise in the first place: "I'm confident the folks who did the (Army's) study did a thorough job."[81]

Most of this NAS committee, perhaps even more so than the experts who reviewed the Boston risk assessments, were card-carrying members of the biodefense complex. Donald Henderson had founded the Johns Hopkins think tank which morphed into University of Pittsburgh Medical Center's Center for Biosecurity. The UPMC Center consistently serves as a prime biodefense propagandist (see Chapters 8 and 15 of this book). Henderson had also played a significant official role in the Bush White House's war on bioterror. Jonathan Richmond, a former safety officer for the CDC, now ran a biosafety consulting business, and had evangelized on behalf of the North Carolina pro-NBAF forces. Henry Mathews, another biosafety consultant, was also a former CDC safety officer, involved in designing the CDC's new high-containment laboratories, with their recent total power loss. Barbara Johnson, another consulting entrepreneur, had worked for the U.S. government for 15 years, "managed the design, construction, and commissioning of a BSL-3 Aerosol Pathogen Test Facility, and launched the U.S. Government's first chemical and biological counterterrorism training facility." The committee's chair, Charles Haas, was "co-director of the Center for Advancing Microbial Risk Assessment that is jointly funded by the U.S. Department of Homeland Security and the U.S. Environmental Protection Agency."[82] Nancy Connell directed the Center for Biodefense at the University of Medicine and Dentistry of New Jersey, the same institution responsible for two different recent incidents of missing plague-infected mice. A chapter in Leonard Cole's book *The Anthrax Letters* describes Connell's transformation from military skeptic into grant-induced bioterror enthusiast.[83]

On March 4, 2010, the NRC panel released its report, and accepted the irrelevancy the Army had already consigned it to. The panel had found multiple problems with the EIS:

> The analyses of the risks and the mitigation measures to address them were not comprehensive and there was insufficient documentation for an independent assessment of the risks to the community posed by biologic agents. The problem was compounded by the fact that the MCE scenarios were not reasonably foreseeable accidents.
>
> The epidemiologic characteristics, including transmission pathways, natural reservoirs, geographic distributions, and clinical outcomes of the pathogens, were not systematically documented.
>
> There was incomplete consideration of some of the possible routes through which the general public might be exposed to pathogens.[84]

For all its expert knowledge, the panel was unable to independently verify the EIS's risk analysis. Consequently, "The information and documentation provided in the EIS are insufficient for an independent assessment of the risks to the community posed by biologic agents."[85] But when the committee did attempt "to independently verify the calculations of doses of infectious agents delivered off site following puff releases," it found "significantly higher concentrations than described in the report."[86]

Despite the EIS's problems, however, "the committee judged that it would not be useful to propose specific revisions to the EIS or supplementary analyses given that the Record of Decision to construct the new USAMRIID was issued and construction has begun on the project."[87] In other words, the Army was going to build the thing—in fact, the Army was *already* building the thing—whatever the EIS said. The Army had already indicated that it regarded the EIS as an irrelevant hoop it had already jumped through, and it was proceeding accordingly. It needed to get on with the business of defending the U.S. from U.S. biodefense.

And it was apparently time for the National Research Council, which had surprisingly embarrassed the NIH at Boston, to get back to its time-honored function of helping the government with its military efforts, something it had been happily doing since the World War II bioweapons days. And so it was abracadabra time again. Even as it declared that the EIS was "insufficient for an independent assessment of the risks to the community," the NRC panel pulled an extra rabbit out of its hat and said it didn't matter, there were no risks, because everything those Army boys and girls did was state-of-the-art: "USAMRIID's current procedures and regulations for its biocontainment facilities meet or exceed the standards of NIH and CDC."[88] The Army had said as much repeatedly over the years; but it at least had admitted that, regretfully, it was sometimes hard to get people to observe those "current procedures and regulations."

As far as all those biolab speeding tickets went, the NRC said, "Measures have been taken to improve safety at USAMRIID when problems have been identified."[89] In other words, Detrick was very good, when things went wrong, at saying "Bad Detrick." It had said "Bad Detrick" so many times and so convincingly in the past, that "the committee has a high degree of confidence that the new USAM-RIID facility will have the appropriate and effective physical security, biosurety program, and biosafety operating practices and procedures in place to protect its workers and the public."[90]

When the panel, seemingly quoting from a Detrick press release, declared that "USAMRIID currently has a robust biosafety program that consistently updates training, SOPs, and other written policies and manuals," the documentation it provided for this statement was insufficient for independent verification. The fact that USAMRIID "has constituted and registered an Institutional Biosafety Committee (IBC)," for instance, seems a bit underwhelming.[91] The claim that recombinant DNA protocols "are closely monitored by the IBC and NIH,"[92] on the other hand, is laughable in light of the Sunshine Project's extensive survey of the failures of the current IBC system.

Unlike the NRC panel, we can't take much comfort from the fact that the CDC will inspect the new USAMRIID facility before it begins work, and that the facility "will be subject to announced and un-announced inspections by CDC, USDA, or both."[93] The CDC's inspections, after all, missed the biosafety train wreck at Texas A&M, and the CDC had trouble with the backup power design for its own new high-containment facilities (unless one thinks state-of-the-art duct tape an appropriate biosafety "enhancement"). The fact that the CDC's inspections "include detailed scrutiny of the receipt, storage, use, and transfer of BSAT"[94] should be positively discouraging, fol-lowing USAMRIID's discovery of a 9000-vial inventory boo-boo in the spring of 2009. The CDC's "detailed scrutiny" apparently missed the fact that, as of early 2009, one-eighth of USAMRIID's existing stock of pathogens wasn't even recorded in inventory.

The panel's perusal of the "list of laboratory-acquired infections that have occurred at USAMRIID since the 1940s" and the "dimin-ishing numbers over the decades since research first began," followed by the confident statement that "These reduced numbers clearly reflect the positive changes in biosafety practices, procedures, use of equipment and better engineering controls,"[95] also seems the sort of propaganda more suited to an Army press release than to an "inde-pendent assessment." In fact, a close look at the figures suggests that the crucial "safety improvement" was termination of the offensive bioweapons program in 1969. Prior to that time, the Army had been manufacturing large quantities of pathogens, merrily infecting "human volunteers" with bioweapons agents, and clandestinely spraying American cities. No doubt stopping all that represented a "positive change in biosafety practices," but maybe the NRC shouldn't take its standards for safe driving from a demolition derby. Especially since the Army, DHS, and NIH, acting like drunk frat boys at the Detrick "biodefense campus," were getting ready to throw the biggest germ-lab block party in history.

The panel was full of praise for USAMRIID's "biosurety" program, declaring that "[t]he Army has taken the lead in establishing a robust biosurety program," and claiming that the Biological Personnel Reliability Program initiated in 2003 "is considered to be a model by other institutions."[96] The panel did not indicate what sort of survey of "other institutions" it conducted to determine the peerless reputation of USAMRIID's "personnel reliability" program. The program had not seemed so superlative after the FBI's statements about Bruce Ivins's mental state and his role in the 2001 anthrax attacks. The panel glossed this over by declaring that "no program can stop all threats of theft or misuse of BSAT. The solution to preventing such incidents is not in stopping all work with BSAT."[97] Certainly, no "biosurety program" can eliminate all threats of BSAT theft or misuse, just as no biodefense program can eliminate all threats of bioterrorism (and certainly not those which it itself poses). *Scaling back* the current zombie-like expansion of biodefense research would at least, however, reduce some of the increased risks that that expansion has brought about.

Having pronounced the appropriate abracadabras, and declared that USAMRIID's procedures and regulations were state-of-the-art, the NRC panel then got down to the real problem. Like Boston University before it, USAMRIID wasn't playing the public relations game very well: "The community has not been made aware of the details of the many safeguards already in place at USAMRIID, the requirements governing the operation of biocontainment facilities, and the Army's ongoing commitment to improving safety and security."[98] And then the committee proceeded to offer USAMRIID a short primer on the arts of germ-lab public relations, encouraging it to provide fact sheets, hold an open house (and presumably, provide certified pathogen-free tea and cookies), create a community advisory board, and "consider strategies that have been used by other containment laboratories" better at the public relations game. Ironically, the committee also suggested that USAMRIID might consider "promptly disclosing laboratory incidents to the public"[99]—something very likely to undo all the hard work of the tea-and-cookie socials—and something at odds with the whole bioweapons/biodefense enterprise since World War II. Keep them posted on our accidents? Who are you kidding?

The best comment on the NRC panel's abracadabras came not from critics of the Detrick expansion, but from reality itself. First was yet another of those "Bad Detrick" incidents. In November 2009, a USAMRIID researcher had been infected by the tularemia she was working with. She had been exposed sometime between November 13 and November 17; she'd called in sick on November 23, and then gone to the Barquist Army Health Care Clinic complaining of flu-like symptoms. She'd not been diagnosed with tularemia until sometime in December.[100]

The NRC panel had been aware of the incident as it prepared its assessment, but had essentially dismissed it with a no-procedures-can-eliminate-all-possibility-of-accidents shrug, something at odds with the panel's mumbo-jumbo elsewhere about state-of-the-art abracadabras. A month after the NRC panel issued its optimistic evaluation of USAMRIID's safety procedures, USAMRIID released its own "Bad Detrick" report on the tularemia incident.

The incident involved several significant ironies. First was the complete lack of suspicion for over a week that the researcher had been infected by the agent she was researching. Just as in the Texas A&M and Boston University incidents, expertise in a pathogen did not seem to include awareness of its clinical symptoms. Or maybe researchers actually believe all the biolab propaganda that says they won't get infected. But as the *Frederick News-Post* noted in its editorial on the incident, "Maybe it's us, but shouldn't those [flulike] symptoms, combined with her job working with infectious bacteria, have raised a red flag for someone—or triggered an automatic investigation?"[101]

Second, "the investigation revealed no clear indication of exactly when or how this individual was exposed." The report could only identify "several lapses in proper laboratory techniques that, taken together, may have increased the individual's risk of exposure." The "lapses" included improper disposal of waste, inadequate respiratory protection given that the individual had not been vaccinated against tularemia, and "improper illness reporting, which appeared to have been caused by confusion regarding the proper procedures for illness reporting at USAMRIID."[102]

The NRC panel's confidence that USAMRIID has the appropriate biosafety practices and procedures "in place"—i.e., written in a manual—may be irrelevant if the procedures aren't followed, whether because of "confusion," overconfidence, or laziness. One would wish that in a $1 billion-plus high-tech playground for dangerous pathogens, somebody would pay some attention to making the procedures for reporting illness *crystal-clear*. But misplaced priorities is what biodefense is all about. The disease being researched—tularemia—affects only about 200 people in the U.S. annually (and is rarely fatal), hardly qualifying as a pandemic threat. Yet it's been researched ad nauseam since World War II and is one of those rare bioweapons agents that the biodefense complex thought we should pay attention to rather than influenza, tuberculosis, and MRSA infections.

The *Frederick News-Post* noted: "Some may wonder—reasonably, we believe—whether this researcher's actions were out of the ordinary, or whether such breaches of procedure are routine at USAMRIID." Well might one wonder. And if one read Richard Preston's paean to brave germ lab researchers, *The Hot Zone*, closely enough, one might have noticed a significant breach of procedure by two former USAMRIID researchers, Peter Jahrling and Tom Geisbert. (Jahrling is directing the NIH's new BSL-4 lab on the Detrick "biodefense campus"; Geisbert has an administrative position with the controversial Boston facility.) Preston describes how, after the two believed they had been exposed to a dangerous strain of Ebola, they rationalized the decision to keep the exposure to themselves.[103] Fortunately, the "Reston" strain of Ebola turned out to be harmful to monkeys but not to humans, so Jahrling's and Geisbert's tempting of fate had no consequence.

Less than two weeks after USAMRIID released the tularemia report, the *Frederick News-Post* reported that numerous building flaws were delaying the opening of the National Biodefense Analysis and Countermeasures Center (NBACC), Homeland Security's state-of-the-art contribution to the Detrick biodefense block party. The problems included valve placement that made it difficult to access the HEPA filters for changing and decontamination, a drainage system that might allow contaminated water to bubble back out of the drains in the lower floors of the building, a ceiling in the BSL-3 area cracked by a decontamination practice, problems with the air pressure sensor, and a leaky roof supposedly resulting from last winter's snowstorms. (Duct tape, anyone?) Needed modifications were going to take several additional months. The facility's director, J. Patrick Fitch, said that "We have a brand-new building, and while we have decades of experience running a containment lab, no one has ever run the NBACC building."[104]

At Detrick and around the country, there are lots of new buildings "no one has ever run before." But hey, they haven't killed anyone yet.

WHAT HAPPENED TO THE ANTI-NBAF SIGN?

JUST AS THE **DEIS** HEARINGS WERE WRAPPING UP, THE **ASSOCIATED** Press obtained a leaked copy of DHS Undersecretary Jay Cohen's July 2007 Final Selection Memorandum. The headline in *USA Today* for the AP story read: "AP Exclusive: US Disregarded Experts Over Biolab." The story revealed that DHS had chosen Mississippi—"home to powerful U.S. lawmakers with sway over the agency"—as a finalist despite a low rating. Selection committees had rated each site numerically on each of the four main selection criteria; those ratings had then been combined into a numerical composite for each site.[1]

Mississippi had come in 14th among 17 sites. The AP story suggested that Mississippi's political connections might have been the reason. Democratic Rep. Bennie Thompson was chairman of the House Homeland Security Committee; Senator Thad Cochran was the top Republican on the Senate Appropriations Committee and the subcommittee dealing with homeland security appropriations.[2] (And something the AP story didn't mention—Mississippi's Governor Haley Barbour was former chairman of the Republican National Committee.)

Mississippi's composite rating was 81; the other finalists had composite ratings ranging from 90 to 94. Cohen had disagreed with the low scores the selection committees had assigned Mississippi for research and workforce; he said the consortium's proposal to have Battelle establish related research programs near the site and to relocate or train a skilled workforce overcame "what would otherwise have been a significant weakness." He referred to this strategy as a "workaround" and said "when built, they come."[3]

The discovery brought both complaints from consortiums with higher scores who'd been eliminated from consideration and concern from some of the other finalists. The plain, hard fact was that all the finalist consortiums were spending public relations and lobbying money lavishly and pulling all the political strings they could. As the *Atlanta Journal-Constitution* noted, "Politicians only lament pork when it's being served to someone else." The *AJC* did conclude, however, that "it appears a whopping platter could be headed to our hungry neighbors in Mississippi."[4]

Cohen's conclusions about the different sites made for interesting reading. At the Kentucky site, for instance: "Seemingly organized demonstrators chanting 'No Bio Lab'; 'Save Our Children' and 'Go Home' were observed picketing at the site by the site visit team." The protest we knew about; what we didn't know about—it wasn't part of the local propaganda—was this: "Some of the opposition group's concerns understandably relate to the consortium's proposal to exercise eminent domain over approximately 10 family farms to create an access road to the proposed site which appears to be 'land-locked' by small, family farms."[5]

Vocal opposition had also helped remove Wisconsin, California, and Missouri from consideration, and it was about to do the same for North Carolina and Georgia.

It was interesting to read the "spin" placed on the withdrawal of the Hinds, Mississippi site—which, if you remember, the *Jackson Clarion-Ledger* had attributed to the spontaneous combustion of ignorant, gossiping opposition. Apparently, the Gulf States Consortium had fibbed a bit to DHS, telling them it had removed the site "due to other development opportunities."[6] The ploy apparently worked, since Cohen concluded: "There has been no observed organized opposition to NBAF in Mississippi at this time."[7] Out of sight, out of mind. Is there no honor at all among thieves?

The memorandum reeked of dramatic irony by the time it was leaked. Cohen had praised the "extensive community acceptance" and the "active, broad, and largely successful . . . public relations, stakeholder outreach and media strategies" of the North Carolina consortium,[8] the highest rated of the five sites by the screening committee. Ditto for the assumption of "strong community support" and "comprehensive outreach and communications plans" of the Georgia consortium.[9]

Close reading would suggest how important a good propaganda apparatus was in Cohen's assessment of the "community acceptance" criterion. As the *Indy Weekly* pointed out in an article titled "First, the Dog, Then the Pony," in gauging the "level of public reaction," the memorandum posed such questions and discussion points as:

"How much of the media coverage is consortium generated versus generated by others?"

"Describe any known public opposition to the NBAF and plans for the consortium to *manage* the opposition." [Emphasis added]

"Describe the consortium's short- and long-term plans to engage and inform local, state and national stakeholders/community."[10]

Quite obviously, community acceptance was never what concerned DHS, only community acquiescence. If people weren't out with protest signs, raising a stink, DHS didn't care whether the "community" wanted NBAF or not. North Carolina and Georgia hadn't developed opposition at the time of Cohen's memorandum, but they would make up for past inattention in a hurry.

Of the Manhattan, Kansas site, Cohen said: "Support is strong and broad from political, business, agricultural, and academic stakeholders." Read "stakeholders" as "bigwigs." And read this as an endorsement of the quantity and persistence of the Kansas propaganda: "The coherent and persistent messaging and continuous outreach from the consortium on the benefits of placing the NBAF at this location continues to foster public support."[11]

The Kansas Pork Railroad would soon demonstrate just how hard they played the propaganda game.

In mid-August, not long after the Kansas DEIS hearing, some of those who had spoken in opposition to NBAF at the DEIS hearing decided to organize an opposition group. *Manhattan Mercury* executive editor Bill Felber showed up at the organizational meeting, and in a "news story" the next day depicted the group as leaving "without a clear idea of what to do next" and as divided about their ultimate goal. Felber then devoted at least a fourth of the article to some serious editorializing, declaring that "Opponents also offered several criticisms of NBAF that fell into the category of 'wrong.'" He offered three examples, and spent three paragraphs "refuting" them.[12]

First, someone claimed—Felber said—that in the event of an outbreak, pets would be killed to prevent the spread of a virus. That was wrong, Felber said, because according to unidentified "authorities," "the killing of pets is not part of either the protocol or practice" for control of an outbreak.

Second, Felber said, someone asserted wrongly that FMD research cannot legally be done on the mainland, when Congress had just passed a law making such research legal.

Third, another speaker wrongly asserted that DHS had estimated the economic damage to Riley County alone from an FMD outbreak to be $4.2 billion. That wasn't right, Felber said; the DHS figure referenced losses for the whole country.

And then, to put the frosting on his editorializing, Felber included some quick reaction from Tom Thornton of the Kansas Bioscience Authority. Thornton pointed to the "late arrival of the opposing forces" (as opposed to the "early," $250,000 taxpayer-funded lobbying effort of the KBA and its well-heeled political insider companions), and alleged "overwhelming public and private support" for NBAF.

The *Mercury* added an official editorial, "NBAF's opponents," just in case Felber's own editorializing hadn't done the trick, attacking the opponents for "presenting"—i.e., speaking in the context of an organizational meeting—statements that were "both inflammatory and incorrect."[13]

The *Mercury* had certainly never applied this sort of scrutiny to the press releases it turned over for the Kansas Bioscience Authority. And it had certainly never sat in on any of the "organizational" meetings of the Kansas Bioscience Authority or the Heartland BioAgro Consortium, alert for any tidbit of spontaneous discussion that might be a wee bit inaccurate. That is because the consortiums have their discussions behind closed doors and, when they go public, do so with talking points and press releases all laid out, like all good public relations prevaricators.

As Tom Manney, leader of the new "No NBAF in Kansas" group, pointed out in responding to the *Mercury's* attacks:

> [I]t is important for the public to know that we are a grassroots group and that no one among us is an expert on this topic. Collectively, we represent a wealth of expertise and information. We are not paid experts or professionals and are doing what we can without the benefit of paid professional staff such as KSU's media relations office or the Kansas NBAF Task Force press room.[14]

Or, he might have added, a biased local newspaper ready to help squelch the first signs of public rumbling.

Two other people wrote letters to the *Mercury* attacking its in-article editorializing and its own inaccuracies. Whereas Felber reported that "about 50 people" attended the meeting, the sign-in sheet indicated 76.[15] And another writer quoted the DEIS on the FMD loss question: "A May 2008 modeling study evaluated the potential economic loss from a foot and mouth virus release in each COUNTY [emphasis added] where the alternatives are located. The economic losses would be between $2.8 billion [Suffolk County, New York] and $4.2 billion [Riley County, Kansas]."

While the reference of county is probably to "release" rather than "loss," that is not perfectly clear, and like so many "statistics" related to NBAF, left open to interpretation and manipulation. Manney's letter noted that the *Mercury* had published the figure $4.2 *million* in a July 20 article, so the *Mercury* obviously had some "accuracy" problems of its own.

On the matter of the legality of FMD research, Manney recognized that Congress had recently passed a law directing the Secretary of Agriculture to issue a permit for FMD research *at a Plum Island replacement facility*—i.e., NBAF. And he went on to add some perspective to the discussion of "legality" at the meeting:

> What we stated at our meeting was that biosafety facilities that could be used for foot-and-mouth disease research already exist on the mainland (such as the KSU Biosafety Research Institute), but that the law would have to be changed. Our statement was not that it would be against the law, but that it has been against the law to study foot-and-mouth disease on the mainland.

The *Mercury* was certainly bending over backward to discredit what it called the "eleventh-hour" opposition. It added a bizarre and false argument in its editorial: "Where opponents have tended—here and elsewhere—to drive their truck off the rhetorical pier has been to extrapolate that because GAO faults DHS, therefore accidents with disastrous economic consequences are essentially inevitable. Their remedy is to apply a 'fail-safe' operational threshold that is logically unattainable."[16]

Where the *Mercury* drives off the logical pier is in asserting that because lab opponents assert that accidents are possible, they are claiming that disastrous accidents are inevitable, or that a fail-safe standard is necessary. It is in fact the lab proponents who insist that the facilities are, if not fail-safe, as close as they can get this side of heaven. Having to pierce this obviously false fairy tale of invulnerability is what drives lab opponents to distraction sometimes.

Two days later, the *Mercury* continued its onslaught on those incautious enough to ruffle Kansas's best-of-all-possible-places-for-a-biolab mantra, giving Tom Thornton and recent KSU germ lab hire Juergen Richt a full article ("Pro-NBAFers dispute claims by opposition") to take a few more whacks, in case the *Mercury* itself hadn't slapped them around enough already. (It was an article, not an op-ed, and the NBAF opponents weren't asked for their reaction.) The article contained the following deceptive and misleading comment: "[Richt] noted that both Canada and Germany already operate level

4 bioscience labs on their mainlands, and pointed out that the Canadian lab is located in Winnipeg, about one hour north of the U.S. border. Foot and mouth disease research is already being conducted there."[17]

This statement, despite the GAO's important caveat that the Winnipeg lab had only recently been established, dealt with much smaller numbers of animals than even Plum Island, and was located in an urban area away from susceptible animals. And while Germany might have a BSL-4 lab on the mainland, it conducted its FMD research on an island.

About a month later, the *Mercury* returned to the propaganda fray. Ned Seaton, in his Q&A column, responded to a question about what would happen in the event of a release from NBAF: "[W]ho would take command of the situation?" the questioner asked. "What local, state and federal authorities would be involved? Would it be like the movies, with the imposition of martial law and FEMA, Homeland Security and the military battling for control?"[18]

Seaton, who apparently found it hard to take anything an NBAF opponent worried about seriously, responded: "Yes, but fortunately Tom Cruise would come and save us from the mutant slimed creatures. So, ummm, let's just dispense with the last part of your question, which makes you look sort of silly."[19]

Apparently Seaton had missed the "Crimson Sky" simulation of an FMD outbreak, which had Kansas Senator Pat Roberts himself playing President, with the National Guard called out to quell riots in the street, execute ten million animals, and dig a 25-mile ditch to hold all the carcasses. That would seem like a bit darker flick than *The NBAF in Wonderland*. But "Crimson Sky" was one of the national cinematic features back when the biodefense complex was hyping bioterror, not trying— as it was now—to squelch questions about mainland FMD research.

About the same time, the Mercury ran a hysterical editorial attacking the recently organized Citizens Coalition of communities living in the shadow of the U.S. biodefense program. It said that "the movement Mom and Pop Farmer attempted to get stirred up"— which it said, condescendingly (and inaccurately, since the local opposition included several retired and current KSU faculty and other diverse elements), had a "certain charm"—"has been hijacked by a broader national coalition pursuing its own agenda. In the process, Mom and Pop have been kicked to the curb, their value as emotional focal points having played out. Meanwhile, this broader coalition—of which the small Manhattan-based group has been allowed to participate as a tag-along partner—last week released a statement making plain its view that the anti-NBAF campaign is no longer, if in their minds it ever really was, about stopping NBAF here or any particular place."[20]

Actually, the coalition was a loose network of groups who had been involved in struggles against major germ lab expansions around the country—some of them groups, like the ones in North Carolina or Missouri, that had been involved in the struggle against NBAF—others facing different threats: the BSL-4 lab in Boston; the huge expansion of BSL-4 facilities at Detrick; the wide range of dangerous research at Lawrence Livermore. The NBAF opponents in Kansas hadn't been "hijacked" by the group, which wasn't an anti-NBAF group per se— but had joined—after being scorned and smeared and ridiculed by their local newspaper and the Kansas NBAF juggernaut—in an effort to share and receive experience and advice.

Of course the *Mercury* was becoming adept at misrepresenting the opposition. The citizens' group's primary concern and focus was the massive, out-of-control proliferation of "biodefense" laboratories (the subject of this book, of course). "One need not read very far into the organization's statement," the *Mercury* screamed, "to discern its non-evidentiary basis."[21] It's true that the *Mercury's* germ lab coverage had more or less ignored the GAO report and the House hearing on the dangers posed by the proliferation of high-containment laboratories, but that's the problem with becoming a project booster instead of a newspaper: you only see the evidence supporting the project you're boosting, and then because you didn't notice the other evidence, assume it doesn't exist.

The *Mercury* proceeded to "sum it up" (the coalition's position) "in one neurotic sentence: All effort at advancement in biosecurity should be taken as an attempt to secretively conduct bioweapons research."[22] This comment came less than two months after the FBI said Bruce Ivins had turned one of our nation's "advancements in biosecurity" against us—an incident the *Mercury* had certainly not wanted to dwell on. Aside from the fact that building over 1200 new high-containment germ labs, and putting—yes, let me use the word—bioweapons agents in the hands of over 15,000 researchers— can hardly be considered an "advancement in biosecurity"—the *Mercury's* summing up of the opposition completely misrepresented the coalition's statement and concerns. Although the historical record amply demonstrates the potential for "secret bioweapons research" in U.S. facilities, the coalition's statement focused upon how the current proliferation may be *perceived* by others: "We join Biological Weapons Convention non-proliferation experts in concluding that we risk creating a biowarfare arms race with those who do not trust and cannot verify our intentions."[23]

I might add that the NBAF-kissing cheap shots and rhetorical dishonesties of hometown newspapers like the *Mercury*—along with the routine deceptions of government agencies like DHS—create multifaceted distrust and suspicion, even paranoia, among ordinary people who have to seek out the truth for themselves.

The Kansas public relations juggernaut's suppression of dissent was nothing if not thorough. Someone wrote in to Seaton's column and asked "What happened to the 'No NBAF in Kansas' billboard that was on Fort Riley Boulevard, near 17th Street? It was there for a couple of weeks and then it was gone."[24]

Seaton responded that the "property owner decided he didn't want to wade into the middle of that controversy, so it came down." The sign had gone up on the property of People's Grocery, which leased the space and "'didn't mind allowing it to go on there,' under the belief that the space for the sign went with the business as part of its lease."[25]

After a couple of weeks, the property owner decided "'he didn't want to get into politics' on that property, and that the NBAF discussion was too much of a hot potato. On further investigation, it turned out that the lease arrangement for People's Grocery restricts the billboard so that it is to be used only to advertise 'on-premise' business."[26]

One wonders which parties from the playful list of NBAF proponents recited by Seaton—"the state's congressional representatives, the Kansas Legislature, Gov. Kathleen Sebelius, K-State, the county commission, the city commission, the Chamber of Commerce, two French hens and a partridge in a pear tree"—helped the property owner get in touch with his true inner political being.

CHAPTER THIRTEEN

Educating the Educated
The Brouhaha in Boston

1

There's gold in these here labs, Mayor Thomas Menino said in January 2003, after Boston University announced it would compete for one of the NIH's new biodefense labs.

But, he told the university, you have to get out there and educate the people; otherwise, you'll have all these not-in-*my*-back-yard people coming at you.[1]

That's right, said liberal lion Edward Kennedy a few months later. "We have to get the best people in the country to work with the local community so they understand it."[2]

Remedial education. It had worked in the case of Massachusetts Governor Mitt Romney. He allowed he'd been a bit worried about the safety of a BSL-4 facility himself, but Mayor Menino had hooked him up with some bona fide safety specialists; his concerns had been "addressed"; and he "came away from those meetings feeling that I and my family and my children and grandchildren are even safer in Boston following construction of this facility than perhaps we are today." Romney then became the university's chief advocate with the Bush administration.[3]

Many Bostonians were more stubborn than Romney. They refused to be properly indoctrinated, to the obvious consternation of the *Boston Herald*'s editorial staff. "The NIMBY crowd rants on," the *Herald* would complain in its 4/22/04 editorial, "A Valued New Neighbor." The poor innocent lab had been the victim of a "raging outbreak" of ignorance.

"We are at war," the *Herald* insisted in June 2005. "Opponents of a Biosafety Level 4 lab in Boston should be ashamed of themselves for playing on phony fears when there are real threats, including bioterror threats, to our safety. . . . Anthrax, Ebola, smallpox, botulism—an attack using any of these is the stuff of nightmares."[4]

Politicians who raised concerns about this "clean, secure research lab" were simply fearmongers,[5] panderers to ignorance.[6]

Look out, shouted the *Herald*. The bioterrorists are coming! Anthrax! Ebola! Smallpox! Botulism!

But what did these fanatical biolab opponents mean playing on phony fears? anthrax ebola smallpox botulism. (To be whispered rapidly, without pausing for breath.) Shhh. Six hundred thousand people in Boston. Poo. Nothing to worry about. Nothing escaping from our $178 million state-of-the-art Starbucks biolab. Thirteen hundred construction jobs. 660 lab positions. $1.7 billion in grants. Nothing to worry about. "Extraordinary measures taken to ensure that lethal agents cannot escape." "Armed guards will monitor checkpoints, labyrinths of hallways will make quick escape impossible, and scientists will wear protective suits." Poor germs will be so scared, they won't know whether they're coming or going. "The lab will be as safe as it is humanly possible to make a facility. . . . [A fact that, even if it were true, wouldn't necessarily reassure.] There are endless layers of redundant systems to assure safety and security."[7] (And endless layers of redundant reassurance to make sure you understand this.)

And such a noble source of filthy lucre. "The research in question is essential on so many levels—surely for national security purposes, but also to keep us all safe from future outbreaks of infectious diseases that are a fact of life, of nature and of our jet-paced society."[8] (But no chance of keeping us safe from future outbreaks of biolab propaganda.)

Education! It's the key to biolab success. And Boston University is an educational institution. So BU takes the mission to heart. So does the National Institute of Allergy and Infectious Diseases (NIAID), the NIH division funding and guiding the new labs. So does the political power structure, Republicans like Menino and Romney, Democrats like Kennedy and Kerry. And so do various newspapers, dutifully printing press releases about the would-be lab's state-of-the-art technologies.

Even the *Bay State Banner*, an African-American newspaper, gets in on the fray. Notable because one of the primary organizations opposing the Boston biolab is Safety Net, "a movement of people of color who have been excluded from the decision-making process to confront power directly and demand fundamental changes in the rules of the game, so that we can achieve our right to a healthy environment."[9] Melvin Miller, publisher of the *Bay State Banner*, is angry with a group called Operation: Over for trying to persuade prospective students to boycott Boston University because of the lab.

"Tell the Truth," insists Miller's editorial. The organization's arguments, Miller says, are based on falsehoods and half-truths. The organization claims that the facility will be located in a "residential, low-income neighborhood." Hold it, Miller says. "Although some affordable housing is in the area, so are some of the city's priciest townhouses and condos."[10] So it's not "exclusively low-income." Nitpicking a bit, maybe? (And a half-truth in its own right? Can't a neighborhood be low-income without being exclusively low-income?) The fact remains that any outbreak would affect a segment of the city's population least able to seek appropriate treatment, and also more likely to be ignored by the political machine.

Also, Miller asserts, the organization inaccurately says the lab is funded by Homeland Security, when it's actually funded by the National Institutes of Health. Oops. He did catch them in an error there. Probably all those newspaper articles that explained how the lab would be a centerpiece of the Bush administration's war on terrorism. And could anyone really tell the NIH and Homeland Security apart in this new world of biodefense? Did Miller know that NIH, Homeland Security, the Department of Defense, and the Department of Agriculture were all shacking up together on a vast new "biodefense campus" at Fort Detrick, ancestral alleged home of the 2001 anthrax attacks? The world's greatest megalopolis of biodefense research, a veritable orgy of state-of-the-art pathogens and allegedly state-of-the-art containment features? (If we needed 1300-plus high-containment facilities to defend ourselves from one mad scientist at Fort Detrick, what will we need to circumvent all these new towers of bio-babble?)

And while it's true, Miller says, that the facility will research dangerous pathogens, it presents no danger to the surrounding community. And here he slips into a truly serious inaccuracy of his own. "Other Level 4 laboratories," he says, "have been operating for years without incident." What he meant to say, we presume, is that the ones in the U.S. hadn't killed anyone yet—if you overlook the small matter of the anthrax letters. But to say Level 4 labs have operated "without incident" just means Miller hasn't read anything but the BUMC's press releases.

Miller's editorial appeared in March 2006. The *Boston Herald*'s accusations about pandering appeared on February 4, 2006, in an editorial titled "Welcome to Our 'Hood." Two days earlier, the NIH had approved its own draft report stating that the Boston project it was funding presented no significant environmental threat. The NIH, BUMC, and the powerful local and state supporters of the planned project—including its media supporters—were at the height of self-assurance and confidence, in a state of mutual self-congratulation. The university said it would begin construction that month and expected to open the facility in summer 2008. Famous last words.

2

Famous first words. The pattern for much that would follow. Boston U, new in the biolab education business, hires a telephone survey company to quiz people living near the lab about their attitudes. The "quiz" is apparently a cover for some sort of Vulcan mind-morph attempt. People who answer the phone are told what a wonderful opportunity the lab will be, helping to create jobs and provide protection in the event of a bioterror attack. Jethro Heiko, a Fenway resident, is told that in a bioterror attack, Boston residents will have "access" because of their proximity.[11]

Think I'll walk down to the biolab and see if they can figure out what's wrong with me. A bizarre bit of hokum. The lab isn't a hospital; sick people showing up at the door of this "highly secure" facility won't be admitted for a tour or given a sick bed (there won't be any). Perhaps a Fenway blood sample might aspire to a luxury accommodation on a state-of-the-art lab slide, but one doubts a Fenway draw will get any special consideration at NIAID's NBL.

Heiko remained unconvinced. He felt the economic benefits were being exaggerated, and he couldn't figure how having the lab next door would make him any safer.[12] Common sense (oops, make that ignorance) 1, Education 0.

Boston U just couldn't get it right. Six years later, NIH's Blue Ribbon Panel would tell them just how badly they'd screwed up the trust thing. Everybody knew the damned lab was going to be built, whatever the people of Roxbury, or Cambridge, or Brookline, or Dorchester, thought about it. The mayor wanted it, the governor wanted it, Kennedy wanted it. God damn it, it was "a critical component of Boston's life sciences infrastructure."[13] What did people think, that this was a damn democracy we live in? Since when do we vote on nuclear plants, or military installations, or genetically modified foods, or biolabs?

Meanwhile, there were these hoops to be gone through. NIH and Boston U had to seek the "input" of the community. They had to act like they cared what the community thought. It was called "public outreach." Like the guy who interrupts a business-trip shagging to call his wife and let her know he's thinking about her. It reassures her; it doesn't keep him from the shagging for all that long; everybody's happy.

Boston U did something wrong. They had to have. Because it was a simple, incontrovertible fact that these labs were absolutely, perfectly safe. The people of Boston just needed to have that explained to them, skillfully. But Boston U had loused it up, embarrassing NIAID, embarrassing the Boston City Council, embarrassing the Massachusetts Department of Environmental Affairs, making it awkward even for the shoot-the-rebels-where-they-stand *Boston Herald*.

It was all about "transparency," which means one thing to biolab opponents and another thing altogether to the biodefense complex. To biolab skeptics, to ordinary folks, transparency means everything is above board. There are no secret experiments; accidents are reported and acknowledged to the public. To the biodefense complex, transparency means something different altogether. It means that laypeople are imbued and comforted with the deep, intuitive faith that there is a biodefense heaven where the people's interests and concerns are taken into account and allowed to adjust the balance of the biodefense universe ever so slightly. Suffer the little children to come unto me, and all will be explained to them.

Boston U's biolab push was transparent, all right, just the wrong way. People saw right through it.

3

It had all started the year after the anthrax attacks. The National Institute of Allergy and Infectious Diseases (NIAID) and its collaborators announced that they would fund a bunch of new high-containment laboratories to research bioterror pathogens.[14] NIAID had first compiled a list of Category A, B, and C pathogens, all of which NIAID felt posed bioterror threats. Influenza and multi-drug resistant tuberculosis had just squeaked through, onto the least-threatening Category C list. Seems these two diseases, which didn't ask for or need the assistance of bioterrorists, nonetheless presented a potential bioterrorist threat. So the NIH would want these diseases to be looked at also, but the funding wouldn't be anything like that for the Category A pathogens.

In February 2002, the agency had worked with a Blue Ribbon Panel to develop a strategic plan for biodefense research.[15] The panel had said "research resources" needed to be expanded; NIAID interpreted this as a need for more high-containment laboratories. The agency decided to create a three-pronged network: regional centers of excellence; regional biocontainment laboratories; and national biocontainment laboratories. The emphasis of all the facilities would be the biodefense research agenda, especially the more dangerous Category A agents: anthrax; smallpox; plague; botulism; tularemia; and a host of viral hemorrhagic fevers such as Ebola, Marburg, Lassa fever, and Rift Valley fever.[16] The regional labs would emphasize BSL-3 research; the national lab or labs would also include BSL-4 facilities, to work with the most dangerous agents for which no vaccine or cure existed.

Later, NIAID and Boston U would try to downplay the biodefense focus and repackage the whole enterprise as "emerging infectious diseases" research. For now, though, the PR was all about bio*defense*.

In January 2003, Boston U announced that it would compete for one of the national facilities.[17] By June, opposition had arisen. A coalition of a dozen community groups, spearheaded by Alternatives for Community and Environment, a ten-year-old organization focused on environmental justice in New England, sent NIH a letter asking that the project be denied until the university provided more information and a meaningful public review process.[18] The Council for Responsible Genetics (CRG), Boston City Councilor Chuck Turner, and State Representative Gloria Fox also signed the letter.

July 27 featured a protest in front of the Medical Center.[19] By August, opponents had gathered several hundred signatures in opposition to the project. In an August 19 *Boston Globe* article, community leaders said the university had done a poor job of advertising the public meetings it had held so far. BU officials retorted that they had held 13 community meetings by that point, attended by a total of 300 people.[20]

Dr. Mark Klempner, BU's provost for research, insisted the facility had strong community support despite "some protestors": "There are always going to be some people who resist change."[21] The *Boston Herald* followed up with an August 4 editorial asking what all the fuss was about ("Some folks always need a cause, don't they?"), and criticizing two City Council opponents as "perennial naysayers."[22] That same month, Councilor Chuck Turner proposed (unsuccessfully) a statute banning BSL-4 research in the city.[23]

Two days before the hearing to consider Turner's measure, the *Herald* published an op-ed from two members of the local biodefense complex. Janice T. Bourque, president of the Massachusetts Biotechnology Council, and Paul Guzzi, president and CEO of the Greater Boston Chamber of Commerce, called the lab a $3 billion booster shot for the state's economy. Their op-ed focused on the economic glories of the project, but included this "note about safety":

> In 72 years of combined operation, none of the six Level 4
> labs in North America has had any environmental releases or

community incidents. Level 4 biosafety facilities are built with multiple containments and safety measures to ensure failsafe levels of protection. The labs work with microscopic amounts of material that pose virtually no threat to the neighboring area.[24]

The op-ed featured the central claim of bioboosters everywhere: that old fail-safe magic will look out for us. That the labs won't deal with truckload lots of pathogens is both obvious and irrelevant. A little goes a long way. The American team that investigated the 1979 Sverdlovsk anthrax release that killed sixty-plus people concluded that a couple of grams could have produced the results. And anthrax is a disease that doesn't spread from person to person; the physical spores must actually be inhaled. All other diseases require for a foothold (or a lunghold) is one infected individual—most likely a researcher infected unawares. One such individual will host millions of pathogens, which he can generously share with anyone outside the lab. Each of the various animals—primates, mice, guinea pigs—deliberately infected for experimental purposes with, one assumes, a tiny injection or an aerosol whiff of microscopic particles—will also act as a breeder reactor for pathogens. Insisting that the lab will only contain microscopic amounts of "material" denies the basic nature of pathogens, which don't travel by way of a Wal-Mart truck.

Citing seventy-two years of combined operation in six level 4 labs in North America (apparently a deliberate effort to include recently built Canadian labs) is an extremely creative way of inflating the alleged trouble-free history of BSL-4 research. In the United States, only facilities at USAMRIID and the CDC have many years of operation at all—a bit more than thirty years each. It is true that, at least until 2001, neither facility had (so far as we know) accidentally killed any member of the American public. Even putting Bruce Ivins aside, however, it is inaccurate to say that none of the labs "has had any environmental releases or incidents." These sorts of genetic truth modifications are exactly why many in Boston and elsewhere refuse to "repeat after me: safe as can be."

The PR also ignores one of the central problems with facilities like the Boston one. The proliferation of new facilities (there will soon be fifteen BSL-4 labs in the U.S., not two) greatly increases the odds of catastrophe, and not just numerically. There are so many new facilities, pathogens, and researchers that no one's really watching over, that theoretically meticulous and highly controlled research is actually more like a car moving down the highway with no driver.

On September 30, 2003, the NIH announced that Boston and Galveston had won the competition for the "national" BSL-4 labs. Newspapers described the Boston facility as a "bioterrorism" or "biodefense" facility, while press releases from the NIH and BUMC pointed out that the facility would also study emerging infectious diseases.[25] City and state leaders gloated that the facility would "put the region on the front lines of the war against terrorism,"[26] and also "promote jobs and economic development" by "securing Boston's role as a center of bio-medical and research activity."[27] BUMC indicated it would call the facility the National Center for Emerging Infectious Diseases and Biodefense (NCEIFDB).[28] Down the road, BU would quietly drop the Biodefense part of the name.

A student group joined the opposition in November 2003, charging that project officials had done an inadequate job of informing the community.[29] Then, in April 2004, the "community activists" (which is what ordinary people are called when they question the decisions made for their communities by the powerful—as though they were some sort of professional subversives. Nobody, however, ever calls people who go along, who don't say anything, "community passivists") gained some important allies. 146 Massachusetts university professors, including two Harvard Nobel laureates, and social "activist" Noam Chomsky, sent the City Council a letter protesting the project. The group included a number of high-powered university biologists, and they raised the same concerns about possibilities of accidents that the community people had raised.[30]

Particularly notable was the opposition of Boston University's own public health professor, Dr. David Ozonoff. He had supported the project early on because he saw a need for work on new and emerging infectious diseases. In a September 2004 presentation to Boston U faculty, staff, and students, however, Ozonoff said he'd concluded that "the proposed laboratory is not meant nor is it suited to pursue a genuine public health agenda." He pointed to the explicit requirement of

NIH that the facility pursue the agency's biodefense research agenda, and to the fact that NIH monies for the project came from the Bush administration's Homeland Security budget. He also raised concerns about the potential for secret research, the creation of "weaponized, new or novel pathogens," lack of effective civilian oversight, environmental justice, and the rift the project had created with the community. ("Community opposition is broad and deep. The reason you had to show your BU ID to get in here tonight is because relationships with the community are at an all-time low.")[31]

None of this opposition forestalled a series of rapid approvals by state and local authorities, each accompanied by a self-congratulating press release from the BUMC. In November, the Massachusetts Secretary of Environmental Affairs approved the Final Environmental Impact Report for the facility. The university said the lab was part of a "national network that will develop drugs, vaccines, and treatments against emerging and re-emerging infectious diseases so that they no longer will pose a threat to public health."[32] In December the Boston Redevelopment Authority granted its approval.[33]

On January 12, 2005, the Boston Zoning Board approved the project. BUMC said this was "the final city regulatory approval needed before construction can begin," and it attributed the various approvals to a "significant public outreach and review process." Senior vice president Richard Towle, quoted for the press release, claimed: "As awareness of the project increased[,] the importance of the laboratory was understood and community support for the project grew."[34]

The world of bioboosting is all about press releases. If you say a facility is failsafe, the facility is failsafe. If you say the community supports it, the community supports it. Just wave the magic PR wand. Abracadabra, shazam, shazam.

4

Meanwhile, the BUMC was concealing evidence that dramatically demonstrated how assuming something's harmless doesn't make it so.

A week after the zoning approval, someone in the media got wind of separate 2004 incidents in which BU researchers had been accidentally infected with tularemia, a biowarfare agent. After the media began asking questions—and after state and local authorities had granted the final approval necessary for construction to begin—BU came clean. Two researchers had become ill in May 2004, and a third in September. One of the May victims had been hospitalized overnight, and the September one had been hospitalized for several days. All had recovered after treatment with antibiotics.[35]

The illnesses had not been diagnosed as tularemia until October. No one suspected tularemia because the researchers believed they were working with a non-virulent strain. (In biodefense heaven, if the label says it's non-virulent, it's non-virulent.) But the incident, like the later Q-fever infection at A&M, highlights a disquieting quirk of biodefense researchers. Their quest for intricate knowledge of deadly pathogens doesn't seem to include knowledge of a disease's basic symptoms. Or perhaps they believe their own propaganda, and don't think it's possible they should be infected by the pathogens they work with.

Because they believed they were working with a non-virulent strain, the researchers had also violated procedures requiring them to work inside an enclosed hood.

The university had reported the cases to city, state, and federal health authorities in November, about the time of public hearings on the lab. But neither the university nor the government agencies (nor the mayor, who had also been informed) disclosed the cases to the public.

The university's (and agency's) explanations for this failure sounded a bit like a man explaining to his latest love interest why he hadn't told her he was married. The university and the agencies said there had been no risk to the public: tularemia isn't transmitted from person to person. They couldn't imagine any other reason why the public might want to know about the incident, such as the ongoing debate over BU's national lab. Among other things, BU had proudly boasted in its NIAID proposal that no BU researcher had ever been infected by a laboratory pathogen. BU didn't bother to update the application after the incident.

BU didn't think the incident relevant to the National Lab debate because the researchers were working in a BSL-2 lab, "which has far less stringent safety standards and is allowed to work only with less dangerous material." The university said this sort of thing just wouldn't happen in a BSL-4 lab.

The university said it hadn't informed the Zoning Commission because the Commission was only concerned with land-development issues.

Of course, one of the land-development issues a zoning commission considers is the proposed "use" of the property, which certainly can and should entail awareness of any risks. A fail-safe facility, of course—if such a thing existed—would present no risks. The tularemia accident reminded Bostonians that BU's existing facility certainly wasn't fail-safe. BU, of course, wanted to argue that the latest, greatest BSL-4 facility would be fail-safe. But a BSL-4 facility depends on researchers following protocols, just as lower-containment facilities do, and about the only thing one *can* guarantee is that researchers won't always do that.

Richard Ebright told *New Scientist* that the Boston mix-up "would be impossible for any competently-run microbiology laboratory." He pointed out that neither the head of Boston U's tularemia research section (dismissed after the incident), nor the four other BU researchers who'd received grants to study tularemia, had published papers on biodefense—and he worried about such groups operating a high-level defense lab in a densely populated city.[36]

He also pointed out the quite different response when a similar incident occurred at a new biodefense lab at Oakland Children's Hospital. In 2004, the same year as the BU incident, the Oakland lab had received live anthrax instead of the dead germs they'd requested, and didn't notice the mix-up until all the experimental animals died. In contrast to Boston's don't-tell-them-anything-they-don't-need-to-know attitude, the Oakland institution had immediately called a press conference and then stopped doing biodefense research.

The incident caused investigations by the Boston Public Health Commission, the CDC, and OSHA. In February 2005, NIAID Director Anthony Fauci felt it necessary to affirm "the confidence we have in BU as an organization." He spun the incident as an "individual event" irrevelant to the "superb training and physical capability" he expected of the BSL-4 facility. "Things like that happen when people are not trained well," he said.[37] Boston U, which had apparently not trained the tularemia victims well, would of course train researchers in the new facility superbly, even though testimony at the October 2007 Energy and Commerce hearing, and publications by members of the biodefense complex, would indicate that training the horde of new researchers was a real problem.

Fauci said BU's critics were being unfair in raising questions about the university's competence to run a BSL-4 facility. "'They' in this case," he said, "is an investigator who did not follow protocol, who did not go under the training, who was not working in a BSL-4." So it was not a matter of poor training after all, but of human nature, which insists on taking shortcuts, not following the protocol to the letter all the time. But everything would be different in the new facility because, well because, well because everyone would be washed in the blood of the BSL-4. So obviously the answer to all our safety problems was just to make everything BSL-4; instead of increasing the number of BSL-4 labs from four to fifteen, just turn all 1300-plus high-containment labs into BSL-4 labs, let everybody research Ebola and smallpox, give the people who were struggling with BSL-2 work a BSL-4 facility to shock and awe them and impress them with the seriousness of what they're about. *Then* they'll follow the protocols.

5

Following the tularemia incident, "[t]he university moved to tighten safety practices in all labs"—just in time for the April 2005 public hearings on NIH's draft environmental impact statement. In responding to the tularemia aftershocks, the university tried to explain why the new lab (as opposed to the other one) would be perfectly safe. It would be "a submarine within a vault," said one BUMC representative.[38] This image would become a popular one for biolab boosters over the next few months; Mississippi governor and former national Republican Party chairman Haley Barbour—who said the DHS-controlled NBAF wouldn't create illegal bioweapons because the CIA and Department of Defense did *that* sort of thing—also liked the submarine-in-a-vault comparison. Perhaps a nation drowning in debt may want to keep a few submarines at the Fed Reserve, but once we go down the road of bizarre and arbitrary comparisons, why not a submarine sandwich in a vault, or the Good Ship Lollipop in a teapot?

NIH's draft environmental impact statement said the lab would pose no substantial environmental risks. NIH was required by law to take public comment on the draft, prior to issuing a final EIS. Barring a major about-face, the EIS would be the final approval the project needed. The *Boston Globe* indicated that the "last NIH approval has been widely assumed to be a given: It was a branch of the NIH, the National Institute of Allergy and Infectious Diseases, that originally selected BU to build the lab."[39]

The NIH lived up to expectations on February 3, 2006, approving its own draft report (and its own project) and ruling that the facility presented no significant environmental threat. Rona Hirschberg of the NIH said there would be "endless layers of redundant systems to assure safety and security."[40]

Opponents sued at both the state and federal level. The state suit challenged the state's approval of the facility's environmental impact statement. In August, State Suffolk Superior Court Judge Ralph Gants ruled that the state's approval was arbitrary and capricious and lacked a rational basis, finding that the earlier environmental impact assessments failed to adequately consider alternative sites or adequately evaluate worst-case scenarios. He ordered BU and the NIH to conduct a new environmental review, and ordered the state environmental affairs office to consider that review with fresh eyes. A bit bizarrely, however, he allowed the construction to proceed.[41]

About a month later, the Boston Public Health Commission announced draft regulations that would require BSL-3 and BSL-4 labs to apply for operating permits; would set training standards and provide for surprise inspections; and would prohibit classified research and research to create bioweapons. The Massachusetts Biotechnology Council, pharmaceutical giant Merck, and Harvard opposed all or part of the rather minor regulations.[42]

In October, federal judge Patty Saris also ruled that construction of the lab could continue while the new environmental review proceeded.[43]

On January 10, 2007, the Cambridge City Council voted unanimously to oppose construction of the BU lab.[44] Brookline and Somerville would also pass ordinances opposing the facility.[45]

In March, BU's existing BSL-3 lab was evacuated after white smoke wafted through the lab.[46] Investigation revealed that medical waste had been absentmindedly left in an autoclave (a sterilizing machine)

all weekend. City health officials, who only learned of the fire incidentally, said the episode showed the need to make sure researchers
understood the "rigorous safety regulations" adopted after the
tularemia incident.[47] Accidents 2, Fail-Safe Labs –2.

The day after the fire, the State Supreme Court announced it would
hear the university's appeal of Judge Gants's order directly, allowing
the appeal to bypass an intermediate appeals court.[48]

In August 2007, BUMC failed to send a representative to a hearing
before the City Council's Public Safety Committee on transportation
plans for the new lab. The executive director of the Boston Public
Health Commission and the Superintendent-in-Chief of the police
department did appear. One councilor—not yet an opponent of the
lab—was disturbed to learn that pathogens would be shipped by
FedEx; another said BU had not yet answered legitimate transportation questions.[49]

On August 23, the NIH issued its latest environmental impact
study, declaring that the facility did not endanger the neighborhood
and would not be safer located in a less populated area ("Indeed, the
numerous safety precautions in the design and operation of a BSL-4
laboratory result in a negligible risk of release of infectious agent into
the surrounding community").[50] Ed Hammond of the Sunshine
Project said this was "the sort of hubris that gets people into trouble
and ends careers." He scoffed at the study's suggestions that such
measures as vaccination of cattle for Rift Valley fever, use of insect
repellents, medical interventions and supportive therapy, could
reduce the [allegedly nonexistent] risk even further:

> [C]attle are not routinely vaccinated for RVF in the US and it
> would take considerable time to do so. People do not nor-
> mally wear insect repellent, and it is laughable on the level of
> duct tape to imply that Bostonians should derive any comfort
> with respect to a BSL-4 lab leak from the availability of insect
> repellent. On medical interventions and supportive therapy:
> As NIAID well knows, BSL-4 diseases are BSL-4 diseases specif-
> ically because they are typically incurable. If generally effec-
> tive "medical interventions" and therapies existed for BSL-4
> diseases, then they would not need to be handled at BSL-4.[51]

In biodefense heaven, of course, any sort of logical contradiction or impossibility is possible: new hordes of deadly pathogen researchers because of an attack by a deadly pathogen researcher; pathogens we're told should frighten us in the hypothetical hands of bioterrorists but not concern us at all in the biodefense lab down the street; curing the incurable. Come forth, Lazarus; take up thy cot and walk.

Other opponents questioned the objectivity of a report from the organization funding the lab's construction. A BU research vice-president said, however: "The only vested interest that NIH has is to maintain its integrity . . . NIH is not going to put a report out there that is going to be reviewed and critiqued . . . if it doesn't meet their standards."[52]

6

Whether the report met NIH's own standards was a question only NIH could answer. But perhaps NIH's standards depended on what it thought it could get by with. Perhaps NIH, which had more or less had its way with a BSL-4 construction out there in Montana, just wasn't used to the sort of scientific scrutiny that would come its way in Boston, where opponents included plenty of scientists, social studies thinkers, and other high-powered academics. It soon became clear that, if the report reflected NIH's standards, then a lot of people didn't think much of NIH's standards.

Jeanne Guillemin, for instance, whose husband Matthew Meselson had led a U.S. investigation into the 1979 Soviet anthrax release, and who had written a rich sociological study of that investigation[53] (and later, a highly readable history of bioweapons),[54] said the scenarios used for the report were highly contrived and selective, because they focused on diseases which weren't really that contagious.[55]

Ed Hammond, not of Boston, but undoubtedly the country's most probing scrutinizer of biodefense bungles and bamboozle, also found NIH's scenarios to be paper tigers. He said the agents selected for the scenarios should have included "both those that are easily transmitted from human to human (for example, pandemic influenza) and agents prepared for certain types of biodefense challenges (for example, dry milled anthrax spores)." He also said some of the scenarios should have included engineering failures, and "malfeasant scenarios" involving deliberate release.[56]

Hundreds turned out for the September 20 public comment hearing on the NIH's new DEIS, including descendants of the late Edward Abbey. Opponents got a little feisty, on occasion interrupting the formal spiel of Dr. Deborah Wilson, Director of Occupational Safety and Health for the NIH. "Get to the point," one person shouted. "When do we get to talk? That's what we came for," said another, prompting loud applause. At 9 P.M., the cutoff point, the line to speak still stretched down the aisle.[57]

On October 14, the state's Office of Energy and Environmental Affairs announced it would pay the National Research Council, an independent body chartered by Congress, $50,000 to evaluate the adequacy of the NIH's new analysis.[58] On October 19, a panel of ten scientists appointed by the Council conducted a hearing in Washington. Wilson and lab director Klempner argued on the lab's behalf; David Ozonoff of BU's School of Public Health and Marc Lipstitch of the Harvard School of Public Health spoke in opposition. Dr. John Ahearne, chair of the panel, asked each speaker "How do you define worst case scenario?" Klempner indicated he would defer to the NIH; the NIH's general counsel, David Lankford, refused to answer the question, asserting that the National Environmental Policy Act restricted what he could say.[59]

The BUMC was quite active on the public relations, oops, education front during this period. On October 3, the BUMC gave three staffers for the *Boston Herald* a tour of the "nearly-completed" facility, focused on its "impressive security measures."[60] On or about October 25 the BUMC provided a similar tour for staff of the *South End News*, focused on the containment features of the BSL-4 lab.[61]

On November 6, the Community Liaison Committee held a two-hour forum designed to reassure locals about the shipping of dangerous pathogens. By and large, the locals who attended were not reassured. Dr. Ara Tasmassian, BU's associate vice-president, dismissed one person's fears of terrorism by declaring that there was an "entire infrastructure of security officers." (One can only note that many people worry about crime, despite the fact there is an entire infrastructure of law enforcement.) Tasmassian also made the bizarre and dubious assertion that there "are many rogue nations that for $150 will ship it [deadly pathogens]"—apparently implying that there is no need to worry about terrorists intercepting a U.S. FedEx shipment because the pathogens can simply be ordered from some unidentified rogue country. Tasmassian did not provide evidence for his claim.[62] On November 26, BUMC wangled an article in *Boston Now* titled "Trying to Stop Deadly Disease: New BU Lab Focused on Fighting Dangerous Ailments."

Run-of-the-mill propaganda became more difficult on November 29, when the NRC panel issued its report concluding that the NIH report's science "was not sound or credible." The federal analysis was "so woefully incomplete that it did not meet basic standards applied to scientific research."[63] Members of the panel told journalists that the study would have been rejected if submitted for publication in a medical journal, or that as a student paper, it would have failed.[64]

The council agreed with Hammond and Guillemin that the agency had focused on the wrong scenarios by not choosing easily transmissible diseases. The council criticized the use of Rift Valley fever in comparing the relative risks of urban and rural sites. Since cows can help spread Rift Valley fever, that rather arbitrary choice made a rural site appear more dangerous than the South End of Boston. Simultaneously, the NIH had failed to consider the health-compromised state of many South End residents already afflicted by the diseases of poverty.[65] And the report lacked sufficient detail and "transparency" about its methodology and conclusions.[66]

Mayor Menino's support was unfazed; he declared that the facility was "too important not to go forward."[67] A few days later, he predicted to the Greater Boston Chamber of Commerce that the facility would open in a year despite the NAS report.[68] Eugene Benson of the Alternatives for Community and Environment noted wryly that there was a Menino Pavilion at BU's Medical Center and that the PI of the proposed lab and others who don't live in Boston (and therefore won't have the facility as their immediate neighbor) had "donated generously to the Mayor's last re-election campaign."[69]

Two days later, the Massachusetts Supreme Court unanimously upheld Judge Gants's 2006 ruling that the state's environmental approval of the project had been arbitrary and capricious.[70]

7

So it was back to the public-relations drawing board for the NIH. In February 2008, the NIH told federal judge Saris, who had retained jurisdiction and the right to intervene in the project even as she allowed construction to proceed, that a new environmental review would not be completed until April 2009.[71] In a March 6 press release, the NIH announced "additional steps in a comprehensive plan to address public safety concerns."[72]

The NIH, of course, didn't acknowledge that there was anything to worry about; it just needed to "address concerns": "All of the analyses conducted to date indicate that the risks posed by this lab are extremely low. We recognize that the community has remaining concerns, however, and we will address those concerns rigorously, objectively, and comprehensively." The press release asserted that "NIH-funded national biocontainment laboratories, such as the one under construction at the Boston University Medical Center, take stringent precautions to protect the laboratory workers' health and to reduce their individual risk, and also to safeguard the surrounding communities and the environment."

The NIH set up an internal committee to guide its "efforts to address safety concerns" and a "Blue Ribbon Panel" to "review current risk assessments and provide independent technical expertise and guidance." The panel included such existing members of the biodefense complex as Jean Patterson of the Southwest Foundation for Biomedical Research, San Antonio's "private" BSL-4 lab, and Samuel Stanley, director of the NIH's Midwest Regional Center of Excellence for Biodefense and Emerging Infectious Diseases, so independence was a relative term, considering members' own reliance on NIH grants to fund research.

The new public relations message from the NIH was: be tough with us, panel. Don't rubber-stamp us.[73] That's been the public relations pattern for the biodefense complex over the last couple of years: proposing "tough new reviews," "tough new regulations" after each new instance of careless and irresponsible behavior. As Eloise Lawrence of the Conservation Law Foundation pointed out, the promised exhaustive review should have taken place "before the first shovel of dirt was turned."[74]

The BUMC Medical Center added its own touch to the public relations effort, declaring that it welcomed additional study and applauded the creation of the independent panel: "We are confident that the lab will be safe, and this third-party examination is an important step in the public process."[75]

In early May, the NRC panel that critiqued the original NIH study addressed a 21-page letter to NIH Director Elias Zerhouni calling for a new, deeper safety review.[76] The Blue Ribbon Panel appointed by the NIH echoed that call a couple of weeks later, and also urged the NIH to do a better job of "involving" the neighborhood: "Make the community part of the discussion, seek their input before you start, continue their input as you are talking." The panel issued its recommendations after a hearing in which over three dozen speakers "overwhelmingly opposed the lab." According to the *Boston Globe*, the few backers were mostly from the building trades and higher-education associations. Several speakers had criticized the panel for holding the hearing on a weekday, preventing many working residents from attending. One speaker called the session a "public health charade. . . . All we need is trapezes and dancing artists."[77]

In the wake of the Ivins revelations, which threw many existing biodefense facilities into an apologetic "aw shucks" mode, the BUMC announced that, by golly, its scientists would face "in-depth screenings" including psychological testing and financial-records scrutiny designed to "spot researchers who might be unstable or whose financial plight could leave them susceptible to stress-fueled mistakes—or even extortion."[78]

The Ivins revelations made even the *Boston Herald* worry a bit. After first declaring that the new building would be "a virtual fortress replete with every safety feature known to man and some unknown, too" (one wonders just how an unknown safety feature is supposed to work: won't the unknown safety feature be afraid of blowing its cover by, well, being safe?), the *Herald* wondered: "[H]ave they prepared sufficiently for a Bruce Ivins?"[79]

Not so coincidentally, the BUMC's announcement of its Ivins-prevention strategies appeared on the day the NIH's Blue Ribbon panel was conducting its first hearing in Boston proper. The panel seemed focused on "bridging gaps," as though the whole debate were simply a matter of public relations. Patricia Hynes, one of BU's own professors of environmental health, pointed out the arrogance of the panel's initial question: "How can institutions most effectively reach out to local communities and educate about these laboratories?" Said Hynes: "The way it's phrased assumes the community is ignorant, and the university is the savant, or the more knowledgeable of the two. . . . Education is a two-way street. I'd like to rephrase the question: How can communities most effectively reach out to local universities to educate them?"[80]

The *Boston Globe* reported that the crowd of more than 300 shouted down the panel at several points.[81] A show of hands indicated that about three-quarters of those present were opposed to the lab.[82] The panel seemed frustrated that the crowd refused to be educated, that some in the community didn't believe statements that biological weapons won't be created in the lab.[83] Panel chair Adel Mahmoud asked, "Do we live in a country in which the laws are respected or not? . . . We are responding according to the laws of the country."[84]

Mahmoud's comment seems oblivious of the last few years of American history. It also denies the vagueness of the line between offense and defense in "biodefense" research, of the loophole (as the U.S. has chosen to interpret it) allowing creation of more deadly pathogens, and even research into delivery systems, so that the new creations can be defended against. To say that this sort of research into biological weapons *agents* (after all, the basic element of a bioweapon is the germ itself) doesn't involve "creating biological weapons" is straight out of the biolab propagandists' playbook, a semantic game that attempts to skirt a thorny issue and characterize the skepticism of locals as some sort of willful ignorance.

WE'RE OFF TO SEE THE WIZARDS, THE WONDERFUL WIZARDS OF OZ

1

In December, articles on the NBAF decision would reveal that KSU's much-vaunted Biosecurity Research Institute, which had "continued to operate" during the June tornado, and which the NBAF propagandizers frequently cited as evidence of Kansas's ability to operate high-containment germ labs "safely," had actually not started working with select agents yet. In point of fact, KSU hadn't operated anything above a BSL-2, which meant it had not worked with any pathogens that might cause serious or lethal disease after inhalation. And the BRI, this "urgent" contribution to "food safety" research, had actually been built on the concept of a high-tech lab apartment complex (build it and they will come, literally), in hopes of leasing space to homeless germ researchers around the country. And now, with all the empty space, they were hoping that DHS would start the NBAF research early. In fact, that had been one of their selling points to DHS: we have this nice empty BSL-3 Ag facility which we aren't using, and wouldn't you like to start the NBAF stuff early?

That wouldn't keep the Kansas politicians from screaming a few months later how absolutely necessary it was to get NBAF running right this minute—no time for an independent safety assessment—our country's entire agriculture was in urgent risk of collapsing if Osama bin Laden even looked at it crosseyed—and this important agricultural research (much of which had already been under way at Plum Island for fifty years, and continued even as the Kansas germ lab porkers squealed hysterically) had to get started *right now*. That nasty Obama should be sending us $150 million this year instead of only $36 million, because after all, the WMD Commission had just looked into its annual crystal ball and declared that *Yes you are getting very sleepy; you are going to experience a bioterror attack by the year 2013*. Only KSU's beloved NBAF could save the country from being devastated by the African swine fever.

Yes, KSU's new BRI, which hadn't really done any work yet, was working quite well.

Out at the University of Georgia, they weren't so lucky. Their new high-tech BSL-3 Ag facility, which was only new because the one that got started in 1999 had been so screwed up that they'd had to gut half of it and start over, actually had animals in it. And the autoclave had malfunctioned, and animal waste had overflowed the drains and leaked into the basement. *Rhett, are you smelling what I'm smelling?* And the Georgia folks had done what germ-labbers usually do and kept the accident from the public and the press, and the university's community liaison committee, until the newspaper got wind of the incident (pun intended) and actually acted like a newspaper. The result was headlines like "Flood at Univ. of Ga. Germ lab revealed," "Vet school leak kept quiet," and an editorial: "Lab's 'minor incidents' stir major doubts."

And out in San Antonio, their existing BSL-4 lab, the Southwestern Foundation for Biomedical Research, got slammed by the GAO for weak perimeter security. Meanwhile, Greg Harman of the *San Antonio Current* continued to blog scornfully about the gee-whiz journalism of the *San Antonio Express-News*:

> While the variety of media in five other states joust and wrestle over the possibility that Homeland Security's proposed National Bio-and-Agro-Defense Facility may pour cement in

their communities, it has been stunning to live through these past few months of silence in dear little San Antonio—where a stranglehold on all things lab-related reigns.

The Express-News' medical writer penned these deeply analytical offerings in 2007: "Scientists are smiling at the city's chances of landing disease lab," "Scientists like S.A.'s chances for lab," and (my favorite) "Biodefense lab needs support of community."

So imagine my surprise when the day after the first public hearing on the topic brought in both an Express-News editorial board endorsement for the lab and a new headline: "Biolab project's hearing has few naysayers."

During my time in the area, I have yet to see an in-depth treatment of the findings of the General Accounting Office, which suggested the proliferation of germ labs was heightening community risks and increasing the potential for accidents. Or any serious treatment of Homeland Security's Draft Environmental Impact Statement: water, power, employment, accidents?[1]

On August 15, just after the hearings, DHS revealed that "new estimates" had raised the cost of the lab by at least $200 million—from $451 million to a range of $648 to $939 million.[2]

In October, Texas Governor Rick Perry, apparently concluding that NBAF might be an auction, kind of, announced that Texas was prepared to increase its incentives by another $56 million, after the legislature met in January 2009.[3] Kansas, which was holding the high bid, screamed foul, too late, auction over. DHS agreed. Seeds for a future lawsuit.

2

NBAF—Got It.[4] *A Monumental Achievement.*[5] *K-State in Hog Heaven.*[6] *'One of Kansas' finest hours': Politicians, educators ecstatic about all-but-sure ag lab at KSU.*[7] *This decision represents very possibly the single greatest event in the history of Kansas State University.*[8] *Years of effort went into Kansas' successful NBAF bid.*[9]

Out near Plum Island, Governor Rell of Connecticut applauded the decision to locate NBAF away from Plum Island.[10] "While the current lab," he said, "has long been a leading location for animal disease research, given the millions of people who live in the area and the unique environmental treasure it represents it is clearly not the place to locate a research facility that handles some of the most dangerous biological substances known to humanity."

Right, governor. Quite a few of us around the country felt that way about it. But out in Kansas they have the wizards, and the $500 million Kansas Bioscience Authority. It will be safe. Jerry Jaax says so. Abracadabra.

Georgia Governor Sonny Perdue blamed a "tiny contingent" of "anti-NBAF activists."[11] UGA President Michael Adams also thought the opposition hurt Georgia's bid, and cited the "somewhat hysterical charges." "I don't think this facility comes anywhere close to a bioterror lab," he complained.[12] Check out the Kansas PR below, Dr. Adams.

Texas and Mississippi made noises about suing DHS over the decision.

Over in North Carolina, a resident wrote in to GNAT: "I can't thank you enough for all your dedication in making the Bio Lab go away. . . . This has been an ongoing silent stress in my life, not knowing if I should stay and continue my work or plan on moving in a few years because of the Bio Lab. You have all been a blessing to my family and many others. It really shows that *right action* can influence the world for good. Our hearts now have to be with the residents of Kansas."[13]

A bit nervously, the Kansas Livestock Association, which, unlike the Kansas Cattlemen's Association, had signed on to the NBAF bandwagon, issued a press release: *KLA Demands NBAF Include Strict Safety Protocols.*[14] Yes, agricultural "stakeholder." It *will* have strict safety protocols. They told you it would; that was good enough for you then; should be good enough for you now. A safety protocol is what they write on a piece of paper. What people do is what people do.

Kind of like a campaign promise. Understand? Except you don't get to vote the rascal out now.

Meanwhile, the politicians had started the propaganda drive for funding. *Economy Won't Stop Biodefense Lab, Senators Say.* "'I have no doubt this will be funded. It's a national priority. It's a national security priority,' said Senator Pat Roberts of the facility, which would conduct research on anthrax and other potential biological terrorism agents."[15]

But they said it wouldn't research anthrax!! Notice how the PR gets "revised" to suit the needs of the lab propagandists? First, they needed the lab for the scary bioterrorism agents. But then they needed people not to be scared of the scary bioterrorism agents, so it was just a few innocent animal diseases, that's all, honest to goodness. Now it's a national security priority, so it's back to the scary bioterrorism agents.

KSU issued back-to-back-to-back press releases on the vacant apartments at the Biosecurity Research Institute: (1) *K-State's Biosecurity Research Institute and the National Bio and Agro-Defense Facility Would Make Powerful Research Team;*[16] (2) *Opinion: K-State's Biosecurity Research Institute Would Enable Research Programs to Transition from Plum Island without Delay During Construction Phase of NBAF;*[17] (3) *K-State's Biosecurity Research Institute in Pat Roberts Hall Positioned to House Federal Researchers Before NBAF Opens.*[18] It was just a matter of getting NBAF to come and fill the empty spaces.

CHAPTER FOURTEEN

Galveston
Trust Us, Part Two

THE UNIVERSITY OF TEXAS MEDICAL BRANCH AT GALVESTON mourned the Sunshine Project's demise with a few crocodile tears, which it said it had not been weaponized.

Galveston, Texas, was the site for the NIAID's *other* National Bio-containment Laboratory, the place where "community relations" had worked the way it was supposed to. In Boston, an assortment of Rox-bury home folks, Harvard Nobel laureates, metro-area community organizations, and municipalities like Cambridge greeted Boston U's and the politicos' pathogenic brainchild with a series of rowdy public meetings, well-stocked Web sites, and lingering lawsuits. The NIH's going-through-the-motions environmental impact assessment proved too embarrassingly superficial even for a National Academies of Science review panel well stocked with card-carrying members of the biodefense complex.

Ditto for the University of California-Davis, where the university's BSL-4 solicitation was embarrassed by a unanimous negative Davis City Council vote in February 2003.[1]

But in Galveston, they had the propaganda machine well oiled, just in time for the anthrax attacks aftermath. When the *Fort Worth Star-Telegram* ran a worshipful article in January 2002 on "Geeks as Gods," two of the three gods it profiled were UTMB's C. J. Peters and Robert Shope. Both were Detrick alumni; Peters had served as USAMRIID director, and Shope had participated in the Q-fever experiments conducted on Seventh Day Adventist servicemen. ("Three brilliant Texas scientists have willingly risked their lives every day in their work against exotic agents like anthrax and Ebola. Now, with our worst fears realized, they're on the front line in the battle to keep America safe.")[2]

And Galveston's public relations strategies were cited with approval in a 2002 article on the siting of BSL-4 facilities appearing, ironically, in the *Journal of Hazardous Materials*.[3] Ironic because the standard public relations strategy for Galveston and similar facilities is simply to deny that they are "hazardous."

As part of being on the front lines in 2002, Galveston was just getting ready to break ground for one BSL-4 facility, America's fifth. But soon they would be seeking a second, larger BSL-4 facility, one of the NIH's National Biocontainment Laboratories. They had the propaganda spiels down by that point, as a May 2005 *Houston Chronicle* article explained:

> Galveston academics faced a formidable task in the late 1990s when they first sought to build an ultrasecure lab to study the world's most infectious diseases.
>
> How best, they wondered, to convince residents of a narrow island, already vulnerable to hurricanes, that they should willingly open themselves up to another—no matter how unlikely an outbreak might be—potential disaster?
>
> The scientists decided that frankness about the risks and rewards as well as transparency about their plans were the best weapons to combat fear and opposition.
>
> It worked. Partly because of their near universal local support, the University of Texas Medical Branch at Galveston landed a whopping $110 million federal grant last year to construct the Galveston National Laboratory on its campus. Construction likely will begin this month, well ahead of schedule.[4]

And in August 2007, the *Fort Worth Star-Telegram*, noting that a recent meeting of "biodefense experts" had recommended keeping "the community informed and reassured," apparently felt the Galveston facility measured up well, because it had an advisory board composed of people in the community, and even listed accidents on its Web site.[5]

But the *Houston Chronicle* article came on the heels of the brouhaha over Boston U's concealment of its tularemia infections, and the *Chronicle* had gotten its hands on some E-mails between UTMB-Galveston administrators and staff suggesting that, despite the institution's self-description as a poster child for transparency, UTMB might have behaved just like Boston U when it came to infected employees. According to a member of the Biodefense Center's leadership team, in a statement not meant for public consumption, "There is a fine balance between unnecessarily alarming the community when it is at no risk, and being transparent to maintain trust."[6]

In other words, if what they don't know won't hurt them, why tell them? But Hammond, consulted for the *Chronicle* article, responded: "He doesn't understand—or too well understands—the issues. The public has a right to know about laboratory accidents not only because of any immediate health threat that they pose, but also because there is a profound public interest in the safe operation of facilities such as UTMB's."[7]

One reason the public wants to know about accidents, even those that don't present "immediate threats," is so it can judge for itself the "safety" of all the biodefense research in its midst. That is especially important as members of the biodefense complex boast about accidents they say don't happen, or the impossibility of killing somebody—which could only be true if biosafety were foolproof. Every accident shows that foolproof biosafety is a myth. Thus, "informing" the community does not necessarily "reassure" the community. Informing is a matter of honesty; reassuring is a matter of public relations. The biodefense complex will almost always choose public relations over honesty.

And UTMB-Galveston's "openness" image would prove to be just that, a form of public relations. Ed Hammond had already discovered that for himself, when he requested minutes from UTMB's IBC meetings in February 2003:

UTMB, an operator of BSL-4 facilities, chose to fight the request. It argued that a state patient privacy law that exempted the minutes of "medical committees" from disclosure trumped federal guidelines that required release [of] the minutes. **Never mind that there is no information about patients in the minutes, because they aren't about medical care. They are about biosafety.** Despite support from the Freedom of Information Foundation of Texas and the ACLU of Texas, we lost the state ruling.

We appealed to the feds, whose rules say that if UTMB doesn't release the minutes, then the school loses grant money. In came the National Institute of Health's Office of Biotechnology Activities (OBA). Faced with blatant violation of its guidelines, OBA blew some wet kisses to UTMB's biodefense bug jockeys and spanked them with a soggy linguini for good measure. **While UTMB now has public IBC minutes, all of them from before 2004 remain secret—despite federal "rules" that "require" UTMB to release the documents.** We still want them. We asked again and UTMB still won't give them up. OBA promised (in writing) a report on its "investigation"; but never provided one [emphasis added].[8]

UTMB's obstreperousness had won it a place on the Sunshine Project's Top 10 Freedom of Information Failures.

What UTMB's obstreperousness did not affect was its funding from the NIH. During the time the NIH was investigating UTMB's failure to comply with the guidelines, it was awarding UTMB one of its two BSL-4 labs, along with a $350 million regional grant. UTMB was awash with money, in those heady days, plopped down amongst a complacent, compliant populace. "[Al Qaeda] couldn't have done more for microbiology if they had been hired as PR," the *New Scientist* reported. (The writer might have given a little credit to the anthrax mailer[s], but perhaps, in October 2004, she still suspected that to be Al Qaeda.) And UTMB was right in the middle of it all, looking forward to researching some "long neglected" exotic agents—something beyond those persistent, but oh so dull, hospital infections.[9] And as late as December 2007, according to an article in the *Galveston Daily News*, UTMB was still basking in warm and fuzzy daydreams of biotech preeminence.[10] A few months before Hurricane Ike, Ed Hammond predicted rough weather for the daydreams: "Ah, hope springs eternal . . . Dream on, Galveston! It is far more likely for Galveston to be wiped out by a hurricane (again) than to turn into a biodefense-driven biotech hub."[11]

And until the next hurricane showed up, Hammond seemed the one gadfly in UTMB's ointment. In 2006, he bounced back from his earlier disappointment with a new request for UTMB records on laboratory mishaps, biosafety oversight, and correspondence involving certain officials. This time, when UTMB resisted, the Texas attorney general went Hammond's way. UTMB and the UT system were so miffed, they sued to overturn the AG's decision in January 2007.[12]

Later that year, UTMB shared the distress of other biodefense facilities about the Texas A&M revelations. That son-of-a-bitch Hammond had made everybody look bad.

Meanwhile, UTMB's own façades were cracking en masse. In January 2008, an internal door separating bird flu and hemorrhagic fever labs failed twice. Not to worry, said university spokesman Chris Comer: "It's containment within containment within containment and negative air pressure throughout."[13]

Good old negative air pressure. For the biodefense PR people, it redeemed a multitude of sins. It was their god, their refuge and strength. Yet they never understood why people got so excited when the negative pressure collapsed, as it did during the Plum Island and—more relevantly—the Galveston hurricanes, or even the CDC lightning storms over in Atlanta.

In September 2008, Hurricane Ike arrived. As hurricanes go, Ike was a modest Category 2, with wind speeds reaching up to 100 miles per hour. UTMB leaders boasted that the new National Biocontainment Laboratory could withstand wind speeds of 140 miles per hour. One might thoughtfully ask why build it someplace where it would definitely need to. As Ike bore down on September 13, and the facility was being evacuated, the editors at *Effect Measure*, an online "forum for progressive public health discussion and argument[,]" were asking just that. "It's not," they said, "like no one ever thought Galveston would be hit by a monster storm. . . . So it seems a bit odd that the geniuses at the Department of Homeland Security and NIH decided that Galveston was a good place for one of the first two high containment biodefense laboratories to be built after 9/11 (the other is situated in a densely populated neighborhood in Boston, another sterling choice)."[14]

One online comment asked: "What, real estate prices were too high on the San Andreas? Or did they lose a bid to combine it with the tornado tracking center at the University of Oklahoma?" Another, apparently influenced by the illogic of building more biodefense facilities to defend ourselves from attacks by biodefense facilities, suggested: "Why build it there? Think of all the money to be made cleaning up the mess and rebuilding it. Again and again and again."[15]

The hurricane apparently disrupted plans for a September opening of the National Biocontainment Laboratory.[16] Meanwhile, the existing BSL-4 lab took precautions by shutting down experiments, incinerating left-over animals and germs, and adding dry ice to refrigerators and freezers. The Associated Press report on the precautions left a little confusion about precisely who or what was put on ice: "The lab kept emergency medical staff and researchers who had ice in freezers as Ike slammed Galveston. Reyes said the lab has ordered and received 44,000 pounds of dry ice from a Houston vendor on contract."[17]

The researchers may have found the ice preferable to incineration.

The early reports all emphasized how well the various BSL-4 facilities had stood up to the damage: "University spokeswoman Chris Comer said the national lab, which is not yet open for research or specimens, withstood this weekend's storm well. At the campus' other high-security labs, there was minor flooding in an unused basement, and a couple of generators failed, she said, but nothing endangered any research."[18]

Critics would notice several discrepancies and contradictions in the reports of how UTMB battened down. We might pause for a moment here and notice how "nothing endangered any research," yet reportedly there had been no research to be endangered since all experiments had been shut down. Or so we were told.

Spokesmen for Texas Governor Rick Perry and the Department of Homeland Security told CNN that the lab's pathogens were destroyed before the hurricane, yet other sources said they were merely stored away. The CNN story itself carried a comment from a former lab worker, who said

> she would be surprised if all of the pathogens had been destroyed, since some of them are rare and extremely valuable.
>
> The facility is the World Health Organization's center for research on arboviruses, such as ticks and mosquitoes, and tropical disease work, said the student, who asked not to be identified.
>
> It also holds ebola virus and fever-causing lassa virus, sometimes-fatal hantaviruses and anthrax bacteria, she said.[19]

Skeptics wondered early on how the labs would hold up under a Category 3, 4, or 5 storm. (Katrina was a Category 4.) When UTMB spokespeople announced plans to scale back research annually during hurricane season, skeptics wondered about the efficiency of conducting research which one could expect to be regularly cut short by hurricane season. Effect Measure, with a post-Ike blog, "Galveston high-security laboratory: dumb and dumber," asked the common-sense question:

> So if you are in the middle of a complicated experiment and a hurricane might/might not be coming (path and severity still unsure), are you going to (a) hurry to get things done? (b) keep working as long as you can and then hurry to decontaminate everything? (c) gather up as much as you can and put it in a "safe place" until things (literally) blow over (i.e., not do what you are supposed to do because you don't want to lose all the work on the experiment)? (d) Say, "I guess I should just destroy everything according to the plan and start all over when the hurricane issue is resolved?" Remember, some very smart people aren't very good at multiple choice tests.[20]

An anonymous Web site, the Biodefense Barbeque, purporting to be the official Web site of the "Texas Biodefense Alliance," found multiple reasons to be skeptical about the rosy post-hurricane public statements at UTMB:

- Why did it apparently take several days to restore emergency power to the BSL-4s, and how long would it have taken if Ike had been a category 3, 4, or even 5 storm? This is, after all, one of the three most dangerous disease agent collections in the United States.
- What building systems are kept online by emergency power? Does the emergency system keep disease samples in the BSL-3 and BSL-4 labs frozen and is it capable of maintaining negative air pressure in the BSL-3/4 zones (and powering the BSL-4 "space suit" air supply systems)?
- UTMB has brought tens of thousands of pounds of dry ice onto the island to maintain samples frozen. If any of this dry ice is to be destined for BSL-3 and/or BSL-4 labs, can the freezer packs be replenished safely? That is, without loss of negative pressure and without other issues, e.g. carbon dioxide buildup in sealed places.
- What is the actual status of the so-called "Galveston National Lab" BSL-4 terror facility? UTMB's web site last reported on it several days ago and stated that it is intact "as far as we can see", suggesting only an exterior appraisal has been performed.
- What is the status of the Keller Building, which contains a BSL-4 lab (the "Shope Lab"), and whose basement was flooded? What building and labs systems were compromised by the flooding, and did water intrude into higher stories?
- How many experiments were prematurely terminated due to Ike, how many animals sacrificed, and what was the cost of these experiments to taxpayers? If an emergency hurricane shutdown like this is necessary once a year or more, what are the impacts on the scientific and fiscal efficiency of the labs?
- What are the direct costs of this shutdown and recovery effort of the terror labs?[21]

The Barbecue's questions were posted on September 16. An October 17 article in the *Texas Observer*, a biweekly investigative magazine, revealed how pertinent the Barbecue's questions had been. Even as the deputy director for the Galveston National Laboratory claimed that the island's biohazard facilities "came through with flying colors" (or floating colors, depending on the preferred method of hurricane transportation), the *Observer* noted not only the flooding in the Keller Building, with its existing BSL-4 lab, but also the loss of power and water at UTMB, including—shades of Plum Island and the CDC—the failure of backup diesel generators, leaving all the labs without power for at least 36 hours, and some without power for a week or more. Consequently—and this had not been previously reported—negative pressure had been lost for an undisclosed period of time.[22]

The *New York Times* touched on the concerns in an October 28 article, "Bio Lab in Galveston Raises Concerns." After raising the concerns, and noting the possible political reasons why the lab had ended up at Galveston, the *Times* allocated a good deal of the article to UTMB propaganda. "The entire island can wash away and this is still going to be here," James LeDuc, the deputy director, insisted.[23] One assumes future BSL-4 facilities, not needing any sort of foundation, not even an island, will simply hang there smugly, castles erected in thin air.

The *Times* article duly recited the gee-whiz technological reasons why it "is almost impossible for diseases to escape." And there was that old black-magic negative air pressure again: "The air pressure in the laboratories is kept lower than in surrounding hallways. Even if the double doors into the laboratories are opened accidentally [say, by a gentlemanly hurricane: *Excuse me, do you mind if I open your double doors a moment*], air rushes in, carrying pathogens up and away through vents to special filters, which are periodically sterilized with formaldehyde and then incinerated."[24]

Except—someone forgot to tell the *Times*—the negative pressure *failed* during the hurricane. And the *Times* addressed the failure of the emergency power system not as recorded history, but as a hypothetical contingency: "Even if the emergency power system were to fail, the freezers can keep the samples of killer diseases dormant for about four days."[25] Good for the freezers, but some of them apparently went without power for longer than four days, according to the *Texas Observer* article.

Meanwhile, a month after the hurricane, the *Austin American-Statesman* was wondering how many people would show up for the presidential election in a county where 25,000 registered voters had been displaced by Ike, streets were littered with rubble, electrical power was still being restored, and schools were just reopening.[26]

November 11 was dedication day for the National Biocontainment Laboratory, however, and time for a few chirpy articles in the Texas media. "National Lab a 'bright spot' for UTMB."[27] "Galveston biodefense lab was fortress during Ike."[28] Unlike the rest of Galveston, and the UTMB campus, the new lab didn't blow, or wash, away. Marvelous. It was, in fact, completely unfazed: "When other labs on the island were forced to shut down before the storm, the National Lab had two backup generators and tens of thousands of pounds of dry ice to safeguard their specimens."[29]

No, folks, the National Lab didn't have to shut down. Because it hadn't opened yet. This reveals a distinct public relations advantage of biodefense expansion. If your old lab (by old we're talking about a Galveston BSL-4 facility that came on line in 2003) has a few problems—as Galveston's Shope building apparently did—you can just ignore it and talk about your new lab, the way a basketball coach with an 8-12 record talks about a promising new recruit. The new lab's work was unfazed. Because it wasn't doing any.

So it really was irrelevant that it had (if it had) "two backup generators and tens of thousands of pounds of dry ice to safeguard their specimens." Except other articles indicated the dry ice went to the labs that were doing research. Because they had pathogens they needed to put in the freezers. Because their backup generators weren't working so well.

One day after dedicating the NBL, UTMB announced that it was laying off or dismissing about 3800 employees, roughly a third of its work force. It cited financial exigency in the aftermath of Hurricane Ike, which had "ripped through Texas on September 13, killing at least 37 and causing $710 million . . . worth of damage to the medical facilities, only $100 million of which is insured."[30] How reassured should one feel about the safety of research at an institution that insures approximately ten percent of the value of its physical plant and facilities?

On December 2, the Texas Faculty Association filed a lawsuit on behalf of the dismissed employees. Those dismissed included 127 faculty members, 83 of whom were tenured or tenure-track. Tom Johnson, the executive director of the Texas Faculty Association, said Texas officials "had a predetermined agenda that had nothing to do with Ike" but was focused instead on privatizing universities and weakening the tenure process. The lawsuit charged that the nine members of the University of Texas Board of Regents violated the Texas Open Meetings Act when they used closed-door meetings and conference calls to reach their decision of mass layoffs.[31]

One of those laid off was reportedly the attorney who handled biosecurity issues for the new lab (not the same attorney who handled UTMB employment issues, one presumes).[32]

What UTMB-Galveston tried next vividly demonstrated the unreliability of promises and public statements made by members of the biodefense complex. Many Galvestonians would feel they'd been had by their biodefense poster child.

On April 29, the *Dallas Morning News* revealed that Houston Republican Sen. Joan Huffman had offered a bill, SB 4646, which would make most information about select agents in Texas labs confidential. The bill had passed out of a Senate committee with no debate, and quickly passed the whole Senate.[33]

It soon became known that UTMB-Galveston was the bill's true sponsor. Hammond, now living in Colombia, was the only critic cited by the initial *Dallas Morning News* article, but other opponents soon joined him: the Texas Press Association; the Texas Daily Newspaper Association; and the Freedom of Information Foundation of Texas. The Galveston City Council would pass a resolution opposing the legislation, and members of UTMB's community advisory board would criticize not only the legislation, but their unhappiness at not having been consulted or even "informed."

The *Galveston Daily News*, represented by associate editor Michael Smith, editor Heber Taylor, and publisher Dolph Tillotson, would soon home in on the legislation's problems, and their sense of betrayal. They were joined by *Houston Chronicle* columnist Lisa Falkenberg.

"The deal," Falkenberg wrote, "was simple: Galveston residents wouldn't fight a maximum-containment laboratory dealing in agents like anthrax and Ebola on their island if the folks running the place vowed to provide not only solid security but also transparency."[34]

Taylor wrote: "University officials promised extraordinary openness in keeping the public informed. People, including me, were impressed by those promises. Lab supporters argued with neighbors that university officials could be trusted to tell us what was being studied at the lab and to inform us of any problems."[35]

Now, said Smith, UTMB was pushing a bill in the Texas legislature "that would bury a whole body of information under a thick layer of secrecy."[36]

And it was being notably duplicitous in explaining why it sought the legislation. UTMB first claimed the bill didn't expand the pool of information that can legally be withheld from the public, but simply aligned Texas law with federal legislation.

What the bill in fact declared, however, was that all "information that pertains to a biological agent or toxin" defined by federal law as a select agent was "confidential and exempted from the requirements" of the Texas Open Records Act. As Smith complained:

> The bill says nothing about aligning the state law with the federal law, as the author and supporters claim was the point. Pressed to explain how that linkage would work, none, not even Huffman who introduced the bill, could do so.
>
> Pressed to explain why "is confidential and exempted from" doesn't mean what it appears to mean to reasonable members of the public and attorneys specializing in open-government law, none could do so.[37]

Hammond had already argued that federal law protected only detailed, site-specific information about where an agent is located, and "little else."[38] And Smith discovered that state law already exempted "[i]nformation that is confidential or exempted from disclosure under a state or federal constitutional provision, statute or common law." That, he noted, was "about as aligned as laws get."[39]

So then Huffman, Smith scoffed,

said she didn't believe the [new] bill was as broad as its own words. She noted that all laws were open to interpretation and abuse. She said that, if this one were badly interpreted or abused, she would be willing to change it in the next session. Asked to change it now, she declined.

What this bill asks is that the people of Texas trade a right to information for an unenforceable promise that those holding the information will always act in good faith, that the Texas attorney general will always act in good faith, and that politicians will always act in good faith.

That's a bad deal.

Supporters claimed they only wanted a very specific, limited bill. If that's the case, they should write one.[40]

Hammond said if the bill passed, it would be impossible to detect the sort of gross violations that occurred at Texas A&M.[41] Tillotson said it would "allow UTMB to stonewall virtually all requests for information. . . . Under the proposed law, if the newspaper heard a rumor about a suspected leak of deadly germs, as it did just months ago, the university could refuse to release any documents about the alleged accident or security breach. That's dangerous to you and your children."[42]

He noted the emphatic statement of the Texas Open Records Act preamble:

It is the policy of this state that each person is entitled . . . at all times to complete information about the affairs of government and the official acts of public officials and employees. The people, in delegating authority, do not give their public servants the right to decide what is good for the people to know and what is not good for them to know. The people insist on remaining informed so that they may retain control over the instruments they have created.[43]

What UTMB wanted, Tillotson concluded, was "the right to decide what is good for the people to know and what is not good for them to know." He cited UT's recent decision to fire 2800 people after a closed meeting of the school's regents as the most recent example of a bad decision made behind closed doors.[44]

UTMB was all over the map in defending the legislation. Smith dealt summarily with some of this bobbing and weaving in a May 17 commentary, as UTMB was offering a somewhat revised version of the bill:

> Callender [UTMB's president] argued Thursday, as others had argued before, that the change was needed to protect security devices and methods. But those already are explicitly and broadly exempted under state law.
>
> Supporters claim they need to align the state law with the federal law. But the state law already exempts anything exempted under federal law.
>
> Supporters say they want to protect the privacy of researchers. When we pointed out that researchers already were well-known and publish under their own names, the rationale became protecting the identities of janitors and laboratory technicians.
>
> It's interesting, as an aside, that the bill at hand doesn't mention exempting records about janitors or other support-level employees, but instead says the "identity of an individual authorized to possess, use or access a select agent," is exempt. We hope the pool of janitors possessing, accessing or using Ebola and anthrax is rather small.[45]

James LeDuc, deputy director of the Galveston National Laboratory, let slip something of the legislation's true motivation:

> LeDuc believes the bill is necessary to protect sensitive information, like names of lab workers or substances being transferred, but also to protect the medical branch from an individual he said has been filing a harassing volume of requests, including some for sensitive information.
>
> "I'm absolutely convinced that all of this is in reaction to this single jerk that [sic] is abusing the intent of the law," LeDuc said.

The attorney general has ruled against UTMB in litigation stemming from those requests, leading UTMB to sue the attorney general to keep information secret.[46]

The "single jerk" was, of course, Hammond, who had initially harassed UTMB by asking it to comply with the NIH Guidelines, and to provide the minutes of its last two IBC meetings. Around the country, requests by Hammond from other institutions had revealed that few IBCs were functioning as they were meant to do—if they were functioning at all.

Hammond had then followed up with an additional request in 2006. Both requests had been resisted in their entirety by UTMB, on fairly specious grounds that denied the basic intent of the Texas Open Records Act. There was no mention of a "harassing volume of requests" in UTMB's arguments before the Texas attorney general. And it is hard to see why Hammond's requests should have constituted an extraordinary burden for an institution devoted to openness the way UTMB boasted it was.

But the only things transparent about UTMB these days were its lies.

In his commentary "'Security' bill backers forfeit trust," Smith took a close look at one of the charges UTMB had made against Hammond:

> Some supporters of controversial Texas Senate bill 2556 are asking for the public's trust, while acting in ways that inspire the opposite.
>
> For example, James Kelso, charged with keeping the University of Texas Medical Branch and the Galveston National Laboratory in compliance with freedom-of-information laws, told me and a Daily News reporter, and apparently also told a Dallas Morning News reporter, that someone had requested "door access codes" for the laboratory.
>
> That phrase conjures images of a keypad on a secure door and at least implies someone sought information about how to enter the laboratory.
>
> That's scary. It's also not true. What Edward Hammond, formerly of the watchdog group the Sunshine Project, requested was "door access logs."

Not the same thing, of course.

While Kelso implied someone sought to breach the laboratory, what Hammond pretty clearly wanted to do was document who was coming and going from it.

Medical branch officials have used that request as justification for the sweeping change proposed in SB 2556.

It's interesting, then, that the Texas attorney general denied Hammond's request for logs, a mere list of names, using the existing law.

After having got caught in the public relations henhouse with a fat roaster in its mouth, UTMB looked for a hole in the wall, the floor, anything. As the Galveston City Council went on record opposing SB 2556, UTMB offered up a revised, narrower version. It still had one eye on Hammond, however:

> Thursday, as the city council attempted in its resolution to offer language narrowing the bill's effect, medical branch president Dr. David Callender warned that it needed to be broad enough to cover "creative" requests for information.
>
> That was an admission the bill is, was designed to be, what critics have argued all along—too broad.
>
> At least now we can all appear to be arguing about the same legislation.
>
> As best we can tell, however, creative requests are any the medical branch deems to be troublesome in any of several broad ways at any given time.[47]

The bill finally passed by the legislature exempted the medical branch from disclosing information about: (1) the specific location of a specific agent; (2) information identifying an individual whose name appears in documentation of the chain of custody of select agents, including a materials transfer agreement; (3) the identity of an individual authorized to possess, use, or access a select agent. The new bill said if a resident of another state was present in Texas and authorized to possess, use, or access a select agent at a Texas facility, the national lab has to disclose information "that would be subject to disclosure under the laws of the state in which the person resides."[48]

Opponents, especially from the press, said the legislation was still too broad and created an unnecessary layer of secrecy.[49] They had repeatedly challenged UTMB to show why the current law was inadequate, and felt they'd never received a satisfactory answer from UTMB.

UTMB sought to regain some sort of propaganda footing. Dr. Stanley Lemon, principal investigator for the Galveston National Laboratory, issued a prepared feel-good statement: "Thanks to the valuable input [read that opposition] of the Galveston community and our local legislators, this bill strikes an effective balance between the safety and security of UTMB's researchers, staff and facilities; helps safeguard vital research collaboration; and enables our ongoing commitment to transparency and preserves our ability to share information with the public."[50] Of course, the only one threatening UTMB's "ability to share information with the public" was UTMB itself. UTMB might still have big pull with the state legislature, but its status with the community of Galveston was in serious doubt. "A trust has been broken," said advisory board member Jackie Cole.[51]

A TEXAS-KANSAS "PUSS WAR," AND THE PROPAGANDA SHIFTS TO CONGRESS

RIGHT AFTER DHS REVEALED THAT KANSAS WAS ITS "PREFERRED alternative," the *San Antonio Express-News* groused that San Antonio had lost mainly because Kansas offered more dollars, and that:

> On March 1, the department pulled a big surprise on the finalists.
>
> Make us a cash offer, Homeland Security said. Send the offer by March 31.
>
> The state legislatures in at least two finalist sites happened to be in session. The Kansas Legislature offered $105 million. Mississippi sweetened its bid by $88 million.
>
> Texas could offer no bid by the deadline because its Legislature meets only every other year and did not have a 2008 session.[1]

John Kerr, chairman of the Texas Biological and Agro-Defense Consortium, "said Homeland Security violated federal bidding regulations when it allowed contenders to modify their bids so late in the process."[2] John Dublin, chairman of the non-profit foundation that owned the land for the proposed Texas site, said the process "had been completely changed in midstream" and "turned into an auction."[3]

A spokeswoman for Kansas Gov. Kathleen Sebelius said: "We're not sure why this caught them off guard as DHS made it clear that it 'strongly encourages cost sharing, including cost sharing in kind from state and local jurisdictions, that could be applied toward construction and operations of the NBAF.'"[4]

Both Texas and Kansas were being way too cute about the March events. Back then, Sebelius had been "surprised" and, apparently, disturbed by DHS's "new" request that the consortium fund the costs of utilities. Texas had said: don't know why she's surprised: it's been part of the plan all along.

But Kansas had gone to the legislature and upped its bid. Now it was Texas complaining that it had been surprised and Kansas saying: don't know what on earth they had to be surprised about.

Once again, is there no honor among thieves?

The answer is no. A press release from Senator Brownback's office in March 2009 congratulated the senator for getting $100 million for Kansas projects into the omnibus appropriations bill. Brownback had voted against the bill itself, however, because of its "bloated spending."[5]

In February, new DHS Secretary Janet Napolitano toured the Biosecurity Research Institute to consider KSU's offer to advance the date Plum Island germs could invade Manhattan. Napolitano said all the things that would be expected in a good PR event: "In reality, this is the best place in the United States to have this type of facility"; and safety would be "built into every square inch."[6] Using, of course, the patented secret safety ingredient out of which such facilities are always constructed. *Lab Plans Gain Traction: Napolitano Says Obama Won't Reverse Course*, the *Topeka Capital-Journal* reported.[7]

About forty protesters picketed outside the BRI during Napolitano's visit. Two days later, the *Manhattan Free Press* reported on the concerns of the 1500 members of the Kansas Cattlemen's Association, who had met the previous week at the Kansas Cattlemen's Convention. They were skeptical about KSU Dean of Agriculture Fred Cholick's assurances that he would not be afraid to live next to NBAF,

or would not worry if his brother-in-law had a beef and sheep operation next to it. They received no answers to their questions about
compensation in the event of an outbreak.[8]

About the same time, KSU chose a new president, Kirk Schulz,
who had been Vice President for Research and Economic Development at Mississippi State University, a member of the Mississippi
NBAF consortium. Apparently, Schulz's NBAF experience didn't hurt
his chances.[9]

In March, KSU announced that it was a finalist to become the new
home of USDA's Arthropod-Borne Animal Diseases Research Facility
(ABADRL), which studies diseases transmitted to livestock by
insects.[10] KSU was on the verge of developing the sort of synergy
which will characterize the Fort Detrick "campus." NBAF could
supply the zoonotic diseases, and ABADRL could supply the insects
to spread them.

Also in March, the Kansas congressional delegation sent a letter to
President Obama, grousing about the fact that his proposed budget
only included $35 million for NBAF. The government should be
spending $150 million next year, the Kansans said: "The need for
accelerated animal disease research has never been stronger [i.e.,
since Kansas was selected for NBAF]. The U.S. government needs to
move aggressively to guard against a bioterror attack."[11] (Notice how
the Kansas delegation just put the bioterror back in bioterrorism?
Perhaps the allegedly "agriculturally-focused" NBAF might try developing a genetically modified cow able to sniff out airborne pathogens
and moo in some unique fashion?)

"Fully fund my pork!!!" one blogger quipped. (And added: "These
people are just shameless.")[12]

In April, the Texas consortium followed through with its lawsuit
threat, filing a complaint for declaratory and injunctive relief in the
United States Court of Federal Claims, alleging numerous irregulari-
ties in the decision-making process, improper political influence, and
major disregard for public safety.[13]

One part of the complaint stated:

In the FEIS, DHS specifically recognized that a moderate to
severe tornado would cause catastrophic consequences to the

NBAF, including the destruction of the exterior and interior walls and the release of deadly pathogens that would potentially decimate the U.S. cattle industry, endanger the health and safety of other livestock, wildlife and humans and result in billions of dollars of damages. Furthermore, in the FEIS, DHS noted that according to its own construction specifications, the NBAF would not be built to withstand wind speeds in excess of 119 miles per hour, despite the fact that only a moderate F-2 tornado, a common occurrence in Kansas, has wind speeds of 113 to 157 miles per hour. *See* FEIS p. 3-431. DHS failed to acknowledge the likelihood that the Manhattan Campus Site may face a much greater threat than an F-2 force tornado.[14]

One blogger noted with a certain wry amusement that "individuals from respective 'losing' states (I use that term loosely) namely Georgia and Texas are now making the 'public safety' argument that consists of the same concerns expressed by citizens during the site selection."[15]

One of the complaint's allegations related to political ties between Thornton and Cohen:

The Kansas Consortium efforts were led by Tom Thornton, a former staffer of Illinois Congressman and Speaker of the House Dennis Hastert, who, with Hastert's intervention obtained an earmark for a project in Illinois in 2005 from an arm of the Defense Department headed by Navy Admiral Jay Cohen. In 2006, Admiral Cohen left the Pentagon to become Under Secretary of DHS with direct responsibility for the NBAF site selection. At about the same time, **Thornton was hired to lead the Kansas NBAF effort, who then retained former Congressman Hastert to lobby Admiral Cohen to select the Kansas site.**[16]

Nothing surprising about influence-peddling on a project like this. Coming from Texas, of course, it's all an intricate pot and kettle dance.

In Kansas, the media response was partly along the pot and kettle lines. Duh, they don't have tornadoes in Texas? The *Lawrence Journal-World* called it "a case of sour grapes," a "Texas tantrum."[17]

The Kansas politicos were serving up pious rhubarb. Kansas Governor Sebelius said "legal action would undermine the NBAF mission and would place 'our national security and food supply [pork] at risk.'" Kansas's two senators said, "Threats of frivolous lawsuits only delay the critical research to be conducted at the lab."[18]

Thornton was serving up slogans. "Kansas offers a solution, not a site," he said.[19]

In July, the judge dismissed the lawsuit without prejudice, judging that it had been filed prematurely because NBAF might never actually be funded or built.[20] That wasn't what Kansas's congressional delegation were telling their constituents, however.

At the end of July, the GAO issued another NBAF report, this more sharply worded than the previous one. The GAO said DHS relied on a rushed, flawed study (now in the EIS) in its decision to locate NBAF on the mainland. It used "unrepresentative accident scenarios"; "outdated modeling"; and "inadequate site information" in its analysis. DHS's analysis was not "scientifically defensible." The report again attacked a main propaganda argument DHS and the Kansas consortium had made for NBAF: that FMD research was being done at Winnipeg. But the GAO said NBAF would have a less sophisticated method for containing releases than the Winnipeg lab, while handling as many as 10 times the animals.[21]

Reps. Stupak and Dingell called the NBAF decision a "foolish tempting of fate"; Stupak announced his intention to schedule a hearing. The hearing would eventually be postponed because of conflicts with health reform legislation, then drop off the map altogether.

The Kansas congressional delegation—no surprise here—attacked the conclusions of their own investigative body, finding the analyses of DHS, which had chosen the Kansas site, a bit more to their liking.[22]

The Texas consortium tried to piggyback onto the GAO Report, even though that report was critical of any mainland location.

Meanwhile a funding battle was in the works, the Senate appropriating $36 million to begin NBAF construction. Senator Brownback stated he had personally added the appropriation to the Senate bill.

The House, on the other hand, wouldn't appropriate anything until an independent risk assessment had been conducted. The House's caution was largely the work of Rep. David Price, D-NC, who had been roundly criticized for his support of NBAF by some of his North Carolina constituents, and who had thrown them as a bone the assurance that the facility would not be funded until an independent risk assessment had been conducted.

A TRIP TO KANSAS, A LETTER TO CONGRESS

IN SEPTEMBER 2009, IN THE MIDST OF THE ONGOING NBAF BATTLES, I made a trip to Manhattan, hoping to meet with some of the NBAF opponents and get some history of the situation there. The Manhattan Alliance for Peace and Justice was screening a film, *Anthrax Wars* (with an accompanying or complementary book, *Dead Silence*), that had some relevance for this book. I decided a trip to see the film would accomplish two things at once.

The book and film both focus on a string of suspicious deaths of bioweapons/biodefense researchers, including Bruce Ivins; David Kelly, the British researcher and WMD inspector; Vladimir Pasechnik, the Soviet defector; and Frank Olson, the USAMRIID researcher who allegedly committed suicide as the result of a CIA experiment slipping LSD into the drinks of bioweapons researchers, as mentioned here in Chapter Four. Olson supposedly started having hallucinations afterward and jumped out a hotel window. That's the official story. The book explores reasons to suggest that Olson may have been murdered because he wanted out of the bioweapons dirty-tricks business and was starting to talk.

Much of *Dead Silence* is in a similar vein. Proving any conspiracy is difficult, of course; proving several is all the more difficult. But it's not hard to be suspicious when the FBI's case against Ivins for sending the anthrax letters is still shadowy; when we lived for eight years under an administration which claimed dictatorial powers for itself in secret documents, and fabricated evidence of weapons of mass destruction in Iraq; and when the official story for a guy's death is that a CIA agent slipped LSD into his drink.

There were two different sessions; I went to both. Eric Nadler, one of the book/film's author/producers, was there both to talk and to answer questions. Most of the attendees seemed to be opponents of the Kansas NBAF effort, there to see if there was any useful information, or sparks that might be struck, or advice that might be offered. But the first night I found myself sitting beside Tom Thornton of the Kansas Bioscience Authority, whose name I take in vain at various points in this book. Linda Weiss, a realtor and the current president of the Manhattan Chamber of Commerce, was also there. Ms. Weiss had sent out an E-mail to members of the Kansas NBAF consortium warning of the "misinformation" that was about to be presented, and asking for all good NBAF propagandists to come to the aid of their germ lab. Apparently, despite having conducted the most thorough and successful propaganda campaign in germ-lab history (except possibly for the one in Galveston), the Kansas germ-lab honchos were frightened some sudden spontaneous revolt might occur and jeopardize their NBAF money. Curiously, neither Weiss nor Thornton bothered to speak at the film screenings. Apparently they noticed no one in the audience whose opinion mattered, who might suddenly be swayed by the anti-bioweapons slant of the evening. Manhattan Broadcasting, however, did cancel a planned interview with Nadler after receiving a concerned E-mail from Weiss. As I've said, they know how to squelch dissent in Manhattan, which is one of the reasons, in addition to the generous incentives and the empty BSL-3-Ag lab, why Homeland Security liked Manhattan so much.

I had sent an E-mail to members of the Kansas No-NBAF mailing list before I arrived, telling them what I was up to and hoping that I could talk to some of them. What I hadn't realized is that, not knowing me, people couldn't be sure I was who I said I was, rather than a spy or infiltrator, and so I had to reach out again after I arrived.

I talked at length to several people, and for shorter periods to others. In my travels around the campus and town, I also talked to several people who hadn't made their opposition to NBAF public, and I'm convinced there's more opposition in Manhattan than DHS and the pro-NBAF consortium acknowledged. In the absence of a formal canvass or survey, however, it's still hard to know what the majority view is. The opponents I spoke to recognize that they started organizing too late, perhaps thinking rational discussion would be enough, perhaps hoping NBAF would go someplace else without their efforts, not recognizing the more-or-less unprincipled juggernaut they were facing.

Manhattan is by and large a company town, however, with KSU being the company. Many people I spoke to, and written comments on the DEIS, attest to the fact of people feeling intimidated. And the parking situation at KSU during the scoping and DEIS hearings probably worked to the advantage of the pro-NBAFers.

Then there was the *Manhattan Mercury*, running propaganda interference on a regular basis.

At the time of my visit, the Kansas opponents were focusing most of their attention on writing Congress (not their own congressmen, of course, which at this point would have been a waste of time), just after the GAO's latest critique and questions about NBAF funding.

The startling thing I discovered was the proximity of the proposed NBAF site to the existing KSU livestock farm and the football stadium. Some crucial extra precautions that Plum Island took after the 1978 FMD outbreak are gone in Manhattan. Large quantities of livestock are located in fields adjoining the NBAF site, and those fields are located next to football stadium parking. Of course it wouldn't matter, if there were such a thing as a failsafe facility.

After returning to Kentucky, I wrote my own letter to the relevant appropriations committees, mentioning the book in progress and what I had discovered in Manhattan. I didn't hear back from any congressmen, but I did hear from Sushil Sharma of the GAO, expressing interest in the book and letting me know of a new GAO report just being released.

I also heard from Bill Felber of the *Manhattan Mercury*, advising me that my open letter had been forwarded to him by a "friend." He wanted to know who was publishing my book and when ("we might want to talk with you when it appears"). I was suspicious of his motives, suspecting that Felber was trying to get info about dates for the pro-NBAF coalition. I wondered, too, how he had learned of my letter: I had carefully avoided sending it to members of the Kansas delegation. I had sent a copy to members of the No NBAF in Kansas list, however. Either, I figured, a chummy, presumably pro-NBAF congressman or staffer had passed it on to the Kansas delegation, who had passed it on to Felber—or maybe some acquaintance of Thornton had passed it on to that former staffer—or there was a "spy" on the No NBAF in Kansas list.

The Senate had passed a bill providing immediate funding for the start of NBAF construction; the House had withheld funding, pending an independent safety assessment. In October, a compromise version came out of conference, appropriating $32 million, $27 million of which could be used immediately for design, planning, and site preparation, but withholding $5 million designated for construction until a safety assessment is completed by DHS and reviewed by the National Academy of Sciences. It wasn't much of a cautionary move, in light of the GAO reports, but it was a teaspoon-sized quantity of caution better than nothing.

CHAPTER FIFTEEN

You Can't Make a Vaccine for Everything
Some Ways Forward

IN THE SUMMER OF 2009, AS I WAS WRITING THIS BOOK, MY father, just recovering his strength from a stem-cell transplant several months earlier, began suffering from nausea, diarrhea, and loss of appetite. Over the next two months, he slowly starved to death, as a result of medical inattention and confident misdiagnosis, after being batted back and forth between two hospitals. By the time the UK Medical Center finally assumed responsibility for their own transplant patient, hospitalized him, and began giving him nutrition, it may already have been too late. He eventually succumbed to a combination of pneumonia, multiple antibiotic-resistant infections, and a rare disease called HLH. At the end, shrunken and bruised all over his body from the incessant blood draws and IVs, breathing from a ventilator, blistered with bed sores, and bleeding through his skin, he looked like a cross between a famine and an Ebola victim.

I may tell the story of that summer in a future book, after I've reviewed the two boxes of medical records our family received—and compared what the doctors wrote in their notes to what they told us. If so, I expect to have a few things to say about medical bureaucracy, assembly-line medicine, the hubris of "experts," and disrespect for the elderly. I mention the matter here for good reason: antibiotic resistance and the deaths it causes are not a hypothetical threat like that of bioterrorism, but an ongoing crisis which gives every indication of getting worse. And antibiotic resistance is just one of several *existing* public health problems slighted in the post-2001 bioterror obsession.

The CDC itself estimated, for instance, that in 2002, in the midst of the panic caused by five anthrax deaths—*inflicted by America's own biodefense complex*—1.7 million Americans acquired hospital infections after being hospitalized for other reasons—and 99,000 of those patients died of the infections. Brad Spielberg, the author of *Rising Plague: The Global Threat from Deadly Bacteria and Our Dwindling Arsenal to Fight Them*, believes more than 300,000 Americans die from infections each year. (He arrives at those figures by noting that sepsis, influenza, and pneumonia alone kill 250,000 to 300,000 people per year, then adjusting upwards to acknowledge other infection-related causes of death.)[1]

A 2007 article in the *Journal of the American Medical Association* estimated that 90,000 invasive methicillin-resistant *staphylococcus aureus* (MRSA) infections—infections of the bloodstream caused by just a single bacterium—occur each year in the United States, causing around 20,000 deaths.[2] Other studies suggest that, if other illnesses MRSA can cause are taken into account (urinary tract infections, pneumonia, skin infections), MRSA probably causes as many as a million U.S. infections per year.[3]

Ed Hammond of the Sunshine Project posted statistics based on examination of the CDC's Aug. 5, 2004 *MMMR Weekly*, showing that of six priority "biodefense" bacterial pathogens—tularemia, brucellosis, plague, anthrax, meliodosis, and glanders—there were *no* cases of the latter three in the U.S. in 2003, only one case of plague, 104 cases of brucellosis, and 129 cases of tularemia. By contrast, there were 335,104 reported cases of gonorrhea, 44,232 of AIDS, 34,270 of syphilis, 21,273 of lyme disease, 16,281 of acute viral hepatitis, 14,883 of tuberculosis, 11,647 of whooping cough, 2,356 of drug-resistant streptococcus, 2,232 cases of Legionnaire's Disease, and 2,077 cases of Rocky Mountain Spotted Fever.[4]

The 758 NIH-funded microbiologists who signed an open letter in 2005 protesting NIH's new bioterror obsession took a slightly different approach, averaging the U.S. cases for 1996-2003, finding an average of 122 cases of tularemia, 103 of brucellosis, 5 of plague, and 3 of anthrax (representing an "average" incorporating the 2001 letter attacks), and no cases of glanders or meliodosis. In contrast, there were 17,403 cases per year of tuberculosis, 42,957 of salmonellosis, 38,007 of syphilis, 346,765 of gonorrhea, 685,508 of Chlamydia, and 4,371 invasive streptococcal infections, 3,083 of which were by a drug-resistant strain of pneumonia.[5]

Worldwide, TB, malaria, and AIDS kill over five million people annually;[6] 36,000 Americans die yearly from garden-variety influenza; the number of American gonorrhea cases has risen to 700,000 annually, and 13% of these are now drug-resistant. Then there are the various sickenings arising in the U.S. from our industrial, global, poorly inspected food system: 110,000 E. coli infections,[7] 1.4 million salmonella infections. (A single incident, the 2007 E. coli spinach contamination, infected 206 Americans, hospitalizing 100, and killing 3.)[8] According to the Trust for America's Health, 76 million Americans—1 in 4—are sickened by foodborne illness each year; 325,000 of them are hospitalized; and 5,000 die.[9]

The most dangerous result of the post-2002 bioterror obsession may ultimately prove to be this distraction of attention from more serious problems. While war-gamers and biodefense think tanks are scaring government officials with exaggerated bioterror scenarios, hundreds of thousands of Americans die annually, prematurely, from diseases that don't need the assistance of bioterrorists. Discussing the rapid rise of drug-resistant tuberculosis around the world, Spielberg warns that antibiotic resistance alone "is a phenomenon that threatens to send us back to the dark ages of health and medicine."[10]

This is the situation Richard Ebright, and the other 757 microbiologists who signed the 2005 open letter to NIH, had in mind, when they wrote that "[t]he diversion of research funds from projects of high public-health importance to projects of high biodefense but low public-health importance represents a misdirection of NIH priorities and a crisis for NIH-supported microbiological research."[11]

In a recent interview, I asked Dr. Ebright (who heads the prestigious

Waksman Institute at Rutgers University) if the concerns he and the 757 other microbiologists expressed in 2005 were still relevant, and he said yes, that there have been some changes "around the margins" but that in general the misdirection of funding priorities still continues. Ebright said the basic problem is that post-2001, in response to the anthrax mailings, NIAID—the branch of NIH responsible for infectious disease research—assembled a panel "primarily of biodefense and military scientists" to draw up a list of Category A, B, and C pathogens. The pathogens they put on the list were those that had actually been developed or considered for development as bioweapons agents. This was then used to frame priorities not only for biodefense research, but for infectious disease research in general. Ebright says that since 2003, when the list was drawn up, in all areas other than HIV and influenza, "all or essentially all" NIAID research solicitations have specified that proposals responding to the solicitations must address Category A, B, or C agents.[12]

"So we now," Ebright says, "end up in the remarkable position that virtually all infectious disease research other than AIDS research is carried out on agents that cause no or nearly no morbidity in the United States. In contrast, the top bacterial disease killers in the U.S. are not in Categories A, B, and C, receive little in support, and have lost support relative to their previous levels. The money has been redirected to diseases that don't kill Americans.

"Basically a list of priorities that was put in place by a group of biodefense and military scientists immediately after the anthrax attacks has frozen—solidified—as the way the federal government now apportions infectious disease research. The top bacterial killer is called Streptococcus pneumonia: it kills more than twice as many people each year as AIDS, but it's not a priority agent, and it receives little support. Next is Staphylococcus aureus, which also kills more than AIDs in the U.S., and is not a priority agent.

"One has taken what would have been a rationally-driven, public-health-driven set of priorities, and inverted it. We have an upside down pyramid with the agents of least importance receiving the most resources. All of that is new, and all of that is a response to the 2001 anthrax mailings."[13]

The 2005 letter signers supported their concerns with some stark statistics, drawn from NIH's own CRISP database. They compared the number of NIAID grants awarded from 1996 to 2000 for research on the six previously mentioned bioweapons agents to those awarded from 2001 to January 2005. Thirty-three grants had been awarded for the first five-year period; 497 grants had been awarded for the succeeding four-year-and-a-month period—133 of them in January 2005 alone. This represented a 1400% increase.[14]

In contrast, the number of grants awarded under one NIAID review group for non-biodefense-related projects had declined from 490 to 289, a 41% decrease; the number awarded under the other review group had declined from 627 to 457, a 27% decrease.[15]

Meanwhile, the Sunshine Project, using the same NIH CRISP database, had compiled figures focused on actual dollar amounts. Those figures indicated that NIAID grant funding for bioweapons bacterial agents that cause anthrax, glanders, meliodosis, brucella, and plague had increased 2388% during the three-year period 2002-2004 as compared with the prior three-year period (going from $7 million to $185 million). Funding for bioweapons viral agents Ebola, Marburg, Lassa, and smallpox had increased 1900%. Funding for research on HIV, gonorrhea, lyme disease, acute viral hepatitis, and tuberculosis (not generally considered bioweapons agents) had all declined during the same period. These funding differentials were the direct result of some pathogens being considered potential or actual bioweapons agents.[16]

One must remember that the NIH is the country's chief *public health* agency. Some readers may have watched science fiction movies in which invading aliens covertly occupy the bodies or assume the identities of hapless Earthlings. It is as though, around 2003, a bunch of generals and military bioweapons researchers had quietly assumed the identities of public health professionals at the NIH and CDC and subverted both agencies to their bioweapons-related research goals.

And as residents around Dugway, Utah and Frederick, Maryland already knew, it wasn't like the military didn't already have biodefense labs able to produce and disseminate killer pathogens to frighten the public. Why should biodefense hijack the country's main public health agencies as well?

But thanks to the co-option of public health by biodefense, we now have a huge proliferation of new facilities, some of them merrily plugging away at creating dangerous new versions of influenza, anthrax, and other pathogens—greatly increasing the likelihood both of a serious accident affecting the public, and of another incident like the 2001 letter attacks, involving a "diversion" from our own facilities. But just say no, mad scientists.

The basic goal of this research—to develop a "countermeasure," a magic bullet, for each of fifty-plus select agents—is ultimately futile. Because if terrorists did at some point in the future obtain the advanced technological expertise and facilities to produce killer pathogens which the WMD Commission, the UPMC, and other current prophets of inevitable disaster attribute to them, these same Afghan cave magicians would presumably have the ability to genetically engineer their way around any existing countermeasures. And the researchers developing defenses against hypothetical genetically engineered pathogens have no way of knowing what specific "engineerings" the cave magicians might choose, something even bio-Cassandras like former Soviet bioweaponeer Ken Alibek acknowledge. And if—as is more likely—a would-be terrorist simply infiltrated one of our own facilities—(s)he would presumably steal or divert something (perhaps a new genetically tweaked pathogen) that didn't have a countermeasure yet.

In her recent history of biological weapons, Jeanne Guillemin says of the current U.S. biodefense approach, with its emphasis on "ingenious technological solutions": "The assumption on which this technologically-driven policy was based is that, with only general reference to an actual bioterrorist threat, scientific research can defeat each established biological agent or their permutations or perhaps achieve defenses applicable to entire categories of diseases or toxins."[17]

Guillemin contrasts this magic-bullet approach to one centered on strengthening the basic public health infrastructure:

> Although largely ignored in current policy, the importance of basic, accessible, and good-quality health services, a combination of public health programs and clinical care, would be the best protection against an unusual deadly outbreak . . . Epidemics are opportunistic, targeting the more vulnerable populations. The better the general health levels in society, so

that disparities are diminished, the less destructive the impact of a large disease outbreak would be.[18]

Guillemin notes that access to "basic, accessible health care" helps people know where to go during an emergency, and that public health education programs are essential to eliciting informed participation at the local level. "Without informed public cooperation," she says, "the American technological investment—in new laboratories and biological research, in biosensors, drug and vaccine stockpiles, and electronic tracking of disease—is nearly useless."[19]

Another recent questioner of the post-2001 biodefense expansion, Professor William Clark, also points out that our public health systems "will . . . provide our major defense against the results of a bioterrorist attack."[20] Yet the state of public health is not good. A 2002 article in *The Nation*, published in the opening throes of the biodefense boom, described a "public health 'train wreck,'" the result of "[t]wo decades of managed care and government cuts" that had left "a depleted system with too few hospitals, overburdened staff, declining access for patients, rising emergency-room visits and an increasing number of uninsured." Former CDC Director Jeffrey Koplan warned that dropping wads of cash on biodefense without addressing these basic problems was like "building walls in a bog."[21]

A 2004 article in *Harper's* also described the collapse of public health, focusing on the huge numbers of Americans without health insurance, the breakdown of infectious disease and medical surveillance and prevention programs, the failures of Food and Drug Administration regulation, the long decline in public health expenditures as a proportion of overall health spending, the decline in state and local public health budgets despite the infusion of federal monies for "bioterrorism response."[22]

The $7 billion in "preparedness funds" that the federal government provided to states and localities between 2003 and 2009 is one category of post-2001 spending which did help the basic public health infrastructure. Yet even this funding was in decline, according to the Trust for America's Health (TFAH), which points to a 27 percent cut in preparedness funding since 2005 when adjusted for inflation.[23]

Even as it notes the help provided by federal emergency preparedness funds over the last several years, the TFAH states flatly:

> Public health infrastructure has been underfunded for
> decades, as documented in repeated assessments by CDC,
> IOM, and other experts. The gaps in infrastructure have ham-
> pered the nation's ability to respond to the H1N1 outbreak as
> effectively and quickly as possible. A key lesson learned from
> the outbreak is that until the infrastructure is strengthened,
> shortcomings in the core public health system will always
> leave the country unnecessarily vulnerable to emerging
> threats.[24]

The TFAH says of H1N1 that "if the virus had become more severe,
our nation's public health system could have been strained beyond
the breaking point."[25]

All through 2009, newspaper articles noted the cutbacks in state
and local public health funding around the country, in California,[26]
New York,[27] Washington,[28] Texas,[29] and Georgia,[30] for instance. The
TFAH also noted these "devastating budget and staff cuts," involving
an estimated loss of 15,000 positions in local health departments since
the beginning of 2008, and similar cuts at the state level,[31] the lack of
"an integrated, national approach to biosurveillance that is capable of
responding to catastrophic health threats or to more familiar problems
such as the contamination of food supplies;"[32] and the serious problems with surge response in hospitals and emergency rooms.[33]

Just as the current approach—developed during the Bush administration and continued by the Obama administration—neglects the
basic public health infrastructure in favor of research focused on
developing technological magic bullets, it also scorns any notion of
strengthening the Biological Weapons Convention, purportedly out
of skepticism that any verification scheme can assure perfect compliance. (Other reasons suggested by various commentators as the "real
ones" include concern about endangering the trade secrets of pharmaceutical and biotech companies, and a desire not to have the U.S.'s
own bioweapons-related research scrutinized too closely.)

This rejection of diplomacy and international efforts represents just another example of the biodefense two-step. Whatever threats—remote or otherwise—bioterrorism represents can never be *completely* eliminated by *any* means. Certainly they cannot be eliminated by dog-chasing-its tail efforts to develop a "countermeasure" for every conceivable genetic reformulation of fifty-plus potential bioweapons agents. Any realistic approach to biodefense would recognize that international efforts at control, while they cannot offer perfect reassurance, at least reduce the risk. The same cannot be said for the U.S.'s mindless proliferation of biodefense research, which has made us less safe by increasing the odds of a serious accident and of a bioterrorist diversion from our own facilities, and by undermining efforts at international control—both by our withdrawal from verification efforts and by our setting dangerous moral examples for other countries.

Random bits of this have begun to leach into the subconscious recesses of a few congressional representatives and federal policymakers. But most of the current proposals for regulatory reform have either been window-dressing efforts formulated by the biodefense complex itself, or proposals—like vesting regulatory authority with the Department of Homeland Security—that threaten to make matters worse. None of the current proposals addresses the most serious problem—that we have too many dangerous labs, doing too many dangerous things. As the GAO's Keith Rhodes suggested in 2007, the huge amounts we have spent on biodefense since 2001 have made us less safe than we already were. And as 758 NIH-funded microbiologists warned us in 2005, one way in which we are less safe is that we have been distracted, in our research efforts and our funding decisions, from more serious public health problems—from diseases that do not need a suspect WMD crystal ball to make them a threat, but which already kill hundreds of thousands of Americans prematurely, each year. The crisis in antibiotic resistance alone—or a real influenza pandemic arising from our factory farms or our reckless labs—threaten to decimate our society in a way only the most extravagant bioweapons scenarios could match.

Part of the way forward is simple: simply divert some of the funds in the biodefense spigots to more effective and less dangerous uses. Richard Ebright believes there urgently needs to be a reassessment of the primary pathogens list that drives NIH and federal research spending across multiple agencies—a reassessment by a panel not dominated by military and biodefense researchers—"to identify those infectious disease pathogens that have highest impact in the U.S. and warrant the highest resources." If that were done, he thinks, the issue of spending would solve itself. This is real "biodefense," he says—defending against threats that actually kill large numbers of people. He also thinks if funding were reprioritized, many of the new high-containment facilities would move back to researching major public health diseases, which generally do not require as high a level of "containment" because they represent diseases already prevalent in society.[34]

Reducing the numbers of high-containment facilities researching bioweapons agents is the single most important step in making the whole biodefense enterprise safer. But Ebright says research with "potential bioweapons agents" also needs to be subjected to considerably more stringent safety and security regulation. With respect to safety, Ebright says the NIH's Guidelines and the BMBL—"recommendations that are appropriate and developed on sound scientific and biosafety principles," but that are largely voluntary, should be "mandatory for all institutions, government and non-government, public and private."[35]

Security, which is not an issue for most pathogens, obviously is an issue for potential bioweapons agents. Ebright says there needs to be a substantial increase in oversight and regulation related to physical security, personnel security, and operational security for bioweapons agents, including video monitoring of any storage area and any workplace. "Video monitoring should be comprehensive, it should be around-the-clock, and records should be stored indefinitely. That's not currently the case. If there had been round-the-clock video monitoring, all indications are that the 2001 anthrax mailings would never have taken place—and we wouldn't be having this discussion."[36]

Ebright also thinks a two-person rule should be mandatory for all institutions, public and private, working with bioweapons agents—two persons present in the room at all times. "You need two persons to open a bank vault or a safe deposit box, but you don't need them to work with these agents. Had there been a two-person rule in effect, all indications are that the 2001 anthrax attacks wouldn't have happened. Either one of these [video monitoring or a two-person rule] would have prevented them. But neither was in place then, and neither has been put in place since then at most institutions."[37]

In terms of personnel security, Ebright says the current background checks are merely database checks, fairly cursory and thus inadequate. "They would detect someone whose name is listed on a terror watch, they would detect someone who is in the United States illegally, they would detect someone who had previously been convicted of a felony, they would detect someone who had previously been declared mentally incompetent or who had been institutionalized involuntarily, but they wouldn't detect much of anything else. They wouldn't have kept off, for instance, the shoe bomber—they wouldn't have kept off Mohammed Atta—they didn't keep off the mentally disturbed individual apparently responsible for the 2001 anthrax attacks. The security evaluations need to become more comprehensive, and they need to include mental health information."[38]

Ebright reiterates that real security requires reducing the numbers of institutions and individuals with access to bioweapons agents—numbers which he says have exploded by a factor of 20 to a factor of 50 since the 2001 anthrax mailings. "That means that even if security had been increased, risks would still be higher, just from the larger base of institutions and individuals performing the work. Unfortunately, there was no significant increase in security, so risks are now very substantially higher."[39]

"Those numbers need to come down. Of course, those numbers would come down by themselves if we set a rational prioritization of research support."[40]

Ebright also talks about the need to address the "dual use" implications of biodefense research—the fact that some research produces or uncovers new areas of vulnerability. "In some cases, that's the express purpose of the research; in some cases, it just happens as the research goes on." But "there needs to be put in place some form of prior oversight of research projects to identify those that have a high risk of uncovering vulnerabilities, particularly those involving bioweapons agents, and that considers whether there needs to be special restrictions on such research projects. And that needs to be mandatory as well."[41]

Ebright believes this could be done through the IBCs, but only if the IBCs are made mandatory and the IBC rules are enforced by the NIH.[42]

With respect to the Lieberman-Collins bill still before Congress at the time this book was being completed, Ebright agrees that the Select Agent list could be streamlined or stratified. He says, however, that "the devil is in the details," and worries about the number of people— primarily recipients of bioweapons agents research support—who've been pushing to divide the list into high-threat agents and low-threat agents, keep the high-threat agents subject to the current Select Agent regulations, and drop the others from the list. "That," he says, "doesn't improve security in any way. The current regulations are inadequate; if you just remove from the Select Agent list things that might not have needed to be on it, you still would have the same inadequacies for the high-threat agents." The problem with the Lieberman-Collins bill is that "while the bill is clear on the idea that the list needs to be reviewed, it's very unclear on what happens after that occurs."[43]

Ebright is not enthusiastic about making DHS the regulatory agency for the high-threat agents, particularly given the identity of the current Director of Science and Technology. "The kind of thing that I get concerned about is that if the Collins-Lieberman bill would be passed—a decision might be made to draw the line at smallpox as the only high-level agent, to remove all other agents from the list, and to have the smallpox rules administered by Tara O'Toole. That would be an example of a very bad outcome in my opinion."[44]

Unfortunately, bad outcomes seem more likely than good ones in the present environment, with the biodefense-complex-infested "Commission on Prevention of Weapons of Mass Destruction Proliferation and Terrorism" shouting hysterically that the bioterrorists are going to get us by the year 2013 at the latest, and assigning "F grades" to an Obama administration that it claims is "not doing enough"— even though that administration has carried forward intact the reckless biodefense policies of the Bush administration.

The Commission first predicted that terrorists would use WMDs "somewhere" by 2013 in its December 2008 *World at Risk* report.

While it alluded to the 250 interviews it said it had conducted with "governmental officials and nongovernmental experts," and to eight commission meetings and one public hearing,[45] it provided little concrete evidence to support its claim that such an attack could be expected.

The Commission followed up *World at Risk* with an October 2009 document apparently intended to put pressure on the Obama administration to ratchet up the biodefense obsession even further. It cited three areas of concern that it claimed "suggest that the Obama administration does not agree with the Commission's assessment of the biological threat."[46] (Since the Commission had managed to acquire a crystal ball—at a bargain price—from an itinerant bioterror gypsy, anyone who questioned their own degree of alarm would of course be wrong.) Of the three "areas of concern," one seemed partly symbolic: Obama, who had appointed a WMD Coordinator, had not appointed "a senior official whose sole responsibility is to improve America's capability for biodefense." The other two concerns involved complaints about inadequate funding: despite sevenfold increases in such funding since 2001, the biodefense complex still wanted more, more, more.

And who did the Commission rely on for their assessment of the funds needed? Our old friends at the University of Pittsburgh Center for Biosecurity, who "recently estimated that $3.39 billion per year in medical countermeasure development support would be required to achieve a 90 percent probability of developing an FDA-licensed countermeasure" for each item on a Health and Human Services wish list focused mainly on anthrax, filoviruses like Ebola, Junin, and smallpox.[47]

The Commission lay on the propaganda horn again with its January 2010 "Report Card," taking the Obama administration to the woodshed for not having ratcheted up the bioterror hysteria even further. The Commission issued the Report Card just before the President's 2010 State of the Union address and proposed budget. The Commission gave the administration and congress an "F" for failing to sufficiently "enhance the nation's capabilities for rapid response to prevent biological attacks from inflicting mass casualties." And what did it find especially troubling? "[T]he lack of priority given to the development of medical countermeasures—the vaccines and medicines that would be required to mitigate the consequences of an attack." And whose figures did the Commission rely on, once again, in proposing a tenfold increase in funding for bioterror countermeasures? The University of Pittsburgh's Center for Biosecurity, though the Center's role in arriving at the figures was no longer cited in this particular Commission release.[48] Mysteriously, the figures had entered the realm of bioterror folklore.

And what staff members of the UPMC Center for Biosecurity served as members of the Commission's Staff for 2009-2010? Colonel Randall Larsen, Executive Director for the Commission, and Gigi Kwik Gronvall, Science Advisor. And what biodefense funding plum is UPMC presently in hot pursuit of? A facility for production of vaccines and "medical countermeasures."

A huge irony here: the Commission noted, in its first report, "the unbridled growth in the number of high-containment laboratories since 2001, which has occurred without effective and coordinated federal oversight."[49] It also noted the increased security problems that this proliferation created, stating that "it would generally be easier for terrorists to steal or divert well-characterized hot strains from a research laboratory or culture collection," and recognizing the difficulty of terrorists acquiring such pathogens otherwise.[50]

Yet this huge proliferation of new facilities, with all the dangers they present, occurred precisely because of the hyping of bioterror after 2001. The Commission's exaggerated predictions—their push to devote more resources to biodefense research—help create the situation they warned about in December 2008. Perhaps that's why concerns about theft or diversion have been muted in later publications of the Commission; instead, *The Clock Is Ticking* emphasized the "ease" of creating biological weapons.[51]

The Graham-Talent Commission's "Report Card" was a propaganda device that received considerable coverage in the mass media. A rebuttal by the Center for Arms Control and Non-Proliferation's Scientists Working Group on Biological and Chemical Weapons received considerably less attention—which is unfortunate. The rebuttal systematically critiques the unrealistic and exaggerated assumptions of fictional bioterrorism exercises and threat scenarios used to hype the bioterror threat, pointing out that "Offensive, including terrorist, use of biological agents presents major technical problems. That is why the Soviet Union, United States, United Kingdom and others needed to spend vast sums for decades in order to research and develop biological weapons. Even then the results were considered an unreliable form of warfare."[52]

An even more detailed critique of the propaganda disseminated by groups like the Graham-Talent Commission was provided in a 2005 publication prepared for and published by the Strategic Studies Institute of the U.S. Army War College. Milton Leitenberg's *Assessing the Biological Weapons and Bioterrorism Threat* systematically debunked claims by people like former Senate Majority Leader William Frist, who said "The greatest existential threat that we have in the world today is biological," and predicted a bioterror attack in the next ten years. Leitenberg ticked off a quick list of concerns more pressing than bioterrorism: ongoing ecological destruction and global climate change; the interaction of global population and waste product growth with diminishing resources; war; deaths due to poverty; major diseases like malaria, tuberculosis, AIDS, diarrheal diseases, measles, and ordinary flus, which together kill millions of people annually. As the biodefense complex shows some current signs of trying to associate itself with and hitch a ride on new pandemic flu fears, it is instructive to read Leitenberg's figures on 2006 NIH funding: $1.76 billion for biodefense; only $120 million for all influenza research.[53]

Leitenberg proceeds to contrast multiple examples of "thoughtless, ill-considered, counterproductive, and extravagant rhetoric" with what available intelligence shows of actual terrorist capabilities. In light of the Graham-Talent Commission's references to Al Qaeda's "interest" in bioterrorism, Leitenberg's review of Al Qaeda's minimal "achievements" (and apparently minimal interest) in bioterrorism in Afghanistan between 1999 and 2001 are especially relevant. Leitenberg's review notes several ironies. First, information on computer discs apparently belonging to Dr. Ayman al-Zawahiri indicated expenditures of only a few thousand dollars on some rudimentary equipment, which al-Zawahiri considered "wasted effort and money"; and in an April 15 memorandum, al-Zawahiri indicated, ". . . we only became aware of them [biological weapons] when the enemy drew our attention to them by repeatedly expressing concerns that they can be produced simply with easily available materials."[54] Apparently Al Qaeda had not found this to be the case.

A second irony was that the U.S. Department of Defense, through the Defense Reutilization and Marketing Service, had been selling surplus biological equipment (at discount prices) to middlemen who then resold it to buyers in the Philippines, Malaysia, Egypt, Canada, Dubai, ands the United Arab Emirates. In 3½ years, the DRMS had sold "18 safety cabinets, 199 incubators, 521 centrifuges, 65 evaporators, and 286,000 full-body protective suits. One can compare this to the reports of the few pieces of elementary equipment found in the al-Qaida site in Afghanistan, and the significance that was given to those finds."[55] There are certain dark parallels here to the biodefense complex's ongoing effort to discover—and in some cases publish the details of—new ways to make killer germs more deadly. With friends/defenders like this working for us, who needs enemies?

Leitenberg offers a detailed critique of the Atlantic Storm January 2005 exercise, conducted by Colonel Randall Larsen and Dr. Tara O'Toole of the UPMC Center for Biosecurity. O'Toole is now the controversial Undersecretary of Science and Technology for the Department of Homeland Security; Larsen just finished serving as Executive Director of the Graham-Talent WMD Commission. The exercise involved a "Radical al-Qaida Splinter Group" obtaining, producing, and distributing a dry-powder preparation of smallpox. Leitenberg examines why the assumptions associated with each stage— obtaining, producing, and distributing—are implausible in the extreme, along with the assumptions about the spread of the ensuing fictitious epidemic.[56] Leitenberg notes that the exercise received significant favorable media attention, and that "As best [as] is known, not a single report of the exercise raised any questions whatsoever regarding the plausibility of the basic assumptions of the scenario." Yet—the enthusiasm of the *Washington Post* to the contrary—"the exercise *was* science fiction because the scenario antecedents are not 'now,' and they were not [then,] in the least plausible."[57]

Leitenberg lists the characterics of bioterror hype between 1995 and 2000 (characteristics that became even more pronounced after 2001) as follows:

- spurious statistics (hoaxes counted as "biological" events);
- unknowable predictions;
- greatly exaggerated consequence estimates;
- gross exaggeration of the feasibility of successfully producing biological agents by nonstate actors, except in the case of recruitment of highly experienced professionals, for which there was still no evidence as of 2000;
- the apparent continued absence of a thorough threat assessment; and,
- thoughtless, ill-considered, counterproductive, and extravagant rhetoric.[58]

The dramatics of the Graham-Talent WMD Commission, carefully timed in an effort to affect the 2011 biodefense budget, are just the latest version of the hype Leitenberg critiqued in 2005. In fact, some of the same people and groups implicated in the 2005 phony scenarios have been providing their helpful advice to the Graham-Talent Commission.

The huge biodefense expansion which has been created and is now being frantically defended by the biodefense public relations machine is almost certainly a greater danger to our country than any current would-be bioterrorists. It is our own biodefense complex, after all, which attacked us in 2001. It is that very complex which is recklessly producing and proliferating both the knowledge and the pathogens which might be used to attack us again. It is that complex which, proudly assuring us it is safe because it hasn't killed anybody yet, raises killer flus from the grave, genetically engineers forms of the bird flu which are both deadlier than nature's versions and are easily spread between people, and spends most of its energies trying to develop "countermeasures" for diseases that rarely kill anyone outside the minds of bioterror fictionists. Meanwhile, really serious health problems like antibiotic resistance are shrugged off.

This situation is what The Scientists Working Group had in mind when it issued its response to the latest Graham-Talent propaganda. The group's summary is worth quoting in its entirety:

- The bioterrorist threat has been greatly exaggerated.
- New bioweapons assessments are needed that take into account the complex set of social and technical issues that shape bioweapons development and use by state and non-state actors, and that focus on more plausible threats than the worst-case scenarios that have largely driven discussion to date.
- Continuing to emphasize and spend billions of dollars on measures to specifically counter bioterrorist threat scenarios distorts our national understanding of the important issues in public health, and diverts scarce scientific talent and resources away from more pressing public health and natural disease threats.
- While it has been argued that spin-offs from biodefense programs contribute to countering natural diseases, the converse is more likely: direct targeting of effort and expenditure on natural disease threats would provide much greater public health benefit, and spin-offs from these programs would significantly strengthen resistance to bioterrorism.
- Bioterrorist threats need to be seen and addressed within a wider public health context—as just one of the many possible ways in which infectious agents may harm human, animal, and plant health.[59]

This is all basic common sense, a quality shared by many scientists, infectious disease doctors, and analysts not blinded by and beholden to the almost $60 billion wrongly and wastefully spent on the wrong "biodefense" strategies. When a pasture on Mark-Welborn Road in Pulaski County, Kentucky is a semifinalist for the country's second biggest biodefense facility, simply because the county's congressman is chairman of the U.S. House's Homeland Security Appropriations subcommittee, when ordinary people around the country have to spend years of their lives rebutting the lies of politicians, universities, and economic development interests, just to keep bioweapons agents out of their back yards, something is terribly wrong. Yet now this same rampant, out-of-control complex wants even more money, more facilities, so it can do whatever reckless research it can get a grant for, so it can then spend its time and our money worrying about "countermeasures" for it created in the first place, in an endless, but profitable (for a few), cycle of dogs-chasing-their-own-tails. Enough is enough. And too much is more than enough.

Just Trust Us

TWO EVENTS OCCURRED JUST AS THIS BOOK WAS GOING TO THE printer that underlined ironically the dangers of an out-of-control biodefense complex. On May 11, 2010, the *Wisconsin State Journal* revealed that Dr. Gary Splitter, a tenured professor in the University of Wisconsin-Madison School of Veterinary Medicine, had been suspended from lab research for five years after conducting unauthorized experiments "that could have posed a risk to human health." The experiments, which brought the university a fairly insignificant $40,000 fine, involved our old Texas A&M friend brucella.[1] Brucella is a disease that, under natural circumstances, affects few Americans, and does not pose a public health threat outside the fevered imaginations of bioweapons and biodefense researchers. Perhaps that explains why Splitter decided to create a more dangerous antibiotic-resistant version.

That Splitter did such dangerous research is not unusual in the twisted labyrinths of biodefense. He didn't seek the appropriate authorizations, however. UW–Madison Provost Paul DeLuca suggested that Splitter's evading of approvals was the only real problem, that the "experiments didn't pose a 'different level of danger or concern' than other disease studies." But the dangers involved in creating antibiotic-resistant pathogens are exactly why approvals are required in the first place. Incidentally, one person working in Splitter's lab contracted brucellosis; university officials said they didn't know whether the antibiotic-resistant strain was involved.[2]

The Splitter incident raises two additional concerns. The CDC and NIH had apparently been "involved" since 2008; as is typical in the secretive world of biodefense research, however, they had kept the situation from the public. The residents of Madison, Wisconsin—and, indeed, the neighbors of more than 1300 "high-containment" germ labs around the country—might have preferred a little less concern about the sensitive psyches of germ lab researchers and a little more frankness with the public.

Perhaps even more troubling were Splitter's efforts to blame the situation on graduate students, and on the university itself for failing to "properly educate researchers about guidelines for working with antibiotic-resistant strains." Troubling because Splitter served on the university's Institutional Biosafety Committee—the closest thing to a germ lab police force that UW-Madison or most universities have— between 2004 and 2006. During this period he acted as a "secondary reviewer" for some of Yoshihiro Kawaoka's reckless "extreme sport" experiments with the 1918 flu, and made motions at IBC meetings to approve that research.[3] With Splitter pleading ignorance of basic safety protocols related to his own dangerous research, we have to wonder what level of scrutiny he and others applied to their germ lab colleague—who just this year announced (proudly) his creation of the world's most dangerous pathogen: deadly forms of the bird flu easily transmissible between humans.

The university's $40,000 fine can be put into context by comparing it to the $50,000 in legal fees incurred by Robert Ferrell in defending himself from *criminal prosecution* for having provided an almost completely harmless pathogen for an anti-genetic-engineering art exhibit—an act that created no risk for the public whatsoever.

Even as the *Wisconsin State Journal* was revealing one scientist's recklessness and blatant disregard for existing regulations, others in the scientific community were being equally reckless in declaring that even those inadequate regulations were hobbling select agent research. The pretext for their complaints was a modest little study just published in the *Proceedings of the National Academy of Sciences*, examining the pace of publications about anthrax and Ebola after enactment of the Patriot Act and 2002 Bioterrorism Preparedness Act, in an effort to determine whether those laws had caused "negative consequences" for select agent research.[4] The study had been conducted in the context of pervasive bellyaching by members of the biodefense complex, including grief by members of the National Science Advisory Board for Biosecurity, lamenting "the unmeasurable cost of select agent research that was not done, [and] suggesting that unnecessary inhibition of this science amounts to a national security and public health threat."[5] The *PNAS* article found to the contrary that "Indicators of the health of the field, such as numbers of papers published per year, number of researchers authoring papers, and influx rate of new authors, indicated an overall stimulus to the field after 2002." The article did conclude—on thin grounds—that there had been a "loss of research efficiency" during this period, because it had found "an approximate 2- to 5-fold increase in the cost of doing select agent research as measured by the number of research papers published per millions of U.S. research dollars awarded." More funds had been awarded, and more papers had been published, but the average "cost" of the papers had risen.[6]

Articles in both *The Scientist* and *Nature* dove head first into this latter statistic to draw unwarranted conclusions that alleged practitioners of the "scientific method" should be ashamed of. The *PNAS* article had warned: "We can make no claims that the trends detected were caused by the anti-bioterrorism laws."[7] The *Nature* and *Scientist* authors, afflicted with the omniscience and omnipotence that inhabitants of biodefense heaven assume as their birthright, however, felt no such compunction. *Nature* led with the headline: "Regulations increase cost of dangerous-pathogen research."[8] *The Scientist* drew a similar conclusion, and led with the headline "Biosecurity Laws Hobble Research"[9]—even though the *PNAS* study found that the "hobbling" hadn't discouraged hordes of new researchers.

To conclude that research is "less efficient" because the amount of funding per paper has risen—without a careful examination of the research and papers involved—is simply irresponsible. Likewise with attributing the increased "costs" to increased regulation. Why not attribute it to the explosion in overall funding, in the suspicion that the costs of a research project—the projected expenses associated with a grant application—expand to absorb the funds that are offered? Diners at an all-you-can-eat buffet don't necessarily take their calories elegantly or efficiently. A horde of researchers new to the field of bioweapons agents were inevitably going to spend a lot of time learning the ropes and spinning their wheels. And their successive efforts to plow the same fields as their forebears probably mean more and more dead ends.

The biodefense propagandists had been shouting loudly for some months that regulations were driving researchers away. Now that a study suggested this wasn't the case—the government had thrown the biodefense money out there and the researchers had flocked to it—they were shifting their tactics, grasping at evidentiary straws to argue that regulation was making their research less "efficient."

It's quite possible that some of the regulations, the recordkeeping requirements, for instance, might actually make select agents' research more "efficient," not to mention more safe. I really don't want our select agent labs run like my household ("Do you remember where we put the bird flu?") With USAMRIID finding 9,000 uninventoried germ vials (many or most of them, presumably, from the "unregulated" days) in its nooks and crannies, with 350 or so U.S. facilities playing around with anthrax—the agent that started the bioterror craziness in the first place—I would prefer to err on the side of too much regulation, an error we're in no danger of making any time soon.

Out in the Wild West, they sometimes hobbled the horses at night to keep them from running off, "pursuing their own agendas." I'd like to hobble Bruce Ivins or whoever it was really mailed the anthrax letters. I'd like to hobble Gary Splitter and his fun and games with antibiotic resistant brucella. I for sure would like to hobble Yoshihiro Kawaoka in his preferred pastime of creating pandemic flu monsters.

What is it about these dudes and dudesses of Germland? They tell us about all their wonderful standard operating procedures, but when it comes to a simple matter of getting authorization for a dangerous experiment, or reporting an accident, they suffer from a peculiar type of attention deficit disorder, unable to decipher the most basic of regulatory requirements. After working a few years in the biodefense realm, they undergo genetic mutations that destroy their capacity for logical discourse and leave them with the ethics of elixir salesmen in a traveling medicine show.

Could anything be more "inefficient," or crazy, or unsafe, than the rampant, rampaging propagation of biodefense facilities and research after the 2001 anthrax attacks—perpetrated from one of those very facilities? And our biodefense cowboys want this Wild West state of affairs to continue, because they feel frustrated by the extra paperwork and the thought that someone is looking nervously over their pathogenic shoulders?

Get a life, germ labbers. Those of us paying for your reckless research, and feeling threatened by it, feel we ought to have something to say about what you're up to. You want to get out of select agent research? Be our frigging guests. There's TB to worry about, and HIV, and all those antibiotic-resistant infections that don't need you to create them before they kill people. And none of them will get the attention they should while you keep doing the biodefense two-step.

Late-Breaking News from Germland and The War on Terror
Gifts and Givers That Keep on Giving

AT LONG LAST, THE PANDEMIC FLU MONSTER WE'VE ALL BEEN WAITING FOR

The bird flus the CDC created to be transmissible between humans (*see* Chapter Seven) had at least been less virulent than the parent bird flu strain, which kills about half the people it infects. In February 2010, however, a three-country team led by University of Wisconsin-Madison germ lab cowboy Yoshihiro Kawaoka—apparently jealous of the recent attention given to the H1N1 swine flu—proudly announced that *they* had created bird flus that were not only transmissible between humans, but were even deadlier than the original bird flu. The CDC had stopped with only 63 possible combinations. But the Kawaoka team said *they* had kept going; *they* had created all the 254 possible new, mutated offspring of H5N1 bird flu and H3N2 human flu; and *they* had created mutated viruses which were both deadly and transmissible. In fact, 22 of their creations were deadlier (in mice) than the parent H5N1 bird flu strain.[1] The Kawaoka boys had thus taken every step possible to demonstrate the possibility of human/bird flu pandemics except actually releasing their creations into the general public.

That final step might yet be taken, since the research was performed at something called "BSL-3 enhanced."[2] Kawaoka was criticized for moving experiments with the 1918 flu (see Chapter Seven) from a BSL-4 lab at Winnipeg to his BSL-3 lab in Madison. As yet, however, there seems to have been no similar outcry on this occasion. One has to ask what BSL-4 containment is for, if not for what can now be considered the deadliest and most dangerous pathogen in the world, one which our germ lab cowboys just created? I propose something called BSL-4 enhanced for such research: a special BSL-4 containment facility constructed in a desert somewhere (not at Dugway, the state of Utah begs) to serve as a "permanent residence" for Kawaoka and his colleagues, including the CDC regulators who consider BSL-3 a sufficient "level of containment."

WHAT SCIENTIFIC REVIEW? THIS CASE IS CLOSED

On February 19, 2010, the FBI announced it was formally closing the Amerithrax investigation, without waiting for the carefully circumscribed review of its science it was permitting the National Academy of Sciences to conduct.[3] The *New York Times* editorialized that the agency's "voluminous circumstantial evidence" was "largely persuasive," but said "its report leaves too many loose ends to be taken as a definitive verdict." It pointed to the lack of direct evidence tying Ivins to the letters, as well as to the problematic nature of "the investigative work that led the F.B.I. to conclude that only Dr. Ivins among perhaps 100 scientists who had access to the same flask, could have sent the letters."[4]

In a commentary published in the January 24, 2010 *Wall Street Journal* just before the FBI "closed the case," Edward Jay Epstein also cited the lack of direct evidence, and the remaining questions about how the 100-plus others with access to the "telltale flask" were eliminated as suspects. Epstein raised questions as well, however, about the FBI's explanation of the silicon—an additive used to weaponize anthrax in the 1960s. The FBI suggested the silicon had been naturally absorbed from the water and nutrients used to grow the anthrax spores. Epstein noted the extraordinarily high silicon content which the FBI eventually revealed had been present in the Leahy letter, and the 56 futile attempts by scientists at Lawrence Livermore National Labs to replicate anything approaching this high silicon content: "Even though they added increasingly high amounts of silicon to the media, they never even came close to the 1.4% in the attack anthrax. Most results were of an order of magnitude lower, with some as low as .001%."[5]

On February 25, the U.S. House of Representatives passed an amendment offered by Representatives Holt and Bartlett to the Intelligence Authorization Act, requiring an independent review of the evidence by the Inspector General of the Office of the Director of National Intelligence.[6]

In April 2010, a former colleague of Ivins at USAMRIID, microbiologist Dr. Henry S. Heine, told the NAS panel and members of the press that the arithmetic of spore cultivation meant Ivins would have needed to process more than 26 gallons of liquid anthrax to produce the quantity of spores mailed in the anthrax letters, and that "He couldn't have done that without us knowing it."[7] Heine also "told the panel that biological containment measures where Dr. Ivins worked were inadequate to prevent the spores from floating out of the laboratory into animal cages and offices. 'You'd have had dead animals or dead people,' he said."[8] An interesting comment, since Ivins supposedly conducted his work in one of Detrick's BSL-3 labs.[9] BSL-3 is the level of containment normally used for working with aerosolized anthrax. These facilities—Detrick included—are fond of telling worried neighbors how absolutely failsafe their BSL-3 containment protocols are.

A PFIZER WHISTLEBLOWER IS VINDICATED

April 1 has been a popular day for germ lab propagandists to make outrageous claims about lab safety. Ironic, then, that on April 1, 2010, a federal jury awarded $1.37 million in damages to a former Pfizer molecular biologist, Becky McClain.[10] The damages were awarded against Pfizer, a member of the Alliance for Biosecurity, for having retaliated against McClain by firing her after she began to suffer from a potassium deficiency causing intermittent paralysis. She attributed the illness to a co-worker's research with a genetically engineered HIV-derived lentivirus, research performed on her work bench for a month without her knowledge. A member of Pfizer's safety committee, McClain had raised safety concerns both before and after her illness.[11]

McClain's attorneys were advising her not to speak to reporters in April 2010 because punitive damages had yet to be awarded, and because they expected Pfizer to ask the judge to overturn the jury award.[12] But she had spoken and written about her experiences in some detail in the spring of 2009, in connection with the California Coalition for Workers Memorial Day. Her articulate address on that occasion, available at http://laborvideoblip.tv, both reinforces and broadens the concerns of this book.

Ms. McClain reinforces this book's message by warning how the public is being kept in the dark about the unacknowledged risks of genetically engineered pathogens, and by noting the almost complete absence of regulation in this area. McClain broadens the concerns of this book by expanding them beyond biodefense per se to embryonic stem cell research, mostly conducted at the BSL-2 level in thousands of private and university labs.

McClain worked at the Pfizer lab in Groton, Connecticut for nine years. As a member of the Pfizer safety committee, she saw "some rather interesting unsafe work practices and unsafe work conditions": genetically engineered viruses found in the dining area; a micro-centrifuge tube found at the bottom of a coke; problems with human and monkey blood samples. The facility was designed so that researchers' offices were located inside the lab areas; and the department break room was located in a working hallway, so that researchers who had removed "personal protection"—gloves, masks, etc.—were eating and drinking among both biological agents and carcinogens. When she brought this up to a manager, he said, "I was in graduate school, I had my hands in this stuff, I don't have cancer yet."[13]

McClain raised concerns again after a building blew up, the roof came off, and a neighborhood was evacuated. She wrote a letter asking Pfizer to explain its safety budget in relation to all the unnecessary risks. Eventually (four months later) she was told that under OSHA standards, her unsafe work environment was "legal." She concluded that Pfizer's safety budget was based on what is "legal," not what is "safe."[14]

In 2002, she says, over a period of months, several people in the embryonic stem cell laboratory contracted a mystery illness, eventually traced to the exhaust from a biological hood where research was performed on mouse embryonic stem cells and genetically engineered viruses. Pfizer's solution was simply to plug the hood and vent it outside into the Groton air.[15]

Soon after she learned of the co-worker's research on her workbench, McClain experienced the first symptoms of her chronic illness. McClain asked the researcher involved what the virus was; the researcher said it was some sort of lentivirus but didn't know anything other than that. She asked him to check; he came back the next day, said "It's safe," and implied it wasn't infectious to humans.[16]

She eventually took medical leave, was hospitalized, and in 2004 reported the incident to OSHA, after which Pfizer terminated her. OSHA told her they didn't have jurisdiction to go in and do a safety inspection for her, but did help pressure Pfizer to name (two years after her exposure) the virus involved. Given the unique nature of genetically engineered viruses—in which each individual virus is unique—the name alone wasn't very helpful. But McClain's efforts to learn the genetic code, and how the virus had been cloned and produced, were stymied by an ultimate ruling that Pfizer's "trade secrets" superseded a worker's right to exposure records. Faced with a lack of jurisdiction by OSHA, worker's compensation, the Connecticut Department of Public Health, NIH, and the CDC, she was forced to file a federal lawsuit.[17] The judge eventually dismissed that part of her lawsuit which attributed her illness to the virus, declaring that she lacked sufficient evidence of causality and that the claim was one which should have been brought under worker's compensation.[18]

McClain says scientists insist genetically engineered viruses are safe because, theoretically, they're made not to replicate. If they don't replicate, they won't cause communicable disease. But when a virus is produced in a BSL-2 lab, no one checks to see if the virus is replicating or not.[19]

Even if a virus doesn't replicate, she says, it can still infect anyone exposed to it. In so infecting, it can permanently integrate itself into a victim's DNA. A frequent goal of genetic engineering is to change the *way* a virus infects; HIV, for instance, might be modified so it infects not just through exchange of blood or other bodily fluids but through inhalation as well. Genetically engineered viruses may infect other animals who can then serve as a reservoir for human infections. If released into the environment, such artificial pathogens may cause increased chronic illness, new emerging disease, metabolic disorders, and clusters of cancer. A virus that replicates can cause a pandemic; a virus that doesn't replicate can be difficult to detect.[20]

Regulation, McClain says, is sorely needed to protect both workers and the public. There is a special need for whistleblower protections and for "transparency": "you want to be able to see what's going on." Yet the biotech industry resists regulation because "they want the freedom to do any type of research without responsibility or liability"; they don't want to put the money into safety (she says most of the Pfizer problems could have been fixed with a bit of remodeling); and "they don't want public scrutiny. They don't want the public to find out what's going on with these dangerous technologies."[21]

"Even more concerning," she says, "is that academia and certain government agencies are so entrenched in large monetary interests to develop embryonic stem cell research with the pharmaceutical industry that there is no longer a group that is free of conflicts of interest to first protect and advocate for the public health and safety."[22]

Jeremy Gruber, president of the Council for Responsible Genetics, a nonprofit Cambridge, Massachusetts organization founded in 1983 to foster "public debate about the social, ethical and environmental implications of genetic technologies,"[23] applauded the jury's damage award, calling Becky McClain "the canary in the coal mine," and warning that regulations had not kept pace with the "explosion of research." Ralph Nader was also taking an interest in the case; he said the field was one which had "been trade-secreted out of the sunlight."[24]

KEEPING THE WORLD SAFE FOR BIOWEAPONS AGENTS: YOUR ANTITERRORISM DOLLARS AT WORK

Kansas State germ lab propagandist Nancy Jaax once suggested that opposing NBAF was downright criminal.[25] Some of America's law enforcement authorities must have harbored similar feelings about biodefense opponents. The NIH, CDC, and DHS might be indifferent to the proliferation of dangerous germ research around the country, and it might take the FBI seven years to get its story straight on who used some of that research to set off the bioterror alarms, but federal and Maryland authorities thought they knew terrorists when they saw them.

First in their sights was Steven Kurtz, art professor at SUNY-Buffalo (State University of New York), and a founding member of the Critical Art Ensemble. Kurtz specialized in visual, kinetic art, aimed at illuminating issues that concerned him, like the dangers of genetically modified food, or his "belief that the U.S. government has deliberately scared Americans with the specter of bioterrorism to make them more controllable." On at least one occasion, Kurtz participated in a protest of the Detrick expansion. In May 2004, he was preparing a museum exhibit for a multimedia project commissioned by a British-based arts-science group, The Arts Catalyst. The exhibit would have allowed museum visitors to see if their store-bought food contained genetically modified organisms. It included a food-testing lab and petri dishes containing the relatively harmless bacteria *Bacillus subtilis* and *Serratia marascens*.[26] I say relatively harmless because the bacteria are not "select agents." They were used as tracers by the Army in the 1950s, who sprayed them on millions of Americans in large quantities for germ warfare dispersion experiments. As Leonard Cole's book *Clouds of Secrecy* points out, such indiscriminate, involuntary exposures presented some dangers to immuno-compromised and other especially sensitive individuals.[27] Kurtz's exhibit posed no similar danger. A geneticist friend, Dr. Robert Ferrell, obtained the samples for Kurtz from the American Type Culture Collection.

While Kurtz was preparing the exhibit in May 2004, his wife died suddenly in her sleep. The police who responded to the 911 call saw the petri dishes and called the FBI, who cordoned off Kurtz's house, searched his house, and confiscated his art projects, computer, reference books, notes, and an anti-germ warfare manuscript Kurtz was working on. (He didn't get them back.) Kurtz was "detained" under the Patriot Act for suspicion of bioterrorism. Politicians, including New York Governor George Pataki, and federal prosecutors, including U.S. Attorney Michael Battle, rushed to congratulate the Buffalo Joint Terrorism Task Force for having disrupted a major bioterrorism threat.[28]

FBI tests revealed within a few days that there were no biological agents in the house, and that Kurtz's wife had died of natural causes. After all the self-generated hoopla, however, the feds apparently felt they had to do something, so several months later they charged Kurtz and Ferrell with wire and mail fraud under the Patriot Act. The "fraud" as they saw it was that in ordering the bacteria from the ATCC, Ferrell hadn't revealed the true destination of the bacteria, Kurtz's politically oriented exhibit. Bringing the charges under the Patriot Act allowed for a maximum sentence of 20 years.[29]

The feds would continue this vital struggle against bioterrorism for four years, however. Ferrell, who was preparing to undergo a second stem cell transplant for lymphoma, and who had three strokes during the legal wrangling, pled guilty to a lesser misdemeanor charge after spending $50,000 in legal fees. Kurtz would keep battling. On April 21, 2008, federal judge Richard Arcara dismissed the indictment as "insufficient on its face." What the feds alleged had happened did not constitute fraud.[30] No one had been defrauded, other than the public who had been told that Big Brother was hard at work protecting them from the ravages of bioterror.

Later in 2008, Barry Kissin, a Frederick, Maryland attorney and prominent opponent of the Fort Detrick expansion, learned that he was one of 53 people who the Maryland State Police had labeled as terrorists, kept files on, and subjected to undercover surveillance. The surveillance operation, launched in 2005, targeted such potential Al Qaeda operatives as Catholic nuns, environmental organizations devoted to establishing bike lanes, PETA (People for the Ethical Treatment of Animals), Amnesty International, and death penalty and anti-war groups. They were suspected of such crimes as "Terrorism-anti-government," "Terrorism-Anti-War-Protesters," and "civil rights."[31] In February 2009, the *Washington Post* would disclose that the Maryland State Police had been assisted in their surveillance by the US Department of Homeland Security.[32]

Notice the full circle we have made in this book. DHS—whose interest in my rural county first drew my attention to the subject of biodefense—is caught engaging in yet another instance of the sort of police state, secretive activity which makes people distrust anything it says about its germ labs. And back in 2001, someone had sent anthrax-laced letters to two Democratic senators resisting the Bush-proposed Patriot Act. Then, the same FBI which had so much trouble pinpointing the perpetrator or perpetrators of that real "bioterrorism incident" had no difficulty or qualms about disrupting the political protest of a biodefense critic. Meanwhile, in the name of "biodefense," research with and stocks of *real* bioweapons agents, posing *real* dangers, were proliferating about the country helter-skelter, with no end in sight.

SWINE FLU SUSPICIONS

In January 2010, the Council of Europe launched an investigation into whether the World Health Organization had deliberately hyped the H1N1 situation, among other means by redefining the definition of a pandemic, under the pressure or influence of pharmaceutical companies. The companies had profited from large, non-revocable, contractual sales to European governments. This investigation followed a 2009 Danish parliamentary review of links between WHO flu expert Albert Osterhaus and the pharmaceutical companies.[33] In November 2009, a team led by Adrian Gibbs had published a study suggesting that H1N1 might have emerged from a lab.[34] Also in 2009, Austrian journalist Jane Burgermeister sued Baxter, alleging that its shipment of seasonal flu vaccines contaminated with bird flu was a deliberate effort to promote a pandemic.[35]

In the aftermath of the anthrax attacks, and in the midst of reckless flu experiments with bird flu and the 1918 flu, suspicions of this sort arise easily.

THE BIOTERROR DUCKS KEEP QUACKING

The congressional mandate of the Graham-Talent WMD Commission expired on Feb. 26, 2010. The commission had found pulling the bioterror fire alarm so entertaining, however, it decided it would not fade quietly into history. Graham, Talent, and former UPMC staffer and commission executive director Colonel Randall Larsen announced that they were forming a new "Bipartisan WMD Research Center" so they could keep pulling the bioterror fire alarm on a regular basis. "We're not a lame duck. We're not going away," Larsen said.[36]

Acknowledgments

Deb Bledsoe of Appalachian Science in the Public Interest deserves the special gratitude of all Kentuckians for tactical advice and practical suggestions to the Kentucky anti-NBAF group. I thank Deb also for encouragement and suggestions during the time the book was being written.

Thanks to Winifred Golden of the Castiglia Agency for digging my query letter out of the slush pile, and for having the perceptiveness to recognize competent writing and a worthy project. Thanks to my editor at Pegasus, Jessica Case, for working with me during the difficult period following my father's death, and for bringing a fresh set of eyes to the manuscript.

Thanks to "C," who, because of institutional affiliation, prefers to remain anonymous, but who read major portions of the manuscript in progress, offered practical suggestions, and most importantly, assured me that the writing was compelling and interesting.

Thanks to my friends Richard and Atsuko Krause, and Jeff and Linda Worley, for providing refuge and consolation and human connection during a difficult period of my life, and for their comments on portions of the manuscript.

455

Thanks to my mother, who is an outdoors person, not a reader, but who, with the intuition of a mother, understood my need to finish this book during a difficult time for our family.

Thanks to Debra Champion of the Agape Agency in Bowling Green for having counseled me and helped me stay focused on this project through the two most difficult periods of my life.

Thanks to Jennifer Mattern and Terri Brown of the University of Kentucky Libraries for their help with book renewals and circulation limitations in place for those who are mere residents of the state, and not UK faculty, staff, or students.

Special thanks to Ed Hammond of the Sunshine Project, whose dedicated and selfless work over several years helped make this book possible. Hammond's research, much of it preserved in the online archives at www.sunshine-project.org, not only informs this book throughout, but helped bring about what little government scrutiny the "germs gone wild" subject has received. Thanks also to the Council for Responsible Genetics, whose online tabulation of lab accidents, at www.councilforresponsiblegenetics.org, and to the Associated Press, whose 2007 summary of recent lab accidents, helped breach the walls of secrecy around this important subject.

I also owe a special thanks to authors whose prior publications and research made the historical parts of this book possible, including Leonard Cole, Jeanne Guillemin, Ed Regis, Michael Carroll, Marilyn Thompson, and Glenn Greenwald, and to those journalists—perhaps not a majority in this book—who chose to act like journalists instead of demented biolab cheerleaders.

Finally, I—and my fellow citizens—all owe an enormous debt of gratitude to our many fellow citizens around the country who took the time to go beyond the public relations hoopla of well-heeled economic development consortiums, to find out the facts for themselves, and to speak up. The spirit of democracy is still alive, no thanks to our greasy politicians, our greedy chambers of commerce, our empire-seeking universities, or our ignorant and duplicitous "experts."

This book was encouraged early on by an Al Smith Fellowship from the Kentucky Arts Council.

Notes

CHAPTER ONE

1. Scott Higham and Robert O'Harrow Jr., "The Quest for Hometown Security," *Washington Post*, Dec. 25, 2005.
2. Ibid.
3. Robert O'Harrow Jr. and Scott Higham, "Post-9/11 Rush Mixed Politics with Security," *Washington Post*, Dec. 25, 2005, A01.
4. John Stamper, Greg Kocher and Bill Estep, "Kentucky Vies for Bioterrorism Lab," *Lexington Herald-Leader*, Feb. 2, 2006, A1.
5. "We Should Support Effort to Bring Bio-Lab to Pulaski," *Somerset Commonwealth Journal*, March 22, 2006.
6. Jeff Neal, "Expert Proclaims Bio Lab 'Safe,'" *Somerset Commonwealth Journal*, April 1, 2006.
7. Jeff Neal, "How Safe Will National Lab Be?" *Somerset Commonwealth Journal*, Feb. 21, 2006.
8. Chris Harris, "Fletcher: Labs are Traditionally Very Safe," *Somerset Commonwealth Journal*, April 25, 2007.
9. Ibid.
10. Blake Aued, "Some Sure, Some Not, of Lab's Safety," *Athens Banner-Herald*, Sept. 16, 2007.
11. Don McAlpin, "Bio Lab Is Big Opportunity for Pulaski," *Somerset Commonwealth Journal*, March 24, 2006.
12. Stephen Smith, "Foes Vow to Fight Bioterror Lab," *Boston Globe*, Oct. 1, 2003.
13. House Subcommittee on Oversight and Investigations of the Committee on Energy and Commerce, *Germs, Viruses, and Secrets: The Silent Proliferation of Bio-Laboratories in the United States*, 110th Cong., 1st sess., 2007, 110.
14. Ibid., 10, 148.
15. Ibid., 24. The more than 1356 BSL-3 Labs are distributed among 400 or so institutions.
16. The facility at Georgia was a small "glove box" facility.
17. Ibid., 22-23.
18. Starbucks Company Fact Sheet, Feb. 2008.
19. Janine DeFao, "Live Anthrax Samples Sent to Institute," *San Francisco Chronicle*, June 11, 2004, B1.
20. "An Open Letter to Elias Zerhouni," *Science* 307 (2005): 1409-1410; Debora MacKenzie, "Top US Biologists Oppose Biodefence Boom," *New Scientist*, March 1, 2005, 38.
21. "Germs, Viruses, and Secrets," 172.
22. Ibid., 147-148.
23. Ibid., 134-135.
24. Ibid., 11.
25. "The NIAID Newbies: How Expert Are They?" in *Some Statistics about the US Biodefense Program and Public Health*, The Sunshine Project, http://www.sunshine-project.org.
26. Ibid., 37.

27. Ibid., 65.
28. Jason Winders, "Maybe We Can Mess with Texas for Lab Site," *Athens Banner-Herald*, July 14, 2007.
29. "Biodefense Bid," *Lawrence Journal-World*, Feb. 18, 2007.
30. Mark Fagan, "Kansas Makes Run for Lab," *Lawrence Journal-World*, Feb. 21, 2007.
31. "Georgia to Bid on Bio- and Agro-Defense Research Facility," *Atlanta Business Chronicle*, Feb. 27, 2006.
32. Greg Harman, "Banging the Drum for Bio-defense," *San Antonio Current*, Aug. 15, 2007.
33. "State Charging after $450M Agri-Security Facility," *Triangle Business Journal*, Dec. 8, 2006.
34. "Sebelius Pushes to Win Biodefense Facility for State," *Kansas City Business Journal*, Jan. 11, 2007.
35. Associated Press, "Bioscience Authority Budgets $1 Million for NBAF Defense," *Manhattan Mercury*, July 23, 2009.

NBAF 1

1. Michael Carroll, *Lab 257: The Disturbing Story of the Government's Secret Plum Island Germ Laboratory* (New York: William Morrow, 2004), 43-44.
2. Ibid.
3. Ibid., 45.
4. Ibid., 47-49.
5. Cited in Bruce Lambert, "Closely Guarded Secrets: Some Islands You Can't Get to Visit," *New York Times*, May 17, 1998, Section 14LI, 6.
6. General Accounting Office, *Combating Bioterrorism: Actions Needed to Improve Security at Plum Island Animal Disease Center*, GAO-03-847, Sept. 2006, p. 16, FN 15.
7. In Appendix B, "A Review of Biocontainment Lapses and Laboratory-Aquired Infections."
8. Diana Jean Schemo, "Unit for Animal-Disease Study Trims Safeguards," *New York Times*, Nov. 26, 1992, B1.
9. Judith Miller, "Long Island Lab May Do Studies of Bioterrorism," *New York Times*, Sept. 22, 1999, A1.
10. Schemo, "Unit Trims Safeguards."
11. Diana Jean Schemo, "Research Unit Is Postponing Major Projects," *New York Times*, Dec. 24, 1992, B1.
12. Schemo, "Unit Trims Safeguards."
13. *Lab 257*, 191-206.
14. Ralph Ginzburg, "Top-Secret Plum I. Beckons Public," *New York Times*, Aug. 28, 1994, Sec. 13LI, 1.
15. Robert Gearty, "The Secret World of Plum Island," *New York Daily News*, Nov. 21, 1999, Suburban Section, 13.
16. Karen Guzman, "Residents Fear Plan to Study Lethal Viruses," *Hartford Courant*, Jan. 20, 2000, Town News, B1.
17. John Rather, "Congressman Opposes Disease Center Upgrade," *New York Times*, Jan. 30, 2000, Sec. 14LI, 5; Michael Cooper, "The Long Shadow of Science Past: Long Island Labs, On the Defensive, Struggle for Community Confidence," *New York Times*, Sec. 1, 33.
18. John Rather, "Epicenter of Foot-and-Mouth Research," *New York Times*, April 1, 2001, Sec. 14LI, 9.
19. Ianthe Jeanne Dugan, "Anthrax Island? No, Plum Island Is Alcatraz for Animal Illnesses," *Wall Street Journal*, Jan. 8, 2002, A1.
20. Richard Pyle, "Probe of Alleged 'Sabotage' at Federal Lab Facility," *Associated Press*, Aug. 21, 2002.
21. Marc Santora, "Security at U.S. Lab Is Questioned by Senators," *New York Times*, Oct. 24, 2002, B5.
22. Marc Santora, "U.S. Lab Worker Hired During Strike Had Arrest Record," *New York Times*, Nov. 19, 2002, B1.

23. Richard Pyle, "Clinton Urges USDA to Shut Plum Island Facility to Fix Power Problem," *Associated Press*, Dec. 19, 2002; Marc Santora, "Power Fails for 3 Hours at Plum Island Infectious Disease Lab," *New York Times*, Dec. 20, 2002, B1; Associated Press, "Plum Island Uses Backup Generators after Power Fluctuations," Dec. 23, 2002; Associated Press, "Power Problems at Plum Island Continue Over Weekend," Dec. 23, 2002.

24. Santora, "Power Fails for 3 Hours."

25. Ibid.

26. "Future of Animal Research Lab Is Questioned," *New York Times*, Feb. 10, 2003, B5.

27. Ibid.; Frank Eltman, Associated Press, "New Role of Plum Island Raises Concerns," *Albany Times Union*, March 23, 2003, D3.

28. John Rather, "A Peek at the New Plum Island," *New York Times*, June 1, 2003, Sec. 14LI, 1.

29. Ibid.

30. Ibid.

31. Ibid.

32. Ibid.

33. "Plum Island Won't Become Test Site for Deadly Diseases, Lawmakers Say," *Associated Press*, June 5, 2003.

34. GAO, *Combating Bioterrorism*.

35. Ibid., 2.

36. Ibid., 3-4.

37. Ibid., 14.

38. Ibid., 4.

39. Ibid.

40. Ibid., 17-19.

41. Ibid., 20-21.

42. Ibid., 50.

43. John Rather, "Heaping More Dirt on Plum Island," *New York Times*, Feb. 15, 2004, Sec. 14LI, 1.

44. Ibid.

45. Ibid.

46. Patrick Healy, "Plum I. Was Ready for Its Close-Up," *New York Times*, Feb. 22, 2004, Sec. 14LI, 3.

47. Ibid.

48. Ibid.

49. John Rather, "Lab's Research Gets Spread Around," *New York Times*, May 16, 2004, Sec. 14LI, 3.

50. John Rather, "Plum Island Reports Disease Outbreak," *New York Times*, Aug. 22, 2004, Sec. 14LI, 3.

51. Ibid.

52. Jocelyn Kaiser, "Plum Island Breaches Assailed," *Science* 305 (2004): 1225.

CHAPTER TWO

1. Ewen Callaway, "Biodefence Work Halted at US University," News@Nature.com, July 3, 2007, http://www.nature.com/news; "About the FAZD Center," National Center for Foreign Animal and Zoonotic Disease Defense, http://fazd.tamu.edu, accessed Oct. 10, 2008.

2. News release, The Sunshine Project, "Texas A&M University Violates Federal Law in Biodefense Lab Infection, April 12, 2007, http://www.sunshine-project.org/TAMU/; Emily Ramshaw, "CDC Probes A&M Bioweapons Infections," *Dallas Morning News*, June 26, 2007.

3. Ibid.

4. News release, The Sunshine Project, April 12, 2007.

5. "Brucellosis," Centers for Disease Control and Prevention, http://www.cdc.gov/ncidod/dbmd/diseaseinfo/brucellosis_t.htm.

6. Sunshine Project News Release, April 12, 2007.

7. Ibid.

8. Dr. Robbin Weyant, Director, Division of Select Agents and Toxins, Coordinating Office of Terrorism Preparedness and Emergency Response, Centers for Disease Control and Prevention, letter, "Cease and Desist Order from Violations of the Public Health Security and Bioterrorism Preparedness and Response Act of 2002," letter to Dr. Richard Ewing, Responsible Official, Texas A&M University, April 20, 2007, accessed at http://www.sunshine-project.org/TAMU/CeaseBrucellaTAMU.pdf.

9. News release, The Sunshine Project, "Bioweapons Infections Hit Texas A&M Again," June 26, 2007, http://www.sunshine-project.org/TAMU/; Ramshaw, "CDC Probes A&M Bioweapons Infections."

10. "Q Fever," Centers for Disease Control and Prevention, http://www.cdc.gov/ncidod/dvrd/qfever/index.htm.

11. News release, The Sunshine Project, June 26, 2007.

12. Ibid.

13. Ramshaw, "CDC Probes A&M Bioweapons Infections."

14. Dr. Julie L. Gerberding, Director, Centers for Disease Control and Prevention, memorandum, "Suspension of Select Agent Work: Texas A&M University," to Dr. Richard Ewing, Responsible Official, Texas A&M University, June 30, 2007, accessed at http://www.sunshine-project.org/TAMU/TAMUCDCMemo.pdf.

15. "Safety Clause," *Nature* 448 (2007): 105-106.

16. http://www.sunshine-project.org.

17. Ibid.

18. "Federal Agency Extends Toxic Research Ban at Texas A&M," *Associated Press*, Sept. 4, 2007.

19. Emily Ramshaw, "Infected A&M Lab Worker Lacked Proper Approval," *Dallas Morning News*, July 27, 2007.

20. Dr. Robbin Weyant, Director, Division of Select Agents and Toxins, Coordinating Office of Terrorism Preparedness and Emergency Response, Centers for Disease Control and Prevention, letter, "Texas A&M University: Report of Site Visit," to Dr. Richard Ewing, Responsible Official, Texas A&M University, August 31, 2007, accessed at http://www.sunshine-project.org/TAMU/CDCTAMUReport.pdf; Emily Ramshaw, "CDC Reprimands Texas A&M Over Lab Safety," *Dallas Morning News*, Sept. 4, 2007; Jennifer Couzin, "Biodefense Research: Lapses in Biosafety Spark Concern," *Science* 317 (2007): 1417.

21. Ramshaw, "CDC Reprimands Texas A&M Over Lab Safety"; Ramshaw, "Until This Year, CDC Missed Worst of A&M Lab Problems," *Dallas Morning News*, Sept. 25, 2007.

22. Ed Hammond, E-mail, biodefense@lists.sunshine-project.org, Sept. 6, 2007.

23. Ramshaw, "Infected A&M Lab Worker Lacked Proper Approval."

24. Couzin, "Lapses in Biosafety Spark Concern."

25. Emily Ramshaw, "A&M Biosafety Director Resigns in Wake of CDC Report," *Dallas Morning News*, Sept. 5, 2007.

26. Alison Young, "CDC Lab's Backup Power Fails During Storm," *Atlanta Journal-Constitution*, July 7, 2007.

27. Ibid.

28. Ibid.

29. Alison Young, "CDC Lab Flaw Gets Congress' Attention," *Atlanta Journal-Constitution*, Aug. 8, 2007, A1.

30. Young, "CDC Lab's Backup Power Fails."

31. Young, "CDC Lab Flaw Gets Congress' Atttention."

32. Alison Young, "CDC Biolab Not Ready After 2½ Years," *Atlanta Journal-Constitution*, May 15, 2008.

33. Posted at biodefense@lists.sunshine-project.org, March 15, 2007.

34. Alison Young, "E-Mails Outline CDC Backup Power Flaws," *Atlanta Journal-Constitution*, July 24, 2007, A1.

35. Brandon Keim, "Critic: CDC Lab Breakdown Could Have Exposed Workers to Deadly

Diseases," *Wired Science*, July 12, 2007, http://www.wired.com/wiredscience/2007/07/cdc-lab-breakdo/#Replay.

36. Comptroller and Auditor General, *The 2001 Outbreak of Foot and Mouth Disease* (London: National Audit Office, 2002), cited in Government Accountability Office, Testimony Before the Subcommittee on Oversight and Investigations, Committee on Energy and Commerce, House of Representatives, *High-Containment Biosafety Laboratories: DHS Lacks Evidence to Conclude That Foot-and-Mouth-Disease Research Can Be Done Safely on the U.S. Mainland*, May 22, 2008, GAO-08-821T, 26.

37. Dr. Iain Anderson CBE, *Foot and Mouth Disease 2007: A Review and Lessons Learned*, presented to the Prime Minister and the Secretary of State for Environment, Food and Rural Affairs (London: Crown Copyright, March 2008), 8-10.

38. Ibid., 10.

39. Andy Coghlan, "Faulty Pipe Blamed for UK Foot and Mouth Outbreak," *New Scientist Environment*, Sept. 7, 2007, http://www.newscientist.com.

40. Health and Safety Executive, *Final Report of Potential Breaches of Biosecurity at the Pirbright Site 2007*, Sept. 2007, amended Dec. 20, 2007.

41. Prof. Brian G. Spratt, *Independent Review of the Safety of U.K. Facilities Handling Foot-and-Mouth Disease Virus*, presented to the Secretary of State for Environment, Food and Rural Affairs and the Chief Veterinary Officer, Aug. 31, 2007.

42. Health and Safety Executive, *Final Report*, 3.

43. Spratt, 10.

44. HSE, 2, 43.

45. Anderson, 10-11.

46. Ibid., 12.

47. Ibid., 11-12.

48. Ibid., 60, 115-117.

49. "Biosecurity Lapses," *Financial Times*, Aug. 29, 2007, http://www.ft.com.

50. HSE, 3-4.

51. Brendan Montague and Jonathan Leake, "Virus Lab 'Ignored Warning of Leak,'" *Times Online*, Sept. 30, 2007, http://www.timesonline.co.uk.

52. The HSE, Spratt, and Anderson investigations cited above, and Sir Bill Callaghan, *A Review of the Regulatory Framework for Handling Animal Pathogens*, presented to the Secretary of State for Environment, Food and Rural Affairs (London: Crown Copyright 2008, Dec.13, 2007).

53. Sheldon H. Harris critiques claims that American POWs were among the germ experiment victims in *Factories of Death: Japanese Biological Warfare 1932-1945* (London: Routledge, 1994), 113-131.

NBAF 2

1. Department of Homeland Security, *Fact Sheet: National Bio and Agro-Defense Facility*, Aug. 22, 2005, http://www.dhs.gov/xnews/releases/press_release-0719.shtm

2. DHS, "National Bio and Agro-Defense Facility (NBAF); Notice of Request for Expression of Interest for Potential Sites for the NBAF," *Federal Register* 71, no. 12 (Jan. 19, 2006): 3107-3109.

3. DHS, "National Bio and Agro-Facility//Information for Potential Respondents//How to Respond; Frequently Asked Questions," http://www.dhs.gov/dhspublic/display?theme=27&content=5391&print=true, accessed March 25, 2006.

4. Ibid.

5. DHS, "Facility Research and Staffing for the National Bio and Agro-Defense Facility," page last modified June 12, 2007.

6. Government Accountability Office, *Plum Island Animal Disease Center: DHS and USDA Are Successfully Coordinating Current Work, but Long-Term Plans Are Being Assessed*, Dec. 2005, GAO-06-132, 29-30.

7. John Dudley Miller, "US Homeland Security to Build Animal Biolab," *The Scientist*, Feb. 6, 2006, http://www.the-scientist.com/news/display/23091/.

8. Frank Eltman, "Lawmakers Seek Meeting to Discuss Future of Plum Island," *Associated Press*, Aug. 25, 2005.

9. Devlin Barrett, "Lawmakers Fret Over Possibility of Human Disease Work at Site," *Associated Press*, Oct. 5, 2005; Lisa Munoz, "Hil., Rep Make Case for Plum Island Lab," *New York Daily News*, Oct. 6, 2005, Suburban, 4.

CHAPTER THREE

1. House Subcommittee on Oversight and Investigations of the Committee on Energy and Commerce, *Germs, Viruses, and Secrets: The Silent Proliferation of Bio-Laboratories in the United States*, 110th Cong., 1st sess., 2007, 28.

2. Ibid.

3. Ibid., 14.

4. Ibid., 81.

5. Ibid., 65.

6. Ibid., 64.

7. Ibid., 62.

8. Ibid., 143.

9. Ibid., 169.

10. Ibid., 87,

11. Ibid., 100.

12. Ibid., 97-98.

13. Ibid., 101.

14. Ibid., 65-66.

15. Ibid., 83.

16. Ibid., 32.

17. Ibid., 32-33.

18. Ibid., 84.

19. All references to "WMD" commission in this book are to the Graham-Talent "Commission for the Prevention of Weapons of Mass Destruction and Profileration and Terrorism," an important ally for the U.S. biodefense complex (see chapters 8 and 15). Not to be confused with the International Weapons of Mass Destruction Commission, which has a genuine interest in reducing the profileration of bioweapon agents and other "WMDs."

20. Ibid., 171.

21. In fact, the organization in question seems a valid scientific enterprise, certainly a less dangerous one than politicians' pet biodefense projects. "Margaret Race" is Dr. Margaret Race, one of several scientists employed by the institute, whose Web site is at http://www.seti.org. Among other things, her work focuses on trying to ensure "shuttles don't inadvertently bring unknown life back from space"—a sensible precaution, but perhaps not for a biodefense complex merrily concocting and resuscitating lethal pathogens.

22. *Germs, Viruses, and Secrets*, 34.

23. Ibid., 87-88.

24. Alison Young, "CDC Action at Germ Lab Questioned," *Atlanta Journal-Constitution*, June 22, 2008.

25. *Germs, Viruses, and Secrets*, 37.

26. Ibid., 67-78.

27. Ibid., 79-80.

28. Ibid., 80.

29. Ibid., 109.

30. Ibid., 108-109.

31. Gigi Kwik Gronvall, Joe Fitzgerald, Allison Chamberlain, Thomas V. Inglesby, and Tara

O'Toole, "High-Containment Biodefense Research Laboratories: Meeting Report and Center Recommendations," *Biosecurity and Bioterrorism* 5 (2007): 81.

32. *See* comments by Burgess and Davis, *Germs, Viruses, and Secrets*, 89, 98.
33. Ibid., 109.
34. Ibid., 72.
35. Ibid., 87.
36. Ibid., 109.
37. Ibid., 110.
38. Ibid., 106.
39. Ibid., 109.
40. U.S. Govt. Accountability Office, *Biosafety Laboratories: Perimeter Security Asssessment of the Nation's Five BSL-4 Laboratories*, Sept. 2008, GAO-08-1092, "Highlights."
41. Ibid., 7.
42. Ibid., 2.
43. Ibid., 4.
44. Ibid., 5.
45. Ibid., 10.
46. Ibid., 11.
47. Ibid., 18-19.
48. Ibid., 11.
49. Ibid., 18.
50. Ibid., 3.
51. Ibid., 19.
52. Ibid.

NBAF 3

1. Sharon Dodson, "National Lab: 'Safe as Going to Wal-Mart,'" *Somerset Commonwealth Journal*, Feb. 24, 2006, A1.
2. Arnold Lindsay, "Miss. Seeks Bioterror Center," *Jackson Clarion-Ledger*, March 21, 2006, 1A.
3. Scott Higham and Robert O'Harrow, Jr., "The Quest for Hometown Security," *Washington Post*, Dec. 25, 2005.
4. "Bio-Research Center a Bigger Fish," *Atlanta Journal-Constitution*, March 9, 2006, A14.
5. "Leaders Should Come Together to Gain Facility," *Athens Banner-Herald*, March 10, 2006.
6. "Defense Facility Seems a Natural Fit for University," *Athens Banner-Herald*, April 4, 2006.
7. Jim Thompson, "The Future Is Now for This Area's Economic Health," *Athens Banner-Herald*, May 14, 2006.
8. Alisa Marie DeMao, "Local Sites Pushed for Defense Lab," *Athens Banner-Herald*, April 1, 2006.
9. Diana Jean Schemo, "Unit for Animal-Disease Study Trim Safeguards," *New York Times*, Nov. 26, 1992, B1.
10. Scott Waller, "Support Key in Bio Facility Bid," *Jackson Clarion-Ledger*, March 22, 2006, 1C.
11. Andre Fujiusa, "County in Running for Bio Unit," *Madison County Journal*, March 22, 2006.
12. Lindsay, "Miss. Seeks Bioterror Center."
13. Ibid.
14. Ibid.
15. Geoff Pender, "State in Running for Federal Lab," *Biloxi Sun Herald*, March 21, 2006, A8.
16. Terry Ganey, "Biomedical Facilities to Be Discussed," *Columbia Daily Tribune*, March 21, 2006; Alex Braun, "MU to Discuss Adding Biodefense Research Centers," *Columbia Missourian*, March 22, 2006.
17. Josh Flory, "Defense Lab Plan Spooks Neighborhood," *Columbia Daily Tribune*, March 24, 2006.

18. Debrin Foxcroft, "Debate Erupts Over Defense Facility Proposal," *Columbia Missourian*, March 24, 2006.
19. Flory, "Defense Lab Plan Spooks Neighborhood."
20. Ibid.
21. National Institute for Hometown Security, "Guest View: Extraordinary Consortium . . . Uncommon Opportunity," *Somerset Commonwealth Journal*, Feb. 22, 2006.
22. Lawson was acquitted of all charges in January 2010. He will presumably resume his affectionate and profitable relationship with Congress.
23. Lee Mueller, "In Pike, Few Object to Big Plans for I-66," *Lexington Herald-Leader*, Aug. 21, 2006, A1.
24. John Cheves, "Kentucky Congressman Hal Rogers Delivers for His District," *Lexington Herald-Leader*, Feb. 6, 2005.
25. John Cheves, "Highway Builder Wields Wide Political Influence," *Lexington Herald-Leader*, Aug. 21, 2005.
26. Ibid.
27. Cheves, "Kentucky Congressman Hal Rogers Delivers"; "Rogers Helps Get Money for a Private Lot," *Congress Daily*, Jan. 20, 2005, 7-8.
28. Cheves, "Kentucky Congressman Hal Rogers Delivers."
29. Robert O'Harrow, Jr. and Scott Higham, "Post 9/11 Rush Mixed Politics with Security," *Washington Post*, Dec. 25, 2005, A1.
30. "We Should Support Effort to Bring Bio-Lab to Pulaski," *Somerset Commonwealth Journal*, March 22, 2006.
31. Ibid.
32. Jeff Neal, "How Safe Will National Lab Be?" *Somerset Commonwealth Journal*, Feb. 21, 2006, A1.
33. Jeff Neal, "Expert Proclaims Bio Lab 'Safe,'" *Somerset Commonwealth Journal*, April 1, 2006, A1.
34. Susan Wheeldon, "Hazmat Personnel Endorse New Lab," *Somerset Commonwealth Journal*, Feb. 24, 2006, A1.
35. Bill Mardis, "Picking Up Steam," *Somerset Commonwealth Journal*, March 1, 2006.
36. Ibid.
37. Bill Estep, "Bio Lab Plan Stirs Lots of Questions," *Lexington Herald-Leader*, March 3, 2006, A1; Sharon Dodson, "Not in Our Backyard," *Somerset Commonwealth Journal*, March 4, 2006, A1.
38. This comment was reported to me by several members of the No Ky Biolab steering group during the summer of 2006.
39. Sharon Dodson, "Petition Opposes National Bio Lab," *Somerset Commonwealth Journal*, March 13, 2006; Bill Estep, "Petition against Lab Gets 2,817 Signatures," *Lexington Herald-Leader*, March 17, 2006, A1.
40. Sarah Vos, "Congressman Says Support Bioterrorism Lab or Forget it," March 1, 2006, B1.
41. "Proposed Bio-Lab Worth Exploring," *Lexington Herald-Leader*, Feb. 26, 2006.
42. Jeff Neal, "'Lab 257': Fact or Fiction?", *Somerset Commonwealth Journal*, March 20, 2006.
43. Jeff Neal, "'It Will Be Safe': Past Plum Island Chief Says Fears Aren't Factual," *Somerset Commonwealth Journal*, March 20, 2006.
44. March 22, 2006.
45. Chris Harris, "Somerset Will Never Be Target for Terrorists—Even with Bio Lab," *Somerset Commonwealth Journal*, April 5, 2006.
46. Jeff Neal, "Fletcher, Rogers Pitch Bio Lab," *Somerset Commonwealth Journal*, April 4, 2006.

NBAF 4

1. Bill Mardis, "Decision on Bio-Lab 'Short List' May Come as Early as Next Month," *Somerset Commonwealth Journal*, Aug. 2, 2006.

CHAPTER FOUR

1. Jeanne Guillemin, *Biological Weapons: From the Invention of State-Sponsored Programs to Contemporary Bioterrorism* (New York: Columbia University Press, 2005), 10.
2. Guillemin, 58-61; Leonard Cole, *Clouds of Secrecy: The Army's Germ Warfare Tests over Populated Areas* (Totowa, NJ: Rowman and Littlefield, 1988), 33-35; Ed Regis, *The Biology of Doom* (New York: Henry Holt, 1999), 33-39.
3. Cole, 33, quoting Richard M. Clendenin, *Science and Technology at Fort Detrick, 1943-1968* (Frederick, MD: Fort Detrick, Historian, Technical Information Division, April 1968), x-xi.
4. Leo P. Brophy, Wyndham D. Miles, and Rexmond C. Cochrane, *The Chemical Warfare Service: From Laboratory to Field* (Washington, DC: Office of the Chief of Military History, US Army, 1959), 111.
5. Guillemin, 60.
6. Judith Miller, Stephen Engelberg, and William Broad, *Germs: Biological Weapons and America's Secret War* (New York: Touchstone, 2002), 39.
7. Regis, 43, 79.
8. Guillemin, 88.
9. Ibid., 93.
10. Miller et al., 40; Regis, 17-19.
11. Cole, 12-13.
12. Ibid., 12-14.
13. Guillemin, 79.
14. Regis, 126.
15. Cole, 13, citing John W. Powell, "Japan's Germ Warfare: The U.S. Coverup of a War Crime," *Bulletin of Concerned Asian Scholars* 12, no. 4 (1980): 3.
16. Ibid.
17. Regis, 111-112.
18. Ibid., 112.
19. Ibid., 127.
20. Guillemin, 91.
21. Regis, 154, 165.
22. Ibid., 94.
23. Guillemin, 105.
24. Ibid., 106.
25. Regis, 164-176.
26. Miller et al., 35.
27. Cole, 3.
28. Ibid., 6.
29. Ibid.
30. Ibid., 81.
31. Ibid., 52.
32. Ibid.
33. Ibid., 52-53.
34. Ibid., 44-45.
35. Ibid., 53.
36. Ibid., 47.
37. Ibid., 45, 48-49.
38. Ibid., 18.
39. Ibid., 95.
40. Ibid., 7.
41. Guillemin, 107; Regis, 177; Seymour M. Hersh, *Chemical and Biological Warfare* (New York: Bobbs-Merrill, 1968), 23.
42. Hersh, 26.

43. Miller et al., 52.
44. Regis, 193-196; Miller et al., 72-73.
45. Regis, 152-154.
46. Ibid., 68.
47. Ibid., 157.
48. Ibid., 157-161, 178.
49. Ibid., 179-181.
50. Bob Coen and Eric Nadler, *Dead Silence: Fear and Terror on the Anthrax Trail* (Berkeley: Counterpoint, 2009), 83-102.
51. Regis, 182-185.
52. Ibid., 231.
53. Lee Davidson, "Maryland Germ Lab Neighbors Either Love or Loathe It: Similar Lab Protested in Utah." *Deseret News*, Oct. 8, 1989.

NBAF 5
1. Jeff Neal, "We Made the Cut," *Somerset Commonwealth Journal*, August 10, 2006.
2. Rebecca K. Quigley, "UGA Still in Running for Biodefense Facility," *Athens Banner-Herald*, Aug. 10, 2006.
3. Don Finley, "3 S.A. Sites Make Cut for Disease Laboratory," *San Antonio Express-News*, Aug. 11, 2006, 1B.
4. Ian Hoffman, "Groups Seeking to Block New Biodefense Lab," *Alameda Times-Star*, Feb. 19, 2006.
5. Press release, Tri-Valley CARES, Nuclear Watch New Mexico, "Community Groups Hail Victory, Court Grants Demand for Environmental Review Before Bio-Warfare Agent Research Facility Opens at Livermore Lab," Oct. 16, 2006.
6. Ibid.
7. Press release, Tri-Valley CARES, "Tri-Valley CARES Sues on 5 Separate Freedom of Information Act Violations," Nov. 14, 2006.
8. Ian Hoffman, "Bio Lab Proposed for Tracy," *Tri-Valley Herald*, Aug. 10, 2006.
9. John Upton, "Questions Linger About Bio-Lab," *Tracy Press*, Oct. 5, 2006.
10. David Wahlberg, "UW Applies for Federal Disease Lab," *Wisconsin State Journal*, Nov. 29, 2006, A1.
11. Ibid.
12. Matthew DeFour, "Dunn Residents, Board Oppose Disease Lab," *Wisconsin State Journal*, March 8, 2007, A1.
13. Ibid.; Andriy Pazuniak, "Town Debates UW Proposal," *Badger Herald*, Dec. 1, 2006.
14. "Proposed Federal Bio Lab Concerns Some Residents," Channel 3 News, Nov. 30, 2006, http://www.channel3000.com/news/10436680/detail.html.
15. Wahlberg, "UW Applies"; Pazuniak, "Town Debates."
16. DeFour, "Dunn Residents, Board Oppose."

CHAPTER FIVE
1. Francis A. Boyle, *Biowarfare and Terrorism* (Atlanta: Clarity Press, 2005), 47.
2. William Brood, "U.S. Germ Warfare Research Pushes Limits," *New York Times*, Sep. 4, 2001.
3. Glenn Greenwald, "Vital Unresolved Anthrax Questions and ABC News," Aug. 1, 2008, http://www.salon.com.
4. Thompson, 83.
5. Ibid., 95.
6. Ibid., 96.
7. Ibid., 98.
8. Ibid., 105.
9. Ibid., 114.
10. Leonard A. Cole, *The Anthrax Letters: A Medical Detective Story* (Washington, D.C.: Joseph Henry Press, 2003), 30.

11. Thompson, 143.
12. Cole, 77.
13. Ibid., 64-65.
14. Ibid., 66-67.
15. Thompson, 136.
16. Ibid., 128-130.
17. Ibid., 29-30.
18. Ibid., 35.
19. Richard Preston, *The Demon in the Freezer* (New York: Random House, 2002), 172.
20. Thompson, 119.
21. Preston, 192-193.
22. Douglas Beecher, "Forensic Application of Microbiological Culture Analysis to Identify Mail Intentionally Contaminated with Bacillus Anthracis Spores," *Applied and Environmental Microbiology* 72 (2006): 5304-5310.
23. *See*, for instance, K. A. Mereish, "Unsupported Conclusions on the Bacillus Anthracis Spores," *Applied and Environmental Microbiology* 73 (2007): 5074.
24. David Willman, "Scientist Concedes 'Honest Mistake' about Weaponized Anthrax," *Los Angeles Times*, Sept. 17, 2008.
25. Associated Press, "Rep. Bartlett Skeptical that Ivins Sent Anthrax," Sept. 12, 2008.
26. Eric Coen and Bob Nadler, *Dead Silence*, 12-13.
27. Ibid., 13.
28. Ibid.
29. Greenwald, "Vital Unresolved Questions."
30. Thomas E. Ricks, *Fiasco: The American Military Adventure in Iraq* (New York: Penguin, 2006), 90.
31. Cole, 239.
32. James Gordon Meek, "FBI Was Told to Blame Anthrax Scare on Al Qaeda by White House Officials," *New York Daily News*, Aug. 2, 2008.
33. Ron Suskind, *The Way of the World* (New York: HarperCollins, 2008), 361-373.
34. "US Pays $5.8 Million in Anthrax Lawsuit," June 30, 2008, *The Great Beyond*, blog at Nature.com, http://blogs.nature.com; David Willman, "U.S. Settles with Anthrax Mailings Subject Steven Hatfill for $5.82 Million," *Los Angeles Times*, June 28, 2008.
35. Christopher Ketcham, "The Anthrax Files," *The American Conservative*, Aug. 25, 2008.
36. Scott Shane, "Portrait Emerges of Anthrax Suspect's Troubled Life," *New York Times*, Jan. 4, 2009.
37. Ibid.
38. Ibid.
39. Dan Vergano and Steve Sternberg, "FBI Did Not Analyze Anthrax from Biodefense Lab," *USA Today*, Sept. 24, 2008.
40. Ketcham, "The Anthrax Files."
41. Ibid.
42. Catherine Herridge and Ian McCaleb, "FBI Focusing on 'About Four' Suspects in 2001 Anthrax Attacks," FOX News.com, March 28, 2008.
43. Ibid.
44. Greenwald, "Vital Unresolved Questions."
45. Yudhjiit Bhattacharjee, "NAS Study May Fail to Settle Anthrax Case," *Science* 322 (2008): 27.
46. Ketcham, "The Anthrax Files."
47. Glenn Greenwald, "Journalists, Their Lying Sources, and the Anthrax Investigation," Aug. 3, 2008, http://www,salon.com.
48. Scott Shane, "Critics of Anthrax Inquiry Seek an Independent Inquiry," *New York Times*, Sept. 23, 2008.
49. The summaries given below rely on a rich, profound, and detailed representation by Jeanne Guillemin, *Anthrax: The Investigation of a Deadly Outbreak* (Berkeley: U. of Cal. Press, 1999).

50. Ibid., 187.
51. Ibid., 234, 242.
52. Ibid.
53. Ibid.
54. Ibid., 105.
55. Ibid., 132.
56. Ibid., 133.
57. Ibid., 133-134, 234.
58. Ibid., 85.
59. Ibid., 202-204.
60. Ibid., 8.
61. Ibid., 8-10.
62. Ibid., 107.
63. Ibid., 195-196.
64. Ibid., 160-161.
65. Ibid., 241-242.
66. Ibid., 163-164.
67. Janine DeFao, "Live anthrax samples sent to institute," *San Francisco Chronicle*, June 11, 2004, B1.
68. Ibid.
69. Larry Margasak and Elizabeth Davidz, "Mishandled Germs: Problems at U.S. Labs that Research Deadliest Biological Agents," *Associated Press*, October 1, 2007, http://hosted.ap.org/specials/interactives/wdc/biohazards.
70. "UMC student OK after anthrax spill," *Jackson Clarion-Ledger*, Aug. 13, 2007.
71. Margasak and Davidz, "Mishandled Germs."
72. "Apparent Unauthorized Shipment of Anthrax," *The Project on Government Oversight*, Nov. 20, 2001, http://www.pogo.org; "LANL Gets Unauthorized Anthrax Shipment," *Inside Energy/with Federal Lands*, Nov. 26, 2001.
73. "LANL Owes an Explanation," *Albuquerque Journal*, Dec. 2, 2001.
74. U.S. Department of Energy, Office of Inspector General, *Inspection Report: Concerns Regarding a Non-Viable (Dead) "Anthrax Spore" Research Project at the Oak Ridge National Laboratory,* DOE/IG-0681, March 2005.
75. David Hambling, "US Army Plans to Bulk-Buy Anthrax," *New Scientist*, Sept. 24, 2005.
76. Elaine M. Grossman, "U.S. Limits Anthrax Vaccine Legal Liability," *Global Security Newswire*, Oct. 19, 2008, http://gsn.nti.org.
77. Ibid.
78. Press release, "Universal Detection Technology Responds to Declaration of Anthrax Emergency," *Market Wire*, Oct. 21, 2008.
79. Thompson, 188.

CHAPTER SIX

1. Ryan J. Foley, "U. of. Wis. Quietly Scraps Risky Lab Equipment," *Associated Press*, January 9, 2009.
2. Ibid.
3. Ibid.
4. Ibid.
5. Ibid.
6. Ibid.
7. U.S. Govt. Accountability Office, *High-Containment Biosafety Laboratories: DHS Lacks Evidence to Conclude that Foot-and-Mouth-Disease Research Can Be Done Safely on the U.S. Mainland*, May 22, 2008, GAO-08-821T, 18.
8. House Subcommittee on Oversight and Investigations of the Committee on Energy and Commerce, *Germs, Viruses, and Secrets: The Silent Proliferation of Bio-Laboratories in the United States*, 110th Cong., 1st sess., 2007, 14.

9. Angela Delli Santi, "2 Dead, Plague-Infested Mice Lost by NJ Lab," *Newsday*, Feb. 6, 2009.
10. Ted Sherman and Josh Margolin, "UMDNJ Facility Loses Two Plague-Infested Dead Lab Mice," *The Star-Ledger*, Feb. 7, 2009.
11. Ibid.
12. Ibid.
13. Janine DeFao, "Live Anthrax Samples Sent to Institute," *San Francisco Chronicle*, June 11, 2004: B1.
14. Martin Enserink and Jocelyn Kaiser, "Accidental Anthrax Shipment Spurs Debate Over Safety," *Science* 305 (2004): 1726-1727.
15. Martin Enserink, "Tiptoeing around Pandora's Box," *Science* 305 (2004): 594-595.
16. DeFao, "Live Anthrax Samples."
17. Larry Margasak and Elizabeth Davidz, "Mishandled Germs: Problems at U.S. Labs that Research Deadliest Biological Agents," Associated Press, Oct. 1, 2007, http://hosted.ap.org/specials/interactives/wdc/biohazards .
18. Ibid.
19. Sean Hao, "Paperwork on Risky Viruses Mislaid," *Honolulu Advertiser*, Nov. 18, 2007.
20. Ibid.
21. Ibid.
22. Ibid.
23. Ibid.
24. Associated Press, "Small Package Explosion Evacuates FedEx Facility," March 19, 2003.
25. Margasak and Davidz, "Mishandled Germs."
26. Ibid.
27. Pamela Fayerman, "Vancouver Lab Mishap Alerted World to Flu Pandemic Risk; Potentially Deadly Strain Shipped to 4,000 Labs in 18 Countries," *Vancouver Sun*, April 14, 2005, A1.
28. Ibid.
29. Ibid.
30. Helen Branswell, "Officials Investigate How Bird Flu Contaminated Vaccines in Europe," *The Canadian Press*, Feb. 26, 2009, http://www.thestar.com; Helen Branswell, "Baxter: Product Contained Live Bird Flu Virus," *Toronto Sun*, Feb. 27, 2009.
31. Margasak and Davidz, "Mishandled Germs."
32. Ibid.
33. Lloyd J. Dumas, *Lethal Arrogance: Human Fallibility and Dangerous Technologies* (New York: St. Martin's Press, 1999), 30.
34. Charles Perrow, *Normal Accidents: Living with High-Risk Technologies* (Princeton, N.J.: Princeton UP, 1999), 330.
35. James R. Chiles, *Inviting Disaster: Lessons from the Edge of Technology* (New York: HarperCollins, 2001), 136.
36. Ibid., 282-283.
37. Ibid., 279.
38. Ibid., 280.
39. Ibid., 281.
40. Ibid.
41. 282.
42. 307.
43. 308.
44. 314-315.
45. Chiles,134.
46. Dumas, 229.
47. Ibid., 258-281.
48. Ibid., 63.
49. Chiles, 40.
50. Ibid., 39.

51. Perrow, 137-139.
52. Ibid., 50-54.
53. Ibid., 8.
54. Ibid., 152-153.
55. Dumas, 185.
56. Ibid., 301.
57. Mark Nichols and Donald MacGillivray, "The Hot Zones," *Maclean's*, Dec. 19, 1994, 48-49.
58. Brian Bergman, "On the Front Lines of the Search for the Cause of SARS," *Maclean's*, April 14, 2003, 24.
59. Debora McKenzie, "China's SARS Cases Sound Alarm Over Lab Security," *New Scientist*, May 1, 2004, 12.
60. Government Accountability Office, Testimony Before the Subcommittee on Oversight and Investigations, Committee on Energy and Commerce, House of Representatives, *High-Containment Biosafety Laboratories: DHS Lacks Evidence to Conclude That Foot-and-Mouth-Disease Research Can Be Done Safely on the U.S. Mainland*, May 22, 2008, GAO-08-821T, 22.
61. Craig Schneider and Ariel Hart, "Floods at Univ. of Ga. Germ Lab Revealed," *Atlanta Journal-Constitution*, Nov. 2, 2008.
62. Ibid.
63. Lee Shearer, "Vet School Leak Kept Quiet: UGA's Liaison Panel at First Not Told of Mishaps," *Athens Banner-Herald*, Nov. 8, 2008.
64. Ibid.
65. Pat Allen, "Memo," Oct. 7, 2008. Accessed at http://www.online.athens.com/multimedia/pdfs/110908lab_leak.pdf.
66. Tom Jackson, "Memo," Oct. 7, 2008. See Allen for Internet address.
67. Shelia Allen, "Memo," Oct. 7, 2008. See Pat Allen for Internet address.
68. Schneider and Hart, "Floods Revealed."
69. Ibid.
70. Shearer, "Vet School Leak Kept Quiet."
71. Robert Fenton, "Memo," Oct. 30, 2008. See Pat Allen for Internet address.
72. Harry Dickerson, "Memo," Oct. 3, 2008. See Pat Allen for Internet address.
73. Fenton, "Memo."
74. Accessed at http://www.online.athens.com/multimedia/pdfs/110908lab_leak.pdf.
75. Monifa Thomas, "Plague Researcher Dies of Infection," *Chicago Sun-Times*, Sept. 20, 2009.
76. Ibid.; Emma Graves Fitzsimmons, "Researcher Had Bacteria for Plague at His Death," *New York Times*, Sept. 21, 2009.

NBAF 8
1. George Diepenbrock, "Biodefense Proposals Face Deadline Today," *Lawrence Journal-World*, Feb. 16, 2007.
2. Dolph C. Simons, Jr., "Simons: Political Interests Have Shattered Bioscience Authority Dream," *Lawrence Journal-World*, June 16, 2007.
3. Ibid.
4. Ibid.; "Partisan Distraction," *Lawrence Journal-World*, Feb. 23, 2007.
5. "Sebelius Pushes to Win Biodefense Facility for State," *Kansas City Business Journal*, Jan. 11, 2007.
6. Cody Howard, "Leavenworth County Works Ahead on Plans for Federal Biocenter," *Lawrence Journal-World*, Sept. 13, 2006.
7. "Roberts Brings USDA Official to Kansas to See Biosecurity Research Expertise," *Cattle Network*, Jan. 10, 2007, http://www.cattlenetwork.com.
8. Associated Press, "K-State's New Lab on Front Lines of Biosecurity," *Wichita Eagle*, March 18, 2007.

9. Rex Dalton, "Kansas Wins Race to Host Biodefense Research Center," *Nature* 456 (2008): 687.
10. Ibid.
11. Ibid.
12. John Taylor, "Task Force Named to Lure Biodefense Lab to Kansas," *Lawrence Journal-World*, Jan. 26, 2007.
13. George Diepenbrock, "Roberts to Push Lab Effort at Legislature," *Lawrence Journal-World*, Feb. 5, 2007.
14. Diepenbrock, "Biodefense Proposals Face Deadline."
15. Mark Fagan, "Kansas Makes Run for Lab," *Lawrence Journal-World*, Feb. 21, 2007.
16. Eric Weslander, "Kansans Lobby for Defense Facility," *Lawrence Journal-World*, March 14, 2007.
17. "Biodefense Bid," *Lawrence Journal-World*, Feb. 18, 2007.
18. "No Need to Look Beyond Kansas," *Manhattan Mercury*, Feb. 6, 2007.
19. "Citizens Need More Information about Agro-, Bioterror Defense Lab," *Manhattan Mercury*, Feb. 8, 2007.
20. Ibid.
21. Bill Felber, "What the City Gets with NBAF," *Manhattan Mercury*, March 4, 2007.
22. Eric Weslander, "Neighbors Raise Biodefense Lab Concerns," *Lawrence Journal-World*, March 1, 2007.
23. Richard Preston, *The Hot Zone* (New York: Random House, 1994), 64.
24. Leah Rupp, "Some in Hinds Fight Bid for Bioterror Lab," *Jackson Clarion-Ledger*, March 25, 2007, 1B.
25. Editorial, *Jackson Clarion-Ledger*, April 2, 2007, 6A.
26. Editorial, *Jackson Clarion-Ledger*, April 23, 2007, 6A.
27. Julie Goodman, "Little Town That Could," *Jackson Clarion-Ledger*, June 6, 2007, 1C.
28. Editorial, *Jackson Clarion-Ledger*, July 15, 2007, 4G.
29. Ibid.
30. Feb. 11, 2007.
31. Don Finley, Feb. 2, 2007.
32. David Hendricks, Feb. 13, 2007.
33. David Hendricks, Feb. 16, 2007.
34. Melissa Golden, "Focus on Key Poultry Lab as UGA Hatches Big Biodefense Goals," *Athens Banner-Herald*, Feb. 18, 2007.
35. Press release, Tri-Valley CARES, "Northern Cailifornia Communities Defeat Bio-Warfare Agent Research Proposal at Livermore Lab Site 300 Warfare," July 11, 2007; David Perlman, "UC Out of the Running for Controversial Biodefense Lab," *San Francisco Chronicle*, July 12, 2007, B3; Mike Martinez, "Site 300 Doesn't Make Bio-Lab List," *Tri-Valley Herald*, July 12, 2007.
36. Matthew DeFour, "Dunn Residents, Board Oppose Disease Lab," *Wisconsin State Journal*, March 8, 2007, A1.
37. Alec Luhn, "In-Depth: Not in Dunn's Backyard," *Badger Herald*, May 3, 2007.
38. Alec Luhn, "Dunn Says 'No' to Facility," *Badger Herald*, April 25, 2007.

CHAPTER SEVEN

1. Associated Press, April 24, 2009. Found at http://www.npr.org on April 29, 2009 under the erroneous title "S.C. Wildfire May Accelerate after Overnight Call." *See also* Donald G. McNeil, Jr., "Unusual Strain of Swine Flu Is Found in People in 2 States," *New York Times*, April 24, 2009, A13.
2. Laurie Garrett, "The Path of a Pandemic," *Newsweek*, May 11, 2009.
3. Martin Enserink, "Tiptoeing around Pandora's Box," *Science* 305 (2004): 594-595.
4. Rachael Nowak and Michael Le Page, "Superflu Is Being Brewed in the Lab," *New Scientist*, Feb. 28, 2004, 6-7.

5. Ibid.
6. Li-Mei Chen, C. Todd Davis, Hong Zhou, Nancy T. Cox, and Ruben O. Donis, "Genetic Compatibility and Virulence of Reassortants Derived from Contemporary Avian H5N1 and Human H3N2 Influenza A Viruses," *PLoS Pathogens*, May 2008, 1.
7. Ibid., Abstract.
8. Enserink, "Tiptoeing around Pandora's Box."
9. Nowak and Le Page, "Superflu Is Being Brewed."
10. "Recreating the Spanish Flu?" Oct. 9, 2003, The Sunshine Project, http://www.sunshine-project.org.
11. Hana M. Weingartl et al., "Experimental Infection of Pigs with the Human 1918 Pandemic Influenza Virus," *Journal of Virology* 83 (2009): 4288.
12. Jeffery K. Taubenberger et al., "Characterization of the 1918 Influenza Virus Polymerase Genes," *Nature* 437 (2005): 889-893.
13. Terrence M. Tumpey et al., "Characterization of the Reconstructed 1918 Spanish Influenza Pandemic Virus," *Science* 310 (2005): 77-80.
14. Wendy Orent, "Playing with Viruses," *Washington Post*, April 17, 2005, B1.
15. Andreas von Bubnoff, "The 1918 Flu Virus Is Resurrected," *Nature* 437 (2005): 794-795.
16. Jocelyn Kaiser, "1918 Flu Experiments Spark Concerns about Biosafety," *Science* 306 (2004): 591.
17. Helen Branswell, "Scientists Jittery Over 1918 Virus," *Times Colonist* (Victoria, British Columbia), Oct. 22, 2004, A4.
18. Bubnoff, "The 1918 Flu Virus Is Resurrected," 795.
19. Andreas von Bubnoff, "Deadly Flu Virus Can Be Sent Through the Mail," *Nature* 438 (2005): 134-135.
20. "Flu in Circulation," *Nature* 438 (2005): 130.
21. Bubnoff, "Deadly Flu Virus Can Be Sent Through the Mail."
22. Margaret Munro, "Winnipeg Lab Plans to Recreate Deadly 1918 Flu Virus," *Edmonton Journal* (Alberta), Nov. 10, 2005, B11.
23. Bubnoff, "Deadly Flu Virus Can Be Sent Through the Mail."
24. Ibid.
25. Weingartl et al., "Experimental Infection of Pigs."
26. Ibid.
27. LA Perrone et al., "Intranasal Vaccination with 1918 Influenza Virus-Like Particles Protects Mice and Ferrets from Lethal 1918 and H5N1 Influenza Virus Challenge," *Journal of Virology* 83 (2009): 5726-34.
28. Helen Branswell, "Officials Investigate How Bird Flu Contaminated Vaccines in Europe," *The Canadian Press*, Feb. 26, 2009, http://www.thestar.com; Helen Branswell, "Baxter: Product Contained Live Bird Flu Virus," *Toronto Sun*, Feb. 27, 2009.
29. Ibid.
30. Ibid.
31. Ibid.; Michelle Fay Cortez and Jason Gale, "Baxter's Vaccine Research Sent Bird Flu Across European Labs," Feb. 24, 2009, http://www.bloomberg.com.
32. Nicole Gaouette, "U.S. Asks Glaxo, Novartis to Start Swine Flu Vaccine," May 22, 2009, Bloomberg.com, http://www.bloomberg.com.
33. Paul Joseph Watson, "'Accidental' Contamination of Vaccine with Live Avian Flu Virus Virtually Impossible," March 5, 2009, http://www.prisonplanet.com.
34. Ibid.
35. Ibid.
36. "Baxter H5N1 Contamination: Causes Were a Unique Combination of Process, Technical and Human Errors," *LifeGen.de*, March 3, 2009, http://www.lifegen.de/newsip/shownews.php4?getnews=2009-03-02-2412&pc=s02.
37. Ibid.
38. Ibid.
39. Ibid.

40. Ibid.
41. Rob Stein and Shankar Vedantam, "Deadly Flu Strain Shipped Worldwide," *Washington Post*, April 13, 2005, A1; "Cleaning the Fridge of Pandemic Virus," *Effect Measure*, April 12, 2005, http://scienceblogs.com/effectmeasure.
42. Stein and Vedantam.
43. "Renegade Flu Virus," *Washington Post*, April 18, 2005, A16.
44. Stein and Vedantam, "Deadly Flu Strain Shipped."
45. Ibid.
46. Ibid.
47. Anita Manning and Julie Schmitt, "Samples of Pandemic Flu Virus Found in 3 Foreign Warehouses of Shipper," *USA Today*, April 17, 2009.
48. Stein and Vedantam, "Deadly Flu Strain Shipped."
49. Associated Press, "CDC: No Sign of Deadly Flu Strain Circulating," April 13, 2005.
50. Manning and Schmidt, "Samples Found."
51. "Don't Worry, Be Happy," *Effect Measure*, April 14, 2005, http://scienceblogs.com/effectmeasure; "'More Than Overdue.' What Is There to Deliberate About?" *Effect Measure*, April 19, 2005.
52. "Evidence of Risk for H2N2 1957 Pandemic Flu," *Recombinomics*, April 18, 2005, http://www.recombinomics.com.
53. "Don't Worry, Be Happy," *Effect Measure*.
54. "Cleaning the Fridge of Pandemic Virus," *Effect Measure*.
55. Andrew S. Ross.
56. Alexander Haislip, April 24, 2009.
57. Cyrus Sanati, April 27, 2009.
58. Philip Dru, "Flying Pigs, Tamiflu and Factory Farms," *NWO Truth*, April 29, 2009, http://nwotruth.com.
59. Haislip.
60. Bruce Japsen, "Swine Flu: Baxter Seeks Swine Flu Sample to Begin Work on Vaccine," *Chicago Tribune*, April 27, 2009.
61. Gaouette, "U.S. Asks Glaxo, Novartis to Start Swine Flu Vaccine."
62. Ibid.
63. Alan M. Wolf, "Novartis Nets Bird Flu Vaccine Deal," *News & Observer*, Jan. 16, 2009.
64. "U.S. Government Accepts $192 Million of Sanofi Pasteur H5N1 Bulk Vaccine Antigen for Pandemic Stockpile," press release from Sanofi Pasteur, April 28, 2008.
65. Rachel Nowak, "Killer Virus," *New Scientist*, Jan. 10, 2001.
66. Ibid.
67. Ibid.
68. Ibid.
69. Debora MacKenzie, "U.S. Develops Lethal New Viruses," *New Scientist*, Oct. 29, 2003.
70. Michael Scherer, "The Next Worst Thing," *Mother Jones*, March/April 2004.
71. MacKenzie, "U.S. Develops."
72. Ibid.
73. Jonathan B. Tucker, "The Smallpox Destruction Debate: Could a Grand Bargain Settle the Issue?" *Arms Control Today*, March 2009, http://www.armscontrol.org/act/2009_03/tucker.
74. Ibid.
75. Nell Boyce, "Smallpox Mixes Make a Stir," *U.S. News & World Report*, Jan. 19, 2004, 64; "Plan to Engineer Smallpox Virus Causes Alarm," press release by the Council for Responsible Genetics, Nov. 12, 2004, http://www.councilforresponsiblegenetics.org.
76. "Plan to Engineer Smallpox Virus Causes Alarm."
77. Peter Jahrling et al., "Exploring the Potential of Variola Virus Infection of Cynomolgus Macaques as a Model for Human Smallpox," *PNAS* 101 (2004): 15196.
78. Peter Aldhous and Michael Reilly, "Bioterror Special: Friend or Foe?" *New Scientist*, Oct. 14, 2006.

79. Charles Piller and Keith R. Yamamoto, *Gene Wars: Military Control over the New Genetic Technologies* (New York, Beech Tree Books, 1988), 96-102.
80. Ibid., 133.
81. Judith Miller, Stephen Engelberg, and William Broad, *Germs: Biological Weapons and America's Secret War* (New York: Touchstone, 2002), 70.

NBAF 9

1. James Bruggers, "U.S. to Scout Pulaski Site for Lab," *Courier-Journal*, April 14, 2007. I have to point out that others beside a reporter shape the final article which appears in a newspaper. Several people who read the *Courier-Journal* regularly, and whose judgment I respect, say that Jim Bruggers is a good reporter. His newspaper and I just had a bad day.
2. Alec Luhn, "In-Depth: Not in Dunn's Backyard," *Badger Herald*, May 3, 2007.
3. Rae Nudson, "Fears about Bio-Defense Lab Grow," *Columbia Missourian*, June 7, 2007.

CHAPTER EIGHT

1. "Farewell Radio and Television Address to the American People."
2. Walter Isaacson, "GQ Icon: Colin Powell," *GQ*, Oct. 2007.
3. Will Prescott and Nick Tankersley, "Powell Calls for More Oversight," video interview, Nov. 18, 2007, at http://hub.ou.edu/multimedia.
4. Alison Motluk, "Protect and Prosper: Big Spending on Biodefence Has Given Research into New Vaccines and Antivirals a Massive Shot in the Arm," *New Scientist*, Oct. 23, 2004.
5. Nick Rees, "New Technology Allows for Sanitizing of Potential Bio-Infected Mail," *Bio-Prep Watch*, Nov. 21, 2009, http://www.bioprepwatch.com.
6. Marc Songini, "Boston's BioDefense Blocks Anthrax Mail Threats," *Mass High Tech: The Journal of New England Technology*, Jan. 30, 2009, http://www.masshightech.com.
7. Rees, "New Technology Allows"; Songini, "Boston's BioDefense."
8. Press release, "BioDefense Corporation Accelerates Launch of MAIL DEFENDER IT as Mail-Based Terrorism Hits New York Times and Well-Known Banks," *Market Wire*, Oct. 23, 2008.
9. Press release, "MEDIA ADVISORY: BioDefense Corporation Sees Rash of 'White Powder' Incidents at Reuters, New York Times, and Major Banks as Real Bio-Chemical Assaults, Not 'Hoaxes,'" *Market Wire* via *Comtex*, Oct. 28, 2008.
10. Ibid.
11. Press release, *Market Wire* via *Comtex*, Dec. 9, 2008.
12. Press release, "Post-Election Skyrocketing Gun Sales Prompt BioDefense Corporation's Warning of Increased Potential for Bio-Chemical Assaults, Such as Anthrax, or 'White Powder' Hoaxes," *Market Wire* via *Comtex*, Nov. 10, 2008.
13. Press release, "Boston-Based BioDefense Corp. Announces New UK Partnership to Expand Markets for Biohazard Protection," *PRNewswire*, April 22, 2008.
14. Press release, "Universal Technology Completes First Phase of Department of Homeland Security SAFETY Act Application for the BSM-2000," *Market Wire*, April 30, 2008; press release, "Universal Detection Technology Announces Financial Results and 5,000% Revenue Increase Due to Increasing Sales of Biological and Other CBRN Detection Products," *Market Wire*, May 19, 2008.
15. Kendra Marr, "Emergent BioSolutions to Buy Maker of Flu Vaccine," *Washington Post*, May 27, 2008, D1.
16. Ken Silverstein, "Flaws in the BioShield: VaxGen Looks for Another Federal Bailout," Washington Babylon, *Harper's*, Dec. 12, 2006.
17. Ibid.
18. "Emergent BioSolutions Angles for New Anthrax Vaccine Deal," *Gazette.Net*, May 5, 2008, http://www.gazette.net; Arthur Allen, "Anthrax Vaccine Loses to Lobbying," *Washington Independent*, May 19, 2008.
19. Allen, "Anthrax Vaccine Loses to Lobbying."
20. Ibid.

21. "US Awards Contracts in Hunt for Versatile Biodefense Drugs," *CIDRAP*, Sept. 23, 2008, http://www.cidrap.umn.edu.

22. Irene Collins, "Two Biotech Companies Are to Develop Anthrax Vaccines," eFlux Media, Sept. 28, 2008, http://www.eflux.com.

23. Katie Arcieri, "The Profit of Protection: Annapolis Biodefense Firm PharmAthene Buys Development Rights for Anthrax Vaccine as Part of $40 Million Deal," *The Capital*, May 11, 2008.

24. Jill Hamburg Coplan, "Securing Government Money for Security," *Business Week*, Aug. 7, 2009.

25. Ibid.

26. Ibid.

27. Press release, "Ridge Global Names Luke Ritter to Executive Team," *PRNewswire*, April 8, 2008.

28. Laurie Bennett, "Former FBI Chief Louis Freeh Sells His Services to Corporate Elite," *Muckety*, May 26, 2008, http://news.muckety.com.

29. Ibid. It is supposedly illegal to buy a sitting judge, but it is legal to buy a *retired* judge. Of course a certain group of sitting Supreme Court justices have now made it legal for corporations to buy sitting Congressmen, Presidents, etc., through unlimited campaign donations from corporate treasuries, so that we may now look forward, perhaps, to our political representatives bearing such frank corporate logos as the Pepsi-Cola Congressman for the Fifth District, the Wal-Mart Senator for Arkansas—*on sale, today, 24 hours only!*—or the Three Mile Island Nuclear Energy President of the United States.

30. Noah Shachtman, "DHS' New Chief Geek is a Bioterror 'Disaster,' Critics Charge," Danger Room, *Wired*, May 6, 2009, http://www.wired.com.

31. Ibid.

32. Wendy Orent, "The Bioterror Bugaboo" (op-ed), *Los Angeles Times*, July 17, 2009.

33. Ibid.

34. Quoted by George Smith in "A Most Catastrophic Nomination," May 7, 2009, http://dickdestiny.com.

35. Quoted by Smith.

36. Ibid.

37. Marcus Stern, "Experts Divided over Risk of Bio-Terrorist Attack," *ProPublica*, Dec. 5, 2008, http://www.propublica.org.

38. Thompson, *The Killer Strain*, 58.

39. Quoted at Smith.

40. Jim McElhatton, "Exclusive: Obama Nominee Omitted Ties to Biotech," *Washington Times*, Sept. 8, 2009.

41. Ibid.

42. Ibid.

43. Press Release, "Alliance for Biosecurity Testifies Before Congress on Funding for Medical Countermeasure Development," *PRNewswire*, March 18, 2009.

44. Jennifer Washburn, *University, Inc.: The Corporate Corruption of Higher Education* (New York: Basic Books, 2005).

45. Kathleen Burge, "New College Major: Homeland Security," *Boston Globe*, Oct. 9, 2008.

46. Ibid.

47. Ibid.

48. Barney Gimbel, "Education: Homeland Security U," *Newsweek*, Sept. 1, 2003.

49. Ibid.

50. Becky Orr, "LCCC Homeland Security Program Expanding," *Wyoming Tribune Eagle*, July 16, 2008.

51. "Complete the Following Fields to Get More Info about Counter Terrorism Training," *Kaplan University*, accessed at http://www.earnmydegree.com.

52. "Western Kentucky University to Offer Master's Degree in Homeland Security," *Business First of Louisville*, May 23, 2008.

53. "Why You Should Consider a Degree in Homeland Security," Feb. 16, 2009, http://www.top-colleges.com.blog.
54. Washburn, 3.
55. Ibid., 4-5.
56. *Nature* 409 (2001): 119.
57. Washburn, 81-82.
58. Ibid., 83.
59. Ibid., 98-102.
60. Ibid., 130-136.
61. Ibid., 136.
62. Washburn also notes the pivotal role played by the 1980 Bayh-Dole Act in this commercialization and privatization of academic research.
63. Ibid., 193.

NBAF 10

1. Paul Snyder, "Dane County Board of Supervisors Votes to Oppose Construction of NBAF Facility at Dunn," *Daily Reporter*, May 9, 2007.
2. Ibid.
3. Rae Nudson, "Fears about Bio-Defense Lab Grow," *Columbia Missourian*, June 7, 2007.
4. Ibid.
5. Ibid.
6. Ibid.
7. Rae Nudson, "Residents Say No to Bio Lab," *Columbia Missourian*, June 8, 2007.
8. Rae Nudson, "Cattlemen End Support for Biolab," *Columbia Missourian*, July 3, 2007.

NBAF 11

1. Susan Wheeldon and Ken Shmidheiser, "Pulaski Site Doesn't Make Final Five Cut," *Somerset Commonwealth Journal*, July 12, 2007.
2. Ibid.
3. Sharon Dodson, "It's Official: No Ky. Bio-Lab," *Somerset Commonwealth Journal*, July 12, 2007.
4. Chris Harris, "Bio-Lab Land to be Auctioned," *Somerset Commonwealth Journal*, July 20, 2007.
5. David Hendricks, "S.A. a Biolab Finalist for Several Reasons," *San Antonio Express-News*, July 12, 2007.
6. Ibid.
7. Will Klusener, "In Union There Is Strength," *Manhattan Mercury*, July 12, 2007.
8. "Biotechnology Dreams Could Become Reality," *Athens Banner-Herald*, July 14, 2007.
9. Jason Winders, "Maybe We Can Mess with Texas for Lab Site," *Athens Banner-Herald*, July 14, 2007.
10. Ken Foskett, "A Very Dangerous Field: UGA Seeks Bioterror Role," *Atlanta Journal-Constitution*, Sept. 2, 2007.
11. John Huie, "Bio-Terror Buck$: Athens in Final Five for Facility," *Flagpole*, July 25, 2007.
12. Ibid.
13. Ibid.
14. Ibid.
15. Ibid.
16. "Luring the Bio-Lab," *Raleigh News & Observer*, July 24, 2007.
17. Tim Simmons, "Butner May Get Security Lab," *Raleigh News & Observer*, July 12, 2007.
18. Ibid.
19. Ibid.

CHAPTER NINE

1. E-mail, Feb. 12, 2008.

2. Peter Aldhous, "Sunshine Snuffed Out," *New Scientist*, Feb. 21, 2008.
3. Ibid.
4. Martin Enserink, "Activist Throws a Bright Light on Institutes' Biosafety Panels," *Science* 305 (2004): 768-769.
5. "Sunshine So Far: A Brief History of the Project," http://www.sunshine-project.org.
6. Enserink; Edward Hammond bio at Sunshine Project Web site, http://www.sunshine-project.org.
7. News Release, "US Armed Forces Push for Offensive Biological Weapons Development," The Sunshine Project, May 8, 2002, http://www.sunshine-project.org.
8. News Release, "Lethal Virus from 1918 Genetically Reconstructed," Oct. 9, 2003; Briefing Paper, "Recreating the Spanish Flu?", Oct. 9, 2003, http://www.sunshine-project.org.
9. Joint News Release, Citizens Education Project, Los Alamos Study Group, Nuclear Watch of New Mexico, Physicians for Social Responsibility, The Sunshine Project, Tri-Valley CARES, "Loose Monkey Teaches Biodefense Lab a Lesson on the Hazards of Secrecy," Feb. 26, 2004, http://www.sunshine-project.org.
10. Joint News Release, The Sunshine Project, Tri-Valley CARES, Citizen's Education Project, Physicians for Social Responsibility, Council for Responsible Genetics, "Institute Responsible for Anthrax Accident in California, in Charge of Safety and Security at Chicago Biodefense Laboratory," June 22, 2004, http://www.sunshine-project.org.
11. News Release, "Tularemia Outbreak: CDC Refuses Rapid Response: Cold Comfort for Communities Confronting Biodefense Labs," The Sunshine Project, Jan. 31, 2005, http://www.sunshine-project.org.
12. Edward Hammond, "Averting Bioterrorism Begins with US Reforms," Sept. 21, 2001, http://www.sunshine-project.org.
13. Hammond, "Operation Infinite Contract: Biotech's Impossible, Profitable War to Defeat Nature," Oct. 3, 2001, http://www.sunshine-project.org.
14. News Release, "Faulty Aerosol Chamber Infects Three," The Sunshine Project, April 18, 2005, http://www.sunshine-project.org.
15. "Civil Society Laboratory Monitoring; Collaborations to Promote Local Laboratory Oversight," Sunshine Project Web site, http://www.sunshine-project.org.
16. "Transparency and Public Accountability in Biodefense: Freedom of Information and Access to Research Information," Sunshine Project Web site, http://www.sunshine-project.org.
17. "Mandate for Failure: The State of Institutional Biosafety Committees in an Age of Biological Weapons Research," The Sunshine Project, Oct. 4, 2004, 3, http://www.sunshine-project.org.
18. Ibid., 5-6.
19. Ibid., 6.
20. Ibid.
21. Ibid.
22. Ibid., 7.
23. Ibid.
24. Ibid.
25. Ibid., 7-8.
26. Ibid., 10.
27. Ibid.
28. Ibid., 11.
29. Ibid., 12.
30. Ibid., 12-13.
31. Ibid., 12, footnote 23.
32. Ibid., 13.
33. Ibid., 14.
34. Ibid., 12.

35. Ibid., 22.
36. Ibid.
37. "US Army Builds Biodefense Lab, Neglects to Inspect It," Biosafety Bites #1, June 28, 2004, http://www.sunshine-project.org.
38. Ibid.
39. Ibid.
40. Ibid.
41. "Moribund: The Department of Homeland Security's Plum Island Biosafety Committee," Biosafety Bites # 3, July 6, 2004, http://www.sunshine-project.org.
42. "No Biosafety Meetings at Rockefeller University," Biosafety Bites #5, July 19, 2004, http://www.sunshine-project.org.
43. "The Institute for Genomic Research: Genomic Powerhouse, Biosafety Tragedy," Biosafety Bites # 6, July 27, 2004, http://www.sunshine-project.org.
44. "No Functional Biosafety Committee at Battelle Memorial Institute," Biosafety Bites # 7, Aug. 10, 2004, http://www.sunshine-project.org.
45. "No Biosafety Reviews at Emory University," Biosafety Bites # 8, Aug. 12, 2004, http://www.sunshine-project.org.
46. "Tulane University's Broken Biosafety Committee," Biosafety Bites # 10, Aug. 18, 2004, http://www.sunshine-project.org.
47. "Incautious University of Washington Bends Biosafety in 1918 'Spanish' Flu Experiments," Biosafety Bites # 4, July 12, 2004, http://www.sunshine-project.org.
48. Ibid.
49. Ibid.
50. Ibid.
51. *Mandate for Failure*, 46.
52. Ibid.
53. Ibid., 47.
54. Ibid.
55. Ibid., 50.
56. Ibid., 48.
57. Ibid.
58. Ibid., 49.
59. Ibid.
60. Ibid., 50.
61. Ibid., 49.
62. "Top 10 Freedom of Information Failures," The Sunshine Project, last updated Feb. 20, 2006, http://www.sunshine-project.org.
63. Ibid.
64. Ibid.
65. Ibid.
66. "Complaint Filed Against 'Sham' University of South Carolina Institutional Biosafety Committee," Biosafety Bites (v. 2) # 13, June 1, 2006, http://www.sunshine-project.org.
67. Ibid.
68. Ibid.
69. "Dukes of Hazards: Georgia's Rebel Yell Against Biosafety," Biosafety Bites (v. 2) # 16, Aug. 15, 2006, http://www.sunshine-project.org.
70. "AlphaVax IBC: It's Been Years Since We Met," Oct. 17, 2006, http://www.sunshine-project.org.
71. "Al-Qaeda Plot Suspected on University of North Carolina Biolab," Biosafety Bites (v.2) #15, Aug. 15, 2006, http://www.sunshine-project.org.
72. Ibid.
73. Ibid.

74. "FBI Wants Biodefense Work Secret, Says Newark, New Jersey Lab," Biosafety Bites (v. 2) # 17, Sept. 20, 2006, http://www.sunshine-project.org.
75. Ibid.
76. J. J. Hermes, "Sun May Set on a Project That Monitored Biological-Weapons Research," *Journal of Higher Education*, Feb. 8, 2008.
77. Jocelyn Kaiser, "Biodefense Watchdog Project Folds, Leaving a Void," *Science* 319 (2008), 886.
78. Ibid.
79. News Release, "Fake Forms Populate Fraudulent University of Texas Biosafety Site: Fake Texas IBC Operates in Parallel to Secret Safety Committee," Oct. 31, 2007, http://www.sunshine-project.org.
80. "Cowboy BSL-4 in Texas Thumbs Its Nose at the NIH Guidelines and the NIH Office of Biotechnology Activities," Biosafety Bites (v. 2) # 14, June 6, 2006, http://www.sunshine-project.org.
81. "The Bird Flu Accident that Officially Didn't Happen, or How the University of Texas at Austin Could Have Caused the Next Influenza Pandemic, but Everybody Lived to Cover It Up," Biosafety Bites # 21 (v. 2), Jan. 11, 2007, updated Jan. 26, 2007, http://www.sunshine-project.org.
82. Ibid.
83. Ibid.
84. "Fake Forms," Oct. 31, 2007.

NBAF 12
1. "San Antonio Ready, Able to Accommodate Biolab," *San Antonio Express-News*, Sept. 11, 2007.
2. Cindy Tumiel, "Biolab Project's Hearing Has Few Naysayers," *San Antonio Express-News*, Sept. 12, 2007, 5B.
3. "County, State Officials Come Out in Support of Biolab," *Madison County Journal*, Sept. 6, 2007.
4. Scott Rothschild, "Biodefense Lab Draws High Level of Support," *Lawrence Journal-World*, August 29, 2007.
5. "Five Sites, One Message," *Manhattan Mercury*, Sept. 21, 2007.
6. Matthew E. Milliken, "Proposed Animal Disease Facility Debated in Granville," *Henderson Daily Dispatch*, Sept. 19, 2007.
7. Will Klusener, "Your Say on NBAF," *Manhattan Mercury*, Aug. 26, 2007.
8. "Make Your View of NBAF Known," *Manhattan Mercury*, Aug. 26, 2007.
9. Klusener, "Your Say on NBAF."
10. Angela Kreps, "Let's Show Manhattan Is Right Site for NBAF," Aug. 26, 2007, C8.
11. Nancy and Jerry Jaax, "Citizens Need Straight Talk on NBAF Safety," *Manhattan Mercury*, Aug. 27, 2007, A6.
12. Ibid.
13. Ned Seaton, "Would NBAF, If Built Here, Be a Target for Terrorists?" *Manhattan Mercury*, Aug. 24, 2007, A4.
14. "I question the wisdom of locating DHS's only biocontainment facilities within approximately 40 miles of each other from a workforce recruiting perspective as well as from the perspective of Homeland Security continuity of operations planning in the event of a disaster or terrorist attack in the greater National Capital region."— *Final Selection Memorandum: Site Selection for the Second Round Potential Sites for the NBAF*, U.S. Dept. of Homeland Security, Science and Technology Directorate, July 2007.
15. Ibid.
16. Ned Seaton, "What If We *Don't* Want NBAF?" *Manhattan Mercury*, Aug. 17, 2007, A1.
17. Ibid.
18. Ibid.

19. Ibid.
20. Lisa Sorg, "Biotech or Biohazard?" *Independent Weekly*, July 25, 2007.
21. Lisa Sorg, "Feel a Chill Coming On?" *Independent Weekly*, July 25, 2007.
22. *Independent Weekly*, July 25, 2007.
23. Greg Harman, "Banging the Drum for Bio-defense," *San Antonio Current*, Aug. 15, 2007.
24. Pete McCommons, "The Pathogen City?" *Flagpole*, Sept. 26, 2007.
25. Will Klusener, "Scoping Things Out," *Manhattan Mercury*, Aug. 29, 2007, A1.
26. Ibid.
27. Klusener, "Worries about NBAF Fallout," *Manhattan Mercury*, Aug. 29, 2007, A1.
28. Tumiel, "Biolab Project's Hearing Has Few Naysayers."
29. Andrew Ujifusa, "Flora Ponders Next Bio Step," *Madison County Journal*, July 19, 2007.
30. "County, State Officials Come Out in Support."

NBAF 13
1. North Carolina Consortium for the National Bio and Agro-Defense Facility, "Ten Common Misconceptions about the National Bio and Agro Defense Facility," http://www.ncc-nbaf.org/faq.cfm, accessed Oct. 1, 2007.
2. Ibid.
3. Ibid.
4. Ibid.
5. Ibid.
6. Joe Skernt, "Ebola Gene Theft a Shocker," *Winnipeg Free Press*, May 14, 2009.
7. Ibid.
8. Ibid.
9. *Final Selection Memorandum: Site Selection for the Second Round Potential Sites for the NBAF*, U.S. Dept. of Homeland Security, Science and Technology Directorate, July 2007, 30.
10. *Focus*, Aug. 10, 2007, accessed at http://www.ncc-nbaf.org/081007_Focus_Butner_Butner_Biohazard_Research_transcript.cfm, October 1, 2007.
11. Ibid.
12. Ibid.
13. E-mail, Oct. 13, 2007.
14. Scott Carr, interviewer, State Government Radio, accessed at http://www.ncc-nbaf.org/NBAF_stategovernmentradio_transcript.cfm, Oct. 1, 2007.
15. *Focus* interview, Aug. 10, 2007.
16. Eric Johnson, "State Pushes for Bio Facility," *Daily Tar Heel*, Sept. 11, 2007.
17. Tim Simmons, "N.C. Making Pitch to Win Coveted Lab," *Raleigh News & Observer*, Sept. 17, 2007.
18. Rick Smith, "Biotech Center Helps Lead Lobbying Efforts for Bio, Agro Defense Facility," *The Skinny*, Sept. 17, 2007, http://www.localtechwire.com.
19. Lisa Sorg, "The Battle Against the Bio-Lab," *Independent Weekly*, Oct. 10, 2007.
20. Matthew E. Milliken, "Proposed Animal Disease Facility Debated in Granville," *Henderson Daily Dispatch*, Oct. 19, 2007.
21. Ibid.
22. William F. West, "Officials Push for Butner Facility," *Durham Herald-Sun*, Sept. 18, 2007.
23. Sorg, "The Battle Against the Bio-Lab"; Jason Winders, "Facing Down 'Big Dogs' on NBAF Site," *Athens Banner-Herald*, Oct. 14, 2007.
24. Winders, "Facing Down 'Big Dogs.'"
25. Sorg, "The Battle Against the Bio-Lab."
26. "Group Reunites to Oppose Lab Site," *Raleigh News & Observer*, Sept. 26, 2007.
27. Rebecca Quigley, "Biodefense Lab Pitched as No Threat," *Athens Banner-Herald*, Aug. 31, 2007.
28. Ibid.
29. Ibid.
30. Ibid.

31. Ibid.
32. Daniel O'Connor, "Facility Concerns Residents," *The Red and Black*, Sept. 4, 2007.
33. Ben Emanuel, "Homeland Security: Paying Athens a Visit," *Flagpole*, Sept. 12, 2007.
34. Blake Aued, "Some Sure, Some Not, of Lab's Safety," *Athens Banner-Herald*, Sept. 15, 2007.
35. David Lee, Harry Dickerson, Fred Quinn and Corrie Brown, "Forum: National Biodefense Facility Can and Will Be Operated Safely," *Athens Banner-Herald*, Sept. 19, 2007.
36. Blake Aued, "After Public Speaks, Bio Lab in Feds' Hands," *Athens Banner-Herald*, Sept. 24, 2007.
37. Blake Aued, "UGA Not Releasing Biosafety Records," *Athens Banner-Herald*, Oct. 6, 2007; Ed Hammond, Comment, Oct. 7, 2007, at biodefense@lists.sunshine-project.org.
38. "Editorial: UGA Should Disclose Records of Lab Accidents," *Athens Banner-Herald*, Oct. 9, 2007.
39. Blake Aued, "UGA Releases Lab Accident Records," *Athens Banner-Herald*, Oct. 17, 2007; Comment, Ed Hammond, Oct. 17, 2007, at biodefense@lists.sunshine-project.org.
40. "Granville Residents Mobilize to Oppose National Bio Agro Defense Facility," http://www.nobio.org, accessed Nov. 1, 2007.
41. Ben Emanuel, "Athens Investigates NBAF: Locals Bring the U.S. Biodefense Program's Biggest Critic for a Visit," *Flagpole*, Jan. 16, 2008.
42. Matthew E. Milliken, "Granville Seeks Contact on Lab," *Durham Herald-Sun*, Dec. 5, 2007.
43. Ibid.
44. Tim Simmons, "Butner Lab Foes Step Up Efforts," *Raleigh News & Observer*, Dec. 18, 2007; "Doctors Take Different Stances on Biodefense Lab in Butner," Dec. 25, 2007, WRAL, http://www.wral.com/news/.
45. Sloane Heffernan, "Granville Commissioners Pull Support of Bio Lab," Jan. 8, 2008, WRAL, Jan. 8, 2008, http://www.wral.com/news/.
46. "Congressmen Seek Information on Bioterror Lab," Jan. 14, 2008, WRAL, http://www.wral.com/news/.
47. "NBAF Forum Tonight at Georgia Center," *Athens Banner-Herald*, Feb. 18, 2008.
48. John Fischer, Joseph Corn, and Daniel Mead, "There Is Nothing to Fear from NBAF," *Athens Banner-Herald*, Feb. 17, 2008.
49. "NBAF Communication Thread Shows Debate Over Debate Follow Up Questions," *Oconee County, Georgia Politics,* Feb. 28, 2008, http://oconeedemocrat.blogspot.com.
50. Blake Aued, "Lab Backers Tout Defense Role," *Athens Banner-Herald*, Feb. 19, 2008.
51. Lisa Sorg, "NBAF Opponents Confront Homeland Security," *Independent Weekly*, Feb. 20, 2008.
52. Don Nelson, "NBAF Forum Went Well," *Athens Banner-Herald*, Feb. 24, 2008.
53. "Editorial: Anger Apparent at NBAF Meeting," *Durham Herald-Sun*, Feb. 24, 2008.
54. "Raleigh Council Withholds Support for Bio Lab," *Raleigh News-Observer*, Feb. 20, 2008.

CHAPTER TEN
1. Gigi Kwik Gronvall, Joe Fitzgerald, Allison Chamberlain, Thomas V. Inglesby, and Tara O'Toole, "High-Containment Biodefense Research Laboratories: Meeting Report and Center Recommendations," *Biosecurity and Bioterrorism* 5 (2007): 81.
2. Ibid.
3. Ibid., 75.
4. Ibid., 81.
5. John Miller, "Biosafety Labs Urged to Report Accidents and Near Misses," *Nature Online* (April 4, 2007), http://www.nature.com/news/2007/070402/full/070402-4.html.
6. Gronvall et al., 79.
7. Ibid., 82.
8. Ibid., 81-82.
9. "Senate Bill Would Alter Biosafety, Select Agent Rules," *Science* 320 (2008): 1573.
10. Justin M. Palk, "Security at Biodefense Labs to Get Tighter," *Frederick News-Post*, Dec. 18, 2008.

11. Debora MacKenzie, "Convenient US Army Rules for Biosafety?" *New Scientist*, Aug. 5, 2008.
12. "Military Biodefense Labs Try to Look Safe," *Effect Measure*, Aug. 28, 2008.
13. Palk, "Security at Biodefense Labs to Get Tighter"; "Anthrax Attacks Less Likely Today, U.S. Army Officer Says," *Global Security Newswire*, Dec. 19, 2008.
14. David Dishneau, "Army Lab Workers Getting Security Refresher," *Baltimore Sun*, Dec. 2, 2008.
15. "Anthrax Attacks Less Likely Today."
16. Press release, U.S. Dept. of Health & Human Services, Biosafety Task Force, "Public Consultation Meeting of the Trans-Federal Task Force on Optimizing Biosafety and Bio-containment Oversight," Nov. 18, 2008.
17. Jacob Goodwin, "President Bush Issues Executive Order to Beef Up Security at Bio Labs," *Government Security News*, Jan. 15, 2009.
18. Tara O'Toole and Thomas Inglesby, "Strategic Priorities for U.S. Biosecurity," in "Biosecurity Memos to the Obama Administration," *Biosecurity and Bioterrorism* 7 (2009): 2.
19. Ibid.
20. Ibid., 4.
21. Gigi Kwik Gronvall, "Preventing the Development and Use of Biological Weapons," in "Biosecurity Memos," 1.
22. Yudhijit Bhattacharjee, "The Danger Within," *Science* 323 (2009): 1283.
23. Ibid., 1282.
24. Ibid., 1283.
25. Paul Basken, "Universities Seek 'Sensible' Balance in Federal Oversight of Toxins in Research." *Chronicle of Higher Education*, March 13, 2009.
26. Ibid.
27. Martin Matishak, "New Report Warns Against Applying Personnel Security Measures to Bioresearch Labs," *Global Security Newswire*, June 9, 2009.
28. Ibid.
29. Ibid.
30. National Academy of Sciences, "Responsible Research with Biological Select Agents and Toxins: Report in Brief," 2009.
31. Bhattacharjee, "The Danger Within."
32. Yudhijiit Bhattacharjee, "Lawmakers Signal Tougher Controls on Pathogen Research," *Science* 326 (2009): 28-29.
33. "Containing Risk," *Nature* 461 (2009): 569-570.
34. "Lieberman's Biodefense Response—Weak," *Armchair Generalist*, Sept. 10, 2009.

NBAF 14
1. Scott Rothschild, "State Leaders Lobby for Research Facility," *Lawrence Journal-World*, Feb. 26, 2008; Rothschild, "High Hopes for Biodefense Lab," *Lawrence Journal-World*, Feb. 28, 2008; "Ongoing Effort," *Lawrence Journal-World*, March 3, 2008.
2. John Hanna, "Defense Lab Will Need Its Own Power Plant," *Topeka Capital-Journal*, March 15, 2008, 1A.
3. Ibid.
4. Ibid.
5. Ibid.
6. Blake Aued, "NBAF Doesn't Call for Power Plant: Kansas Governor Misunderstood, Officials Say," *Athens Banner-Herald*, March 18, 2008.
7. Ibid.
8. Lisa Sorg, "The Buck Stops Here on NBAF," *Independent Weekly*, April 2, 2008.
9. Ibid.
10. Ibid.
11. "Power Play? New Requirements for a Federal Defense Laboratory Raise a Number of Questions," *Lawrence Journal-World*, March 17, 2008.
12. Kathryn Waller, "Powering Up NBAF No Problem," *Manhattan Mercury*, March 17, 2008.

13. Scott Rothschild, "Infrastructure for Proposed Biosecurity Lab Could Cost Kansans $164 Million," *Lawrence Journal-World*, March 18, 2008; James Carlson, "Biodefense Center Bill Signed," *Topeka Capital-Journal*, March 29, 2008.

14. Carlson, "Biodefense Center Bill Signed."

15. Government Accountability Office, Testimony Before the Subcommittee on Oversight and Investigations, Committee on Energy and Commerce, House of Representatives, *High-Containment Biosafety Laboratories: DHS Lacks Evidence to Conclude That Foot-and-Mouth-Disease Research Can Be Done Safely on the U.S. Mainland*, May 22, 2008, GAO-08-821T, 2-3.

16. Ibid., 2, 7-8.

17. Ibid., 26.

18. Ibid., 4.

19. Ibid.

20. Ibid., 12n24.

21. Ibid., 13.

22. Ibid., 14.

23. Ibid., 15.

24. Ibid.

25. Ibid., 6.

26. Ibid., 6-7, 19.

27. Ibid., 19-20.

28. Ibid., 20-21.

29. Ibid., 21.

30. *See* "Beef Assoc Slams Foot and Mouth Plan," *Weekly Times*, Dec. 29, 2008; Lucy Skuthorp, "Uproar Over Foot and Mouth Import Proposal," *Rural Press*, Dec. 23, 2008.

31. Ibid., 21-22.

32. Ibid., 22.

33. Ibid.

34. Ibid, 23.

35. Ibid., 24.

36. Ibid., 6.

37. Statement of Bruce Knight, Under Secretary, Marketing and Regulatory Programs, Department of Agriculture, Before the Subcommittee on Oversight and Investigations, House Energy and Commerce Committee, May 22, 2008, 13; Kevin Shea, Animal and Plant Health Inspection Service, USDA, letter to Leroy Watson, Legislative Director, National Grange of the Patrons of Husbandry, May 16, 2008, included in Watson's written statement for the Subcommittee, 10-11.

38. GAO, 6.

39. House Subcommittee on Oversight and Investigations of the Committee on Energy and Commerce, *Germs, Viruses, and Secrets: Government Plans to Move Exotic Disease Research to the Mainland United States*, 110th Cong., 2nd sess., May 22, 2008.

40. Ibid.

41. Ibid.

42. Ibid.

43. Written Testimony of Ray L. Wulf, President & CEO, American Farmers and Ranchers, Submitted for the Record to the Subcommittee on Oversight and Investigations, House Energy and Commerce Committee, May 22, 2008.

44. House Subcommittee, May 22, 2008.

45. Testimony of the National Pork Producer Council Before the Subcommittee on Oversight and Investigations, House Energy and Commerce Committee, May 22, 2008, 10.

46. House Subcommittee, May 22, 2008.

47. Ibid.

48. Ibid.

49. Ibid.

50. Ibid.
51. Ibid. *See* GAO Testimony, 13-14, for GAO's phrasing and the list and years of FMD lab releases.
52. Ibid.
53. Ibid.

CHAPTER ELEVEN

1. Christopher Smart, "FBI Believes Anthrax Scientist Killed with Manipulated Spores from Dugway Lab," *Salt Lake Tribune*, Aug. 8, 2008.
2. Leonard Cole, *The Eleventh Plague: The Politics of Biological and Chemical Warfare* (New York: W. H. Freeman, 1997), 60.
3. Ed Regis, *The Biology of Doom* (New York: Henry Holt, 1999), 139-143.
4. Ibid., 154.
5. Ibid., 3-4, 164-175.
6. Ibid., 175.
7. Ibid., 176.
8. Judith Miller, Stephen Engelberg, and William Broad, *Germs: Biological Weapons and America's Secret War* (New York: Touchstone, 2002), 47-48.
9. Ibid., 48.
10. Regis, 168.
11. Lee Davidson, "Like Sheep to the Slaughter?" *Deseret News*, May 30, 1993.
12. Ibid.
13. Ibid.
14. Ibid.
15. Ibid.
16. Regis, 209.
17. Jeanne Guillemin, *Biological Weapons: From the Invention of State-Sponsored Programs to Contemporary Bioterrorism* (New York: Columbia University Press, 2005), 120.
18. Regis, 209.
19. Davidson, "Like Sheep to the Slaughter?"
20. Ibid.
21. Ibid.
22. Ibid.
23. Ibid.
24. Ibid.
25. Ibid.
26. Ibid.
27. Ibid.
28. Cole, *The Eleventh Plague*, 60.
29. Seymour M. Hersh, *Chemical and Biological Warfare* (New York: Bobbs-Merrill, 1968), 140.
30. Ibid., 141.
31. Ibid., 141-142.
32. Leonard Cole, *Clouds of Secrecy: The Army's Germ Warfare Tests over Populated Areas* (Totowa, N.J.: Rowman and Littlefield, 1988), 144.
33. Ibid.
34. Ibid.
35. Ibid.
36. Ibid., 144-145.
37. Ibid., 145.
38. Ibid., 146-147.
39. Ibid., 146.
40. Ibid.
41. Lee Davidson, "S.L. Crowd Vigorously Opposes Plans for a Germ Lab at Dugway," *Deseret News*, March 23, 1988.

42. Ibid.
43. Ibid.
44. Ibid.
45. Ibid.
46. Lee Davidson, "U. Experts Condemn Germ Lab," *Deseret News*, Aug. 23, 1988.
47. Lee Davidson, "Army Embraces Openness Policy," *Deseret News*, April 4, 1988.
48. "Army Protesters Request List of Tests Performed at Dugway," *Deseret News*, April 21, 1988.
49. Davidson, "Army Embraces Openness Policy."
50. "Army Protesters Request List."
51. Lee Davidson, "Dugway Conducted 173 Open-Air Tests Over Past 10 Years," *Deseret News*, May 3, 1988.
52. Davidson, "Army Embraces Openness Policy."
53. Davidson, "'Open' Records on Tests Are Mostly Unobtainable," *Deseret News*, Sept. 19, 1988.
54. Joseph Bauman, "Military Ignores Us, Group Charges," *Deseret News*, Feb. 10, 1989.
55. Gordon Eliot White, "EPA Takes Aim at Proposed Dugway Lab," *Deseret News*, April 12, 1988.
56. "Don't Mail Toxins, California Begs Army," *Deseret News*, May 17, 1988.
57. Lee Davidson, "State Says Army Failed to Get OKs for Open-Air Dugway Tests," *Deseret News*, May 18, 1988.
58. Davidson, "State, Dugway Plan a Settlement," *Deseret News*, June 16, 1988.
59. "Dugway Plagued by Old Hazardous-Waste Sites," *Deseret News*, Sept. 21, 1988.
60. Lee Davidson, "Public Land Likely Contaminated with Dugway Arms," *Deseret News*, Nov. 13, 1988.
61. "Army Still Needs to Answer Questions on Dugway Lab," *Deseret News*, May 13, 1988; "Build Utah Trust of Dugway with Truth, Less Secrecy," *Deseret News*, June 7, 1988.
62. "Army Still Needs to Answer Questions."
63. Charles Piller and Keith R. Yamamoto, *Gene Wars: Military Control over the New Genetic Technologies* (New York, Beech Tree Books, 1988).
64. Ibid., 271.
65. Ibid., 272.
66. Ibid., 272-274.
67. Ibid., 274.
68. Lee Davidson, "Issues Remain Despite Cancellation of Top-Level Germ Lab," *Deseret News*, Sept. 20, 1988.
69. Matthew S. Brown, "Suit Against Dugway Revives Memories of '84," *Deseret News*, April 6, 1993.
70. Lee Davidson, "Germ-Warfare Reports Contradictory," *Deseret News*, April 24, 1989.
71. Ibid.
72. Ibid.
73. Associated Press, "Dugway EIS Draws Fire," *Deseret News*, May 11, 1989.
74. Valerie Schulties, "Panel Says It's Safety Watchdog at Dugway, Not a Public-Relations Committee for Military," *Deseret News*, Oct. 16, 1989.
75. Associated Press, "Professor Assails Change in Safety Rules," *Deseret News*, Feb. 11, 1990.
76. Matthew S. Brown, "State Worries That Germ Lab Could Be Used for Exotic Tests," *Deseret News*, Aug. 1, 1992.
77. Joseph Bauman, "Germ Tests May Resume in 1 Month at Dugway," *Deseret News*, April 17, 1991.
78. Bauman, "Dugway Releases List of Pathogens It Plans to Test," *Deseret News*, April 24, 1991.
79. Bauman, "Dugway Gets Panel's Approval to Resume Pathogen Testing," *Deseret News*, May 1, 1991.
80. Bauman, "Panel Studies Planned Germ, Chemical Testing," *Deseret News*, May 3, 1991.
81. Matthew S. Brown, "Dugway Germ Testing Fails to Get Clean Bill of Health," *Deseret News*, July 31, 1992.

82. "Downwinders File Lawsuit to Halt Dugway Toxin Tests," *Deseret News*, July 3, 1991; Brown, "Suit Against Dugway Revives Memories."

83. Jim Thompson, "Downwinders Gain Support from Utah Physicians Group," *Deseret News*, Dec. 19, 1991.

84. "Army Reviews Dugway EIS," *Deseret News*, Aug. 7, 1991.

85. "Army Says Lab Would Conduct Defensive Tests," *Deseret News*, April 28, 1992.

86. Matthew S. Brown, "Dugway Outweighs Risks for Tester," *Deseret News*, June 7, 1992.

87. Alvin Hatch, "Testing by Dugway Killed Sheep," *Deseret News*, June 30, 1992.

88. Ibid.

89. Ibid.

90. "What Future for Dugway?" *Deseret News*, Feb. 10, 1993.

91. Matthew S. Brown, "Dugway A Step Closer to Getting Germ Lab," *Deseret News*, April 29, 1993.

92. Lee Davidson, "Hansen Attacks Group for Opposing Lab," *Deseret News*, June 24, 1993.

93. Ibid.

94. Matthew S. Brown, "Dugway Test-Monitor Panel Wants to Hold Secret Meetings," *Deseret News*, Sept. 12, 1993.

95. Brown, "Dugway Cited for 22 Violations," *Deseret News*, Sept. 20, 1993.

96. Lee Davidson, "Secret Tests Released Radiation at Dugway," *Deseret News*, Dec. 15, 1993.

97. Davidson, "Distrust: Documents Show 6 Dugway Tests in '50s Spread More Radiation Than 68 Other Trials Combined," *Deseret News*, Dec. 3, 1994.

98. Matthew S. Brown, "Dugway Pursues a Vaccine Facility," *Deseret News*, May 21, 1994; Brent Israelsen, "Leavitt Requests a Full Disclosure," June 1, 1994.

99. Associated Press, "Dugway Adds More Strains to Its List of Test Organisms," *Deseret News*, Feb. 18, 1994.

100. Associated Press, "Judge Dismisses Downwinders' Suit Over Dugway Testing," *Deseret News*, Jan. 5, 1995.

101. Lee Davidson, "Watchdog Group Upset that Dugway Chosen as Test Site for Germ Agents," *Deseret News*, Dec. 9, 1998; "Dugway's Increased Defense Role," *Deseret News*, Dec. 14, 1998.

102. "Dugway's Increased Defense Role."

103. "Dugway Is a Valuable Asset," *Deseret News*, March 16, 2000.

104. Lee Davidson, "Dugway Finds Old Germ-War Bomblets," *Deseret News*, March 6, 1999.

105. "Areas Suspected of Contamination," *Deseret News*, Feb. 12, 2001.

106. Lee Davidson, "Dugway Has All Its Anthrax," *Deseret News*, Oct. 17, 2001.

107. Scott Shane, "Anthrax Army Matches Army Spores," *Baltimore Sun*, Dec. 12, 2001.

108. John Fialka and Gary Fields, "Army Confirms Experiments with Anthrax at Utah Facility," *Deseret News*, Dec. 14, 2001.

109. Ibid.

110. Joe Bauman, "Dugway Security Called Sloppy," *Deseret News*, May 23, 2002.

111. Ibid.

112. Ibid.

113. Joe Bauman, "Whistle-Blower Suing Army," *Deseret News*, Jan. 14, 2002.

114. Bauman, "Judge Lambasts Dugway," *Deseret News*, Aug. 12, 2002.

115. Bauman, "Dugway Handled Anthrax Safely, Says Ex-Lab Chief," *Deseret News*, May 24, 2002.

116. Laura Hancock, "Activist Opposes Bigger Dugway," *Deseret News*, Sept. 24, 2002.

117. Patty Henetz, "Dugway to Get 4 New Germ Labs," *Deseret News*, Feb. 20, 2004.

118. Ibid.

119. Joe Bauman, "Dugway Labs May Be Permanent," *Deseret News*, March 4, 2004.

120. Bauman, "Dugway May Expand Its Size," *Deseret News*, Oct. 29, 2004.

121. Lee Davidson, "Dugway Expansion a Mystery," *Deseret News*, March 31, 2005.

122. Ibid.

123. Lee Davidson, "Dugway's Size Unclear," *Deseret News*, Aug. 1, 2005.

124. Davidson, "Tests Won't Hurt Utah, Army Says," *Deseret News*, May 31, 2006.
125. Ibid.
126. Ibid.
127. Stephen Speckman, "Military Wants to Renovate Dugway," *Deseret News*, June 2, 2007; Matthew D. LaPlante, "Dugway Aims to Revive Cold War Lab," *Salt Lake Tribune*, March 22, 2007; Steve Erickson, "Army to Triple Germ War Tests in UT," March 2007, http://www.resistinc.org/newsletter/issues/2007/03.
128. Erickson, "Army to Triple."
129. Ibid.

NBAF 15
1. Larry Margasak, Associated Press, "Defense Lab Sought by Kansas Carries Risk," *Topeka Capital-Journal*, April 12, 2008, 6A; Associated Press, "Dems Question NBAF Site: Sites Near Livestock Go Against Long-Standing Safety Practices," *Manhattan Mercury*, April 11, 2008, A1.
2. "KSU: Concerns Unfounded," *Manhattan Mercury*, April 11, 2008, A1.
3. "Dems Question NBAF Site."
4. Ibid.
5. "KSU: Concerns Unfounded."
6. Ron Trewyn and Jerry Jaax, "NBAF No Threat to Area Livestock or People," *Manhattan Mercury*, April 17, 2008, A6.
7. Adrianna DeWeese, "6 U.S. Sites Serve as Contenders for Controversial NBAF," *Kansas State Collegian*, May 7, 2008, 1.
8. Ibid.
9. Tim Carpenter, "Economic Injection," *Topeka Capital-Journal*, June 8, 2008, 10.
10. Press release, Kansas State University, "K-State's Biosecurity Research Institute Continues to Operate During Tornado," June 13, 2008; "In Aftermath of Storm, K-State Buildings Patched; Permanent Repairs Begin," *Kansas City infoZine*, June 14, 2008.
11. Pete Cohen, "Could BRI Withstand a More Powerful Tornado?" *Manhattan Mercury*, June 19, 2008, A6.
12. Jason Gertzen, "Kansas' Bid for National Bio and Agro-Defense Facility on Track," *Kansas City Star*, June 20, 2008.
13. Sarah Nightingale, "NBAF's Potential Costs, Benefits," *Manhattan Mercury*, Aug. 1, 2008.
14. Ibid.

NBAF 16
1. Lisa Sorg, "Big Dough Goes to NBAF PR," *Independent Weekly*, July 16, 2008.
2. Jack Betts, "Should Public Money Go for Advocacy?" *Charlotte Observer*, July 16, 2008.
3. John Hood, "A Lab, a Grant and A Bad Odor," *Carolina Journal*, July 17, 2008.
4. Ibid.
5. Lisa Sorg, "NBAF Supporter Says No Thanks to Funding," *Independent Weekly*, Aug. 5, 2008.
6. David Bracken, "Raleigh Recommends against Defense Lab," *Independent Weekly*, July 29, 2008.
7. "Raleigh Opposes Butner Bio-Lab," WRAL.com, Aug. 5, 2008.
8. Lisa Sorg, "Durham Commissioners Nix NBAF," *Independent Weekly*, Aug. 12, 2008.
9. "Butner Says No to Hosting NBAF," *Durham Herald-Sun*, Aug. 8, 2008.
10. Sorg, "Durham Commissioners Nix NBAF"; Sommer Brokaw, "Residents Win Battle Against Bio Lab," *Triangle Tribune*, Aug. 12, 2008; Matthew E. Milliken, "NBAF Gets 'No' Vote in Durham," *Durham Herald-Sun*, Aug. 12, 2008; Milliken, "Anti-Lab Vote Premature, Says Council Member," *Durham Herald-Sun*, Aug. 20, 2008.
11. Bill Felber, "The Choice Is NBAFfling," *Manhattan Mercury*, July 27, 2008, C1.
12. Many municipalities sell dried sewage sludge for use as fertilizer. Even without the presence of germ lab wastes, however, such sludge is often contaminated with cadmium and other toxic heavy metals.

CHAPTER TWELVE

1. Martin Enserink, "On Biowarfare's Frontline," *Science* 296 (2002): 1954.
2. Alison Walker, "Beyond the Breach: An Eye on Safety," *Frederick News-Post*, April 2006.
3. Justin Palk, "Security at Military Biolabs to Get Tighter," *Frederick News-Post*, Dec. 23, 2008.
4. Walker, "Beyond the Breach: What Went Wrong."
5. Walker, "Beyond the Breach: Risky Business."
6. Ibid.
7. Walker, "An Eye on Safety."
8. David Dishneau, "Infected Researcher Broke Safety Rule at Army Lab," *Associated Press*, May 15, 2000; Steve Vogel, "Army Studies Safety at Fort Detrick Lab," *Washington Post*, May 16, 2000, B3.
9. David Dishneau, "Anthrax Army Lab Security Called Lax," *Associated Press*, Dec. 22, 2001.
10. "Assessment of Detrick's Environmental Policies to Begin," *Associated Press*, Oct. 30, 2005.
11. Keith L. Martin, "Army Lab Security Probe Begins," *Gazette.Net*, Aug. 14, 2008.
12. "Fort Detrick Waste Cleanup Cost Grows," *Associated Press*, May 10, 2001.
13. "Chemical Waste Cleanup Plan Proposed at Fort Detrick," *Associated Press*, May 22, 2000.
14. "Fort Detrick Waste Cleanup Cost Grows."
15. "Chemical Dump Site at Fort Detrick Cleaned Up," *Associated Press*, June 8, 2004.
16. "Fort Detrick Waste Cleanup Cost Grows."
17. David Dishneau, "Infectious Germs Halt Cleanup of Fort Detrick Dump," *Associated Press*, April 11, 2002.
18. "Fort Detrick Unearths Hazardous Surprises," *Associated Press*, May 27, 2003; Lois Ember, "Fort Detrick Cleans Up," *Chemical and Engineering News*, June 2, 2003.
19. Joby Warrick, "'No One Asked Questions': Scientists Recount U.S. Biodefense Labs' Security Lapses," *Washington Post*, Feb. 19, 2002, A1.
20. Ibid.
21. "Report: 1992 Audit Found 27 Sets of Specimens Missing," *Associated Press*, Jan. 20, 2002.
22. "FBI Launches Investigation into Missing Lab Specimens," *Associated Press*, Jan. 24, 2002. The GAO has since been renamed the Government Accountability Office.
23. "Report: 1992 Audit."
24. "Fort Detrick Locates Most Missing Pathogens; None Were Dangerous," *Associated Press*, Jan. 31, 2002.
25. Yudhijit Bhattacharjee, "U.S. Army Lab Freezes Research on Dangerous Pathogens," *Science*, Feb. 7, 2009.
26. David Wood, "Biodefense Lab Starts Inventory of Deadly Samples," *Baltimore Sun*, Feb. 10, 2009.
27. "Army Biodefense Lab Shuts Down to Check If Anything Is Missing," *Discover*, Feb. 10, 2009.
28. Bhattacharjee, "U.S. Army Lab Freezes Research."
29. Wood, "Biodefense Lab Starts Inventory."
30. Ibid.
31. Ibid.
32. Ibid.
33. Ibid.
34. Jason Sigger, "Top Army Biowar Lab Suspends Research After Toxin-Tracking Scare (Updated Again)," *Wired: Danger Room*, Feb. 9, 2009, http://www.wired.com/dangerroom/2009/02/army-biolab-und/.
35. Nelson Hernandez, "Most Research Suspended at Fort Detrick," *Washington Post*, Feb. 10, 2009, B2.
36. Justin M. Palk, "Fort Detrick Disease Samples May Be Missing," *Frederick News-Post*, April 22, 2009.
37. Nelson Hernandez, "Inventory Uncovers 9200 More Pathogens," *Washington Post*, June 18, 2009.

38. Enserink, "On Biowarfare's Frontline."
39. David Dishneau, "Army Post Tarnished by 'Devastating' Anthrax Claim," *Associated Press*, Aug. 7, 2008.
40. David Dishneau, "FBI Probes Possibility Anthrax Was Smuggled Out of Fort Detrick, Maryland," *Associated Press*, June 14, 2002.
41. David Dishneau, "Army, NIH Plan New Bioterrorism Labs at Fort Detrick," *Associated Press*, April 10, 2002.
42. Press Release, National Institute of Allergy and Infectious Diseases, "USAMRIID-NIAID Announce Research Partnership at Fort Detrick," Oct. 15, 2002; Scott Shane, "Research Institutes Unveil Plans for Fort Detrick Lab Expansion," *Baltimore Sun*, Oct. 16, 2002.
43. David Dishneau, "Feds See Little Risk from Bio Threat Lab, Inside or Out," *Associated Press*, Dec. 16, 2004.
44. "Proposed NIH Biodefense Lab at Fort Detrick Deemed Safe," *Associated Press*, Feb. 4, 2004.
45. "CDC to Join Frederick Interagency Biodefense Center," *Associated Press*, May 4, 2005.
46. "Proposed NIH Biodefense Lab at Fort Detrick Deemed Safe."
47. Ibid.
48. Department of Homeland Security, "About National Biodefense Analysis and Counter-measures Center," page last modified Dec. 18, 2006.
49. Ibid.
50. Ibid.
51. Press release, "DHS Starts Construction for the National Biodefense Analysis and Coun-termeaasures Center," June 26, 2006.
52. Ibid.
53. "Weapons Labs' Biological Research Raises Concerns," *Arms Control Today*, March 2008.
54. Joby Warrick, "The Secretive Fight Against Bioterror," *Washington Post*, July 30, 2006, A1.
55. "U.S. Biodefense Lab Raises Concerns," *United Press International*, July 31, 2006.
56. Warrick, "The Secretive Fight Against Bioterror."
57. Patrick Fitch, the NBACC's director, in a bit of slipping and sliding reminiscent of the DHS "change in plans" for NBAF, said four years later that the facility "will not create threats in order to study them," according to the Federation of American Scientists' Strategic Security Blog for May 13, 2008. Richard Ebright responded as follows: "The truthfulness of the statement depends on the definition of 'create threats' (and probably also on the definitions of 'in,' 'order,' 'to,' 'study,' and 'them'—all of which presumably will be defined in an equally tendentious, equally dishonest, fashion."
58. Ibid.
59. Ibid.
60. Associated Press, "U.S. Dedicates Biodefense Center in Frederick," *Baltimore Sun*, Oct. 23, 2008.
61. Frederick Kunkle, "Fort Detrick Neighbors Jittery Over Expansion," *Washington Post*, Feb. 27, 2006.
62. Ibid.
63. David Dishneau, "Army Says Biodefense Lab Plan Addresses Terrorist Threat," *Associated Press*, Dec. 27, 2006.
64. Ibid.
65. Alison Walker-Baird, "Detrick Opponents Call for End to Growth," *Frederick News-Post*, Aug. 26, 2007.
66. Katherine Heerbrandt, "Gray Speaks Out," *Frederick News-Post*, Sept. 5, 2007.
67. Heerbrandt, "Breach of Trust," *Frederick News-Post*, Nov. 7, 2007.
68. Timothy B. Wheeler, "Biodefense Lab Causing Qualms," *Baltimore Sun*, Nov. 19, 2007.
69. Justin M. Palk, "Public Discusses Fort Detrick Lab Expansion," *Frederick News-Post*, Nov. 20, 2007.
70. Palk, "Commissioners Split on Detrick Review," *Frederick News-Post*, Nov. 28, 2007.

71. "Detrick FEIS Review," Nov. 12, 2007.
72. Associated Press, "Frederick May Seek Review of Army Biodefense Lab Plan," *Washington Post*, April 8, 2008.
73. "Maryland: Frederick to Seek Biodefense Lab Plan Review," *Associated Press*, April 10, 2008.
74. David Dishneau, "Mikulski to Seek Review of Fort Detrick Biodefense Lab Plans," *Associated Press*, April 10, 2008.
75. Justin M. Palk, "Army Rejects Petition for Study of Biolabs," *Frederick News-Post*, July 1, 2008.
76. "Commissioners Offer Revised Detrick Study," *Frederick News-Post*, Aug. 29, 2008.
77. Justin M. Palk, "Mikulski Puts USAMRIID Study in Funding Bill," *Frederick News-Post*, Sept. 25, 2008.
78. Palk, "NIH Seeks Comment on Transport Program," *Frederick News-Post*, July 8, 2008; Palk, "Skeptics Call for Detailed Study of NIH Proposal," *Frederick News-Post*, July 10, 2008; "NIH Ponders Logistics of Biodefense Lab Accidents," *Associated Press*, July 8, 2008.
79. Palk, "NIH Seeks Comment."
80. National Academies, "Project Information: Evaluation of the Health and Safety Risk Analyses for the Planned Expansion of USAMRIID's Biosafety Level 3 and 4 Laboratories at Fort Detrick, Maryland."
81. Megan Eckstein, "Army Breaks Ground on New Infectious Disease Labs," *Frederick News-Post*, Aug. 28, 2009.
82. National Research Council, *Evaluation of the Health and Safety Risks of the New USAMRIID High Containment Facilities at Fort Detrick, Maryland* (Washington, D.C.: National Academies Press, 2010), 49-51, http://www.nap.edu/catalog/12871.html.
83. Leonard Cole, *The Anthrax Letters* (Washington, D.C.: Joseph Henry Press, 2003), 138-159.
84. National Research Council, 40.
85. Ibid., 33.
86. Ibid., 35.
87. Ibid., 40.
88. Ibid., 6.
89. Ibid.
90. Ibid.
91. Ibid., 18.
92. Ibid., 23.
93. Ibid.
94. Ibid.
95. Ibid., 18-22.
96. Ibid., 24.
97. Ibid., 5-6.
98. Ibid., 8.
99. Ibid., 45.
100. Megan Eckstein, "Army: Broken Procedures Led to Lab Infection," *Frederick News-Post*, April 7, 2010; "Tularemia in a Laboratory Worker at USAMRIID," USAMRIID press release, April 2010.
101. "USAMRIID Report," *Frederick News-Post*, April 9, 2010.
102. "Tularemia in a Laboratory Worker," USAMRIID press release.
103. Richard Preston, *The Hot Zone* (New York: Random House, 1994), 143-146.
104. Megan Eckstein, "Flaws Delay Opening of Detrick Lab," *Frederick News-Post*, April 16, 2010.

NBAF 17

1. Larry Margasak, Associated Press, "AP Exclusive: US Disregarded Experts Over Biolab," *USA Today*, Aug. 11, 2008.
2. Ibid.

3. DHS, Science and Technology Directorate, *Final Selection Memorandum for Site Selection for the Second Round Potential Sites for the National Bio and Agro-defense Facility (NBAF)*, July 2007, 24.
4. "Pork Always Earns a High Score," *Atlanta Journal-Constitution*, Aug. 12, 2008.
5. *Final Selection Memorandum*, 34.
6. Ibid., 2.
7. Ibid., 25.
8. Ibid., 22-23.
9. Ibid., 20.
10. Lisa Sorg, "First the Dog, Then the Pony," *Independent Weekly*, Aug. 25, 2008.
11. *Final Selection Memorandum*, 19.
12. Bill Felber, "Anti-NBAF Group States Its Case," *Manhattan Mercury*, Aug. 20, 2008, A1.
13. "NBAF's Opponents," *Manhattan Mercury*, Aug. 20, 2008, A7.
14. Tom Manney, "Discrediting NBAF Foes Demeans Mercury," *Manhattan Mercury*, Aug. 25, 2008, A6.
15. Sue Cohen, "Mercury's Account of Meeting of NBAF Opponents Was Slanted," *Manhattan Mercury*, Aug. 22, 2008, A6.
16. "NBAF's Opponents."
17. Bill Felber, "Pro-NBAFers Dispute Claims by Opposition," *Manhattan Mercury*, Aug. 22-23, 2008, A1.
18. Ned Seaton, "What Would Happen If a Disease Escaped at the Proposed NBAF?" *Manhattan Mercury*, Sept. 24, 2008, A4.
19. Ibid.
20. "Taking Over for Mom and Pop," *Manhattan Mercury*, Sept. 24, 2008.
21. Ibid.
22. Ibid.
23. "Statement on the U.S. Biodefense Program from Communities Living in Its Shadow," Sept. 21, 2008.
24. Ned Seaton, "What Happened to the Anti-NBAF Sign?" *Manhattan Mercury*, Oct. 23, 2008, A4.
25. Ibid.
26. Ibid.

CHAPTER THIRTEEN
1. Stephen Smith, "Menino Backs Biosafety Lab Plan," *Boston Globe*, Jan. 18, 2003, B1.
2. Stephen Smith, "Foes Vow to Fight Bioterror Lab," *Boston Globe*, Oct. 1, 2003.
3. Ibid.
4. "Biolab Opponents Exploit Phony Fears," *Boston Herald*, June 9, 2005.
5. Ibid.
6. "Welcome to Our 'Hood," *Boston Herald*, Feb. 4, 2006.
7. Stephen Smith, "Biosafety Lab in South End Gets Final OK," *Boston Globe*, Feb. 3, 2006, A1.
8. "A Valued New Neighbor," *Boston Herald*, April 22, 2004.
9. Safety Net, "Stop the Bioterror Lab," http://www.stopthebiolab.org.
10. "Tell the Truth," *Bay State Banner*, March 9, 2006.
11. Stephen Smith, "Menino Backs Biosafety Lab Plan."
12. Ibid.
13. Stephen Smith, "Foes Vow to Fight Bioterror Lab."
14. National Institute of Allergy and Infectious Diseases, *Request for Proposals and Applications, Regional Biocontainment Laboratories (RBLS) and National Biocontainment Laboratories (NBLS)*, October 15, 2002, 4.
15. Ibid.
16. Ibid., 4-5.
17. Stephen Smith, "BU Center May Seek Biosafety Lab Funds," *Boston Globe*, Jan. 16, 2003.

18. Stephen Smith, "Politicians, Groups Seek Rejection of BU Lab Plan," *Boston Globe*, June 27, 2003, B3; Michael Lasalandra, "Bioterror Lab Bid Draws South End Wrath," *Boston Herald*, June 27, 2003, News Section.

19. Jennifer Heldt Powell, "BU Bioterror Lab Bugs Neighbors," *Boston Herald*, July 31, 2003, Business Section.

20. Katherine Lutz, "Opposition Grows to BU's Biodefense Lab Plan," *Boston Globe*, Aug. 19, 2003.

21. Powell, "BU Bioterror Lab Bugs Neighbors."

22. "U.S. Bioterror Lab Belongs in Boston," *Boston Herald*, Aug. 4, 2003.

23. Jennifer Heldt Powell, "Some Oppose South End Bioterrorism Lab," *Boston Herald*, Aug. 5, 2003, Business Section.

24. Janice T. Bourque and Paul Guzzi, "Op-Ed: As You Were Saying . . . Federal Bioterror Lab Is Just What Doctor Ordered for Hub," *Boston Herald*, Aug. 10, 2003.

25. Jennifer Heldt Powell, "South End to Host Bioterrorism Labs," *Boston Herald*, Oct. 1, 2003; Stephen Smith, "BU Is Seen as US Choice to Build, Run Bioterror Lab," *Boston Globe*, Sept. 30, 2003, A1; Stephen Smith, "Foes Vow to Fight Bioterror Lab"; BUMC Press Release, "NIAID Announces that BUMC Is Recipient of $120 Million," Sept. 30, 2003; NIAID Press Release, "NIAID Funds Construction of Biosafety Laboratories," Sept. 30, 2003.

26. Powell, "South End to Host."

27. BUMC Press Release, "NIAID Announces."

28. Ibid.

29. Jennifer Heldt Powell, "BU's Students Join Critics of Bioterrorism Lab Plans," *Boston Herald*, Nov. 27, 2003, Business Section.

30. Christine MacDonald, "Academic Stars Join a Push against Plan for Bioterrorism Lab," *Boston Globe*, April 14, 2004, B3; Christine MacDonald, "Next Up in Biolab Debate: Dueling Scientists at Hall," *Boston Globe*, April 18, 2004, 5.

31. David Ozonoff, "The Top Ten Reasons to Oppose the Biolab," presentation to faculty, staff, and students of Boston University, Sept. 23, 2004. Available at http://www.ace-ej.org.

32. BUMC Press Release, "MA Office of Environmental Affairs Approves Biosafety Lab," Nov. 17, 2004.

33. BUMC Press Release, "Boston Redevelopment Authority Board Approves Biosafety Laboratory," Dec. 15, 2004.

34. BUMC Press Release, "Zoning Commission Board Approves BioSquare II and Biosafety Lab," Jan. 12, 2005.

35. Jeff Hecht and Debora MacKenzie, "Safety Fears Raised Over Biosafety Lapse," *New Scientist*, Jan. 20, 2005; Stephen Smith, "Bacterium Infected 3 at BU Biolab," *Boston Globe*, Jan. 19, 2005.

36. Hecht and MacKenzie, "Safety Fears Raised."

37. BUMC Press Release, "Key US Figure Boosts BU Biolab Plan," Feb. 24, 200[5].

38. Stephen Smith, "Final Public Hearing Set on University Biolab," *Boston Globe*, April 25, 2005.

39. Stephen Smith, "Strong Views Mark Federal Hearing on BU Disease Research Lab," *Boston Globe*, April 26, 2005.

40. Stephen Smith, "Biosafety Lab in South End Gets Final OK," *Boston Globe*, Feb. 3, 2006, A1.

41. Stephen Smith, "Judge Orders New Review of BU Biolab," *Boston Globe*, Aug. 4, 2006.

42. John Dudley Miller, "Boston to Regulate Research," *The Scientist*, Sept. 11, 2006, http://www.the-scientist.com/news/display/24679/.

43. Raja Mishra and John R. Ellement, "Biomedical Lab Evacuated," *Boston Globe*, March 21, 2007.

44. Nicholas K. Tabor, "City Objects to BU Biolab Building," *Harvard Crimson*, Jan. 10, 2007.

45. Town of Brookline, Feb. 27, 2007. Available at "Stop the Bioterror Lab," http://www.stopthebiolab.org.

46. Mishra and Ellement, "Biomedical Lab Evacuated."

47. Stephen Smith, "Smoldering Waste Caused BU Fire," *Boston Globe*, April 26, 2007.
48. Stephen Smith, "Top State Court to Rule on BU Biolab," *Boston Globe*, March 22, 2007.
49. Posted by Chris Lovett, "Unconnected Dots to the Biolab," *Civic Boston*, Aug. 2, 2007, at http://civicboston.blogspot.com/2007/08/unconnected-dots-to-biolab.html.
50. Stephen Smith and Felicia Mello, "Study Says Biolab Not a Threat to S. End," *Boston Globe*, Aug. 24, 2007; Linda Rodriguez, "BioLab Deemed 'Safe' by NIH," *South End News*, Aug. 24, 2007.
51. Comment posted Aug. 24, 2007, at biodefense@lists.sunshine-project.org.
52. Rodriguez, "BioLab Deemed 'Safe'."
53. *Anthrax: The Investigation of a Deadly Outbreak* (Berkeley: Univ. of Calif. Press, 1999).
54. *Biological Weapons: From the Invention of State-Sponsored Programs to Contemporary Bioterrorism* (New York: Columbia Univ. Press, 2005).
55. Smith and Mello, "Study Says Biolab Not a Threat."
56. Comment posted Aug. 24, 2007.
57. Justin Rice, "Hundreds Turn Out for BioLab Hearing," *South End News*, Sept. 21, 2007.
58. Stephen Smith, "Biolab Faces New Scrutiny from State," *Boston Globe*, Oct. 14, 2007.
59. Lou Manzo, "Scientists Spar at BioLab Hearing," *South End News*, Oct. 25, 2007.
60. Jay Fitzgerald, "Biolab's Security Drive," *Boston Herald*, Oct. 4, 2007.
61. Lou Manzo, "BU Medical Center Shows Off Biolab," *South End News*, Oct. 25, 2007.
62. Lou Manzo, "BU Officials Discuss Shipment Protocol to BioLab," *South End News*, Nov. 7, 2007.
63. Stephen Smith, "US Review of BU Biolab Inadequate, Panel Finds," *Boston Globe*, Nov. 30, 2007; Rick Weiss, "Experts 'Fail' Risk Analysis for Boston Bioterror Lab," *Washington Post*, Nov. 30, 2007, A10.
64. Weiss, "Experts 'Fail.'"
65. Ibid.; Smith, "US Review Inadequate."
66. Jeffrey Brainard, "Report Faults NIH Safety Analysis of Biosafety Laboratory at Boston U.," *Chronicle of Higher Education*, Nov. 30, 2007.
67. Smith, "US Review Inadequate."
68. Jay Fitzgerald, "Menino Says Biolab 'Will Go Forward,'" *Boston Herald*, Dec. 12, 2007.
69. Comment posted on Dec. 12, 2007 at biodefense@lists.sunshine-project.org.
70. Stephen Smith, "Ruling May Stall Opening of Biolab," *Boston Globe*, Dec. 14, 2007.
71. Stephen Smith, "Opening of BU Biolab to be Delayed," *Boston Globe*, Feb. 1, 2008.
72. NIH Press Release, "NIH Outlines Next Steps to Address Safety Concerns about Boston-Area Laboratory," March 6, 2008.
73. Stephen Smith, "For Bioterror Lab, A Long Road Seen," *Boston Globe*, March 14, 2008.
74. Ibid.
75. Art Jahnke, "NIH Convenes Expert Panel to Address Biolab Concerns," *BU Today*, March 7, 2008, at http://www.bu.edu/today/.
76. Stephen Smith, "Scientists Call for Biolab Safety Study," *Boston Globe*, May 3, 2008.
77. Stephen Smith, "Panel Urges Review of Possible Lab Threats," *Boston Globe*, May 17, 2008.
78. Stephen Smith, "BU Outlines Biolab Safety Steps," *Boston Globe*, Oct. 14, 2008.
79. "Biolab Is as Safe as Scientists in It," *Boston Herald*, Aug. 13, 2008.
80. Justin A. Rice, "At Biolab Forum, Divides Remain Deep," *Boston Globe*, Oct. 19, 2008.
81. Ibid.
82. Camille Roane, "Residents Voice Biolab Concerns," *Daily Free Press*, Oct. 14, 2008.
83. Rice, "Divides Remain Deep."
84. Roane, "Residents Voice Biolab Concerns."

NBAF 18

1. Greg Harman, "'News' Treads Lightly on Germ-Lab," *Harman on Earth*, Aug. 8, 2008, http://harmanonearth.wordpress.com.
2. Ken Foskett, "Cost of Planned Bio-Lab Increases $200 Million," *Atlanta Journal-Constitution*, Aug. 15, 2008.

3. Bill Felber, "Texas Ups Offer for NBAF," *Manhattan Mercury*, Oct. 10-11, 2008, A1.
4. *Manhattan Mercury*, Dec. 3, 2008, A1.
5. *Manhattan Mercury*, Dec. 3, 2008.
6. *Topeka Capital-Journal*, Dec. 4, 2008, A1.
7. *Topeka Capital-Journal*, Dec. 5, 2008, A1.
8. Jon Wefald, KSU president, quoted in the *Topeka Capital-Journal*, Dec. 5, 2008, A2.
9. *Lawrence Journal-World*, Dec. 6, 2008.
10. "Governor Rell Applauds Decision to Locate New Bio Research Facility Away from Plum Island," *StamfordPlus.com*, Dec. 3, 2008, http://www.stamfordplus.com.
11. Blake Aued, "Perdue Criticizes Anti-NBAF Activists," *Athens Banner-Herald*, Dec. 5, 2008.
12. Lee Shearer, "Adams: Foes of NBAF Likely Hurt UGA's Bid," *Athens Banner-Herald*, Dec. 5, 2008.
13. E-Mail, Jan. 14, 2009.
14. Dec. 10, 2008.
15. "Economy Won't Stop Biodefense Lab, Senators Say," *Global Security Newswire*, Dec. 5, 2008.
16. Dec. 18, 2008.
17. Dec. 19, 2008.
18. Dec. 19, 2008.

CHAPTER FOURTEEN

1. Anna Badkhen, "Fear Follows Plan to Build More Deadly-Disease Labs," *San Francisco Chronicle*, Aug. 22, 2004.
2. Carolyn Poirot, "Geeks as Gods," *Fort Worth Star-Telegram*, Jan. 13, 2002, LIFE, 1. *See also* Martin Enserink, "On Biowarfare's Frontline," *Science* 296 (2002), 1954.
3. Ragnar Lofstedt, "Good and Bad Examples of Siting and Building Biosafety Level 4 Laboratories: A Study of Winnipeg, Galveston and Etobicoke," *Journal of Hazardous Materials*, July 1, 2002, 47-66.
4. Eric Berger, "Accidents Challenge Openness of Biosafety Labs," *Houston Chronicle*, May 9, 2005, A1.
5. Pete Alfano, "As Biodefense Research Booms, Reward Is Weighed Against Risk," *Fort Worth Star-Telegram*, Aug. 23, 2007.
6. Berger, "Accidents Challenge Openness."
7. Ibid.
8. "The Sunshine Project's Top 10 Freedom of Information Failures: University of Texas Medical Branch, Galveston," http://www.sunshine-project.org.
9. Alison Motluck, "Protect and Prosper," *New Scientist*, Oct. 23, 2004, 56.
10. Leigh Jones, "Focus on Galveston for Biotech Growth," *Galveston Daily News*, Dec. 16, 2007.
11. Comment, biodefense@lists.sunshine-project.org, Dec. 18, 2007.
12. "UT Medical Branch Resisting Open-Records Ruling," *Austin American-Statesman*, Sept. 29, 2007.
13. Associated Press, "UTMB Disease Lab Shut Down after Door Fails," *Dallas Morning News*, Jan. 25, 2008.
14. "Why Would Any Sane Person Put a Level 4 Biodefense Lab in Galveston?" *Effect Measure*, Sept. 13, 2008.
15. Ibid.
16. Emily Ramshaw, "Ike Renews Fears for Biodefense Lab on Galveston," *Dallas Morning News*, Sept. 16, 2008.
17. Suzanne Gamboa, Associated Press, "Germ Lab Took Early Precautions as Ike Neared," *Houston Chronicle*, Sept. 16, 2008.
18. Ramshaw, "Ike Renews Fears."
19. Cited in Sue Sturgis, "Ike Coverage: Galveston Biolab Researching Killer Viruses Reportedly Secured Before Hurricane Hit," *Facing South: The Online Magazine of the Institute for*

Southern Studies, Sept. 15, 2008, http://southernstudies.org/2008/09/ike-coverage-galveston-biolab.html.

20. "Galveston High Security Laboratory: Dumb and Dumber," *Effect Measure*, Oct. 31, 2008.
21. Quoted in "UTMB Claims BSL-4 Pathogens Destroyed Before Ike Hit?" *Butner Blogspot*, Sept. 17, 2008, http://butnerblogspot.wordpress.com.
22. Forrest Wilder, "Bugs in the System," *Texas Observer*, Oct. 17, 2008.
23. James C. McKinley, Jr., "Bio Lab in Galveston Raises Concerns," *New York Times*, Oct. 28, 2008.
24. Ibid.
25. Ibid.
26. Laylan Copelin, "Galveston: After the Storm, An Election," *Austin American-Statesman*, Oct. 17, 2008.
27. *Galveston Daily News*, Nov. 11, 2008.
28. Emily Ramshaw, "Galveston Biodefense Lab Was Fortress During Ike," *Dallas Morning News*, Nov. 16, 2008.
29. Ibid.
30. "Texas Med Center to Lay Off 3,800," posted by Jennifer Evans, *The Scientist.com*, Nov. 13, 2008.
31. "Texas Profs Sue University," posted by Elie Dolgin, *The Scientist.com*, Dec. 2, 2008.
32. "Kansas Wins Race to Host Biodefence Research Centre," *Nature* 456 (2008), 687.
33. Emily Ramshaw, "Texas Biolab Bill Would Keep Too Many Secrets, Opponents Say," *Dallas Morning News*, April 29, 2009; Associated Press, "Bill Would Bar Information on Deadly Agents," *Houston Chronicle*, May 6, 2008.
34. Lisa Falkenberg, "Biolab Bill Leaves Much in the Dark," *Houston Chronicle*, May 13, 2009.
35. Heber Taylor, "Help Put an End to a Bad Bill," *Galveston Daily News*, May 10, 2009.
36. Michael A. Smith, "Don't Trade Rights for Mere Promises," *Galveston Daily News*, May 10, 2009.
37. Ibid.
38. Ramshaw, "Texas Biolab Bill Would Keep Too Many Secrets."
39. Michael A. Smith, "'Security' Bill Backers Forfeit Trust," *Galveston Daily News*, May 15, 2009.
40. Smith, "Don't Trade Rights for Mere Promises."
41. Ramshaw, "Texas Biolab Bill Would Keep Too Many Secrets."
42. Dolph Tillotson, "Killing Germ-Lab Bill Vital for All," *Galveston Daily News*, May 14, 2009.
43. Ibid.
44. Ibid.
45. Michael A. Smith, "UTMB Bill Compromise Needs More Give," *Galveston Daily News*, May 17, 2009.
46. Falkenberg, "Biolab Bill Leaves Much in the Dark."
47. Smith, "UTMB Bill Compromise Needs More Give."
48. "Biolab Bill Awaiting Perry's Signature," *Galveston Daily News*, June 2, 2009.
49. Ibid.
50. Ibid.
51. Laura Elder, "UTMB Offers Another Draft of Biolab Bill," *Galveston Daily News*, May 16, 2009.

NBAF 19

1. David Hendricks, "Feds Changed the Bidding Game," *San Antonio Express-News*, Dec. 5, 2008.
2. Cindy Tumiel, "S.A. Shakes Fist Over Decision on Federal Lab," *San Antonio Express-News*, Jan. 9, 2009.
3. John Milburn, "Texas Ponders Suing Over Biothreat Lab Decision," *Associated Press*, Jan. 14, 2009.

4. Ibid.
5. Press release, U.S. Senator Sam Brownback, "Brownback Secures Funds for Kansas: Votes Against Overall Omnibus Bill," March 11, 2009.
6. Bill Felber, "DHS' Napolitano Confirms NBAF During Visit," *Manhattan Mercury*, Feb. 11, 2009, A1.
7. Tim Carpenter, *Topeka Capital-Journal*, Feb. 11, 2009, A1.
8. "Farmers Skeptical of Bio Lab Safety," *Manhattan Free Press*, Feb. 12, 2009.
9. Tim Carpenter, "Academics, Athletics Factors in Accepting KSU Presidency," *Topeka Capital-Journal*, Feb. 12, 2009.
10. "Life Sciences Leaders Shoot for New Asset as K-State Makes Final Four for Federal Lab," *Kansas City Business Journal*, March 13, 2009. KSU "won" that competition in October 2009. The lab's mosquitoes were reportedly quite excited about their new NBAF neighbor.
11. "Tiahrt to Obama: You're Shortchanging Kansas Food-Safety Lab," Prime Buzz, March 26, 2009, KansasCity.com.
12. Ibid.
13. *Texas Bio- and Agro-Defense Consortium v. U.S.*, U.S. Ct. of Fed. Claims, filed April 23, 2009.
14. Ibid., 3.
15. "There's No Place Like . . . Wait, Texas for the NBAF? Read the Complaint," *Butner Blogspot*, April 24, 2009, http://butnerblogspot.wordpress.com.
16. *TBAC v. US*, 13.
17. "Texas Tantrum," *Lawrence Journal-World*, April 27, 2009.
18. "Senators: NBAF Lawsuit 'Frivolous,'" *Topeka Capital-Journal*, April 22, 2009.
19. "Sebelius Defends NBAF Site Selection as Texas Group Threatens Legal Action," *Kansas City Business Journal*, April 22, 2009.
20. Associated Press, "NBAF Suit Axed," *Manhattan Mercury*, July 17-18, 2009, A1.
21. Carol D. Leonnig, "Infectious Diseases Study Site Questioned," *Washington Post*, July 27, 2009; Government Accountability Office, *Biological Research: Observations on DHS's Analyses Concerning Whether FMD Research Can Be Done as Safely on the Mainland as on Plum Island*, July 2009, GAO-09-747.
22. Sally Schuff, *Feedstuffs*, July 28, 2009; "KS Lawmakers Fight Back Against Article on NBAF," KSN.Com, July 27, 2009.

CHAPTER FIFTEEN
1. Brad Spielberg, *Rising Plague: The Global Threat from Deadly Bacteria and Our Dwindling Arsenal to Fight Them* (Amherst, N.Y.: Prometheus, 2009), 35-36.
2. R. M. Klevens et al., "Invasive Methicillin-Resistant *Staphylococcus Aureus* Infections in the United States," *JAMA* 298 (2007): 1763-71.
3. H. F. Chambers, "Community-Associated MSRA—Resistance and Virulence Converge," *New England Journal of Medicine* 352 (2005): 1485-87. (Cited in Spielberg, 56-57.) Stephanie Woodard, "Concerns Over Superbugs in Our Food Supply," *Prevention*, July 15, 2009. Accessed at MSNBC.com.
4. "Some Statistics About the US Biodefense Program and Public Health," http://www.sunshine-project.org.
5. Supporting Online Material for S. Altman et al., "An Open Letter to Elias Zerhouni," *Science* 307 (2005), 1409: "Appendix 1: Public Health Relevance of Prioritized Bioweapons Agents," http://www.sciencemag.org/feature/misc/microbio/Altman_SOM.pdf.
6. John W. Wright et al., ed, *The New York Times 2009 Almanac* (New York: Penguin, 2008), 452.
7. Eric Schlosser, "Bad Meat: The Scandal of Our Food," *The Nation*, Sept. 16, 2002, 6.
8. Allan Turner, "Tomato Salmonella Scare Stirs FDA's Critics," *Houston Chronicle*, June 7, 2008.
9. Trust for America's Health, *Ready or Not? Protecting the Public's Health from Diseases, Disasters, and Bioterrorism* (2009), 28.

10. *Rising Plague*, 71.
11. "An Open Letter to Elias Zerhouni," March 2005.
12. Author's telephone interview of Ebright, February 22, 2010.
13. Ibid.
14. "An Open Letter," "Appendix 2: Increase in Number of Grants for Research on Prioritized Bioweapons Agents."
15. Ibid., "Appendix 3: Decrease in Number of Grants for Research on Non-Biodefense-Related Microbial Physiology, Genetics, and Pathogenesis."
16. "Some Statistics About the US Biodefense Program and Public Health."
17. Jeanne Guillemin, *Biological Weapons: From the Invention of State-Sponsored Programs to Contemporary Bioterrorism* (New York: Columbia Univ. Press, 2005), 199.
18. Ibid., 197-198.
19. Ibid., 198.
20. William R. Clark, *Bracing for Armageddon? The Science and Politics of Bioterrorism in America* (New York: Oxford Univ. Press, 2008), 65.
21. Katherine Eban, "Waiting for Bioterror: Is Our Public Health System Ready?" *The Nation*, Dec. 9, 2002, 12.
22. Ronald J. Glasser, "We Are Not Immune: Influenza, SARS, and the Collapse of Public Health," *Harper's*, July 2004, 35-42.
23. *Ready or Not*, 3.
24. Ibid., 57.
25. Ibid., 49.
26. Barbara Anderson, "Valley Health Departments Forced to Cut Back, *Fresno Bee*, March 17, 2009; Dr. Douglas K. Owens, "Opinion: Who You Gonna Call? The Role of Public Health Departments," *The Mercury News*, May 4, 2009.
27. William Murphy, "Local Concerns Over Plan to Cut Bioterrorism Funding," *Newsday*, March 27, 2009; Carol Hill, "A Necessary County Program Faces Elimination," *Ithaca Journal*, May 15, 2009.
28. Barbara J. Bly, RN, "Public Health Is Going on Life Support," *HeraldNet* (Snohomish County, WA), April 11, 2009.
29. Randy Lee Loftis and Scott Farwell, "Federal Funds That Put Dallas County's Flu Response in Place May Not Keep Coming In," *Dallas Morning News*, May 3, 2009.
30. Alan Judd, "Health Cuts Raise Worry," *Atlanta Journal Constitution*, May 10, 2009.
31. *Ready or Not*, 51.
32. Ibid., 60.
33. Ibid., 67-70.
34. Author's telephone interview of Ebright, Feb. 22, 2010.
35. Ibid.
36. Ibid.
37. Ibid.
38. Ibid.
39. Ibid.
40. Ibid.
41. Ibid.
42. Ibid.
43. Ibid.
44. Ibid.
45. *World at Risk: The Report of the Commission on the Prevention of WMD Proliferation and Terrorism* (New York: Vintage Books, 2008), vi.
46. Commission on the Prevention of Weapons of Mass Destruction Proliferation and Terrorism, *The Clock Is Ticking: A Progress Report on America's Preparedness to Prevent Weapons of Mass Destruction Proliferation and Terrorism*, October 21, 2009, 11-13.
47. *The Clock Is Ticking*, 11-12.
48. Commission on the Prevention of Weapons of Mass Destruction Proliferation and

Terrorism, *Prevention of the WMD Proliferation and Terrorism Report Card: An Assessment of the U.S. Government's Progress in Protecting the United States from Weapons of Mass Destruction Proliferation and Terrorism*, January 2010, 6.

49. *World at Risk*, 29-30.
50. Ibid., 24.
51. *The Clock Is Ticking*, 6.
52. Center for Arms Control and Non-Proliferation, *Biological Threats: A Matter of Balance*, Jan. 26, 2010, http://www.armscontrolcenter.org/policy/biochem/articles/biological_threats_a_matter_of_balance.
53. Milton Leitenberg, *Assessing the Biological Weapons and Bioterrorism Threat* (Carlisle, PA: Strategic Studies Institute, US Army War College, 2005), 1-7.
54. Ibid., 35.
55. Ibid., 41.
56. Ibid., 51-59.
57. Ibid., 59.
58. Ibid., 45.
59. Ibid.

A FINAL POSTSCRIPT

1. Deborah Ziff and Ron Seely, "UW-Madison Professor Barred from Lab for Potentially Dangerous Experiments," *Wisconsin State Journal*, May 11, 2010.
2. Ibid.
3. "UNIVWISCONSINMAD.pdf," records supplied on April 28, 2006, in response to a March 12, 2005 request from the Sunshine Project, located at *The IBC Archive, Sunshine Project/FOI Fund*, www.sunshine-project.org.
4. M. Beatrice Dias et al., "Effects of the USA PATRIOT ACT and the 2002 Bioterrorism Preparedness Act on Select Agent Research in the United States," *PNAS Early Edition*, May 10, 2010, www.pnas.org/cgi/doi/10.1073/pnas.0915002107.
5. Ibid., citing Franz, David. R. et al., "The 'Nuclearization' of Biology Is a Threat to Health and Security," *Biosecurity and Bioterrorism* 7 (2009): 243-244.
6. Dias et al.
7. Ibid.
8. Heidi Ledford, "Regulations Increase Cost of Dangerous-Pathogen Research," *Nature*, May 10, 2010, http://www.nature.com/news.
9. Bob Grant, "Biosecurity Laws Hobble Research," *The Scientist*, May 10, 2010, http://www.the-scientist.com/blog/display/57399/.

EPILOGUE

1. Bryan Walsh, "After H1N1, Researchers Warn of a Potential New Superbug," *Time*, Feb. 22, 2010; "Contagious Hybrid Bird Flu-Human Flu Created in Lab," *Environment News Service*, Feb. 26, 2010; Chengjun Li et al., "Reassortment Between Avian H5N1 and Human H3N2 Influenza Viruses Creates Hybrid Viruses with Substantial Virulence," *PNAS* 107 (2010), 4687-4692.
2. "All experiments with live viruses and with transfectants generated by reverse genetics were performed in an enhanced biosafety level 3 containment laboratory approved for such use by the Centers for Disease Control and Prevention and the US Department of Agriculture." Li et al, 4691.
3. Scott Shane, "F.B.I., Laying Out Evidence, Closes Anthrax Case," *New York Times*, Feb. 19, 2010.
4. "The F.B.I.'s Anthrax Case," *New York Times*, Feb. 28, 2010.
5. Edward Jay Epstein, "The Anthrax Attacks Remain Unsolved," *The Wall Street Journal*, Jan. 24, 2010.
6. "Bill for More Investigation of '01 Anthrax Case Passes House," *Baltimore Sun*, Feb. 26,

2010; Erin Duffy, "Holt: Last Word Not in on Anthrax Case," *New Jersey Times*, Feb. 26, 2010.

7. Gary Matsumoto, "Colleague Says Anthrax Numbers Add Up to Unsolved Case," *ProPublica*, April 23, 2010.

8. Scott Shane, "Colleague Disputes Case Against Anthrax Suspect," *New York Times*, April 22, 2010.

9. Matsumoto, "Colleague Says Anthrax Numbers Add Up to Unsolved Case."

10. Andrew Pollack and Duff Wilson, "A Pfizer Whistle-Blower Is Awarded $1.4 Million," *New York Times*, April 2, 2010.

11. Becky McClain, videotaped address at California Coalition for Workers Memorial Day, available at Labor Video Project, http://laborvideo_blip.tv.

12. Pollack and Wilson, "A Pfizer Whistle-Blower."

13. McClain, videotaped address.

14. Ibid.

15. Ibid.

16. Ibid.

17. Ibid.

18. Pollack and Wilson, "A Pfizer Whistle-Blower."

19. McClain, videotaped address.

20. Ibid.

21. Ibid.

22. McClain, "Dangers in Embryonic Stem Cell Research."

23. "About CRG," *Council for Responsible Genetics* Web site, http://www.councilforresponsiblegenetics.org.

24. Pollack and Wilson, "A Pfizer Whistle-Blower."

25. Sarah Nightingale, "NBAF's Potential Costs, Benefits," *Manhattan Mercury*, Aug. 1, 2008.

26. Patricia J. Williams, "Strange Culture," *The Nation*, Oct. 29, 2007, 9; Andrew Z. Galarneau, "Steve Kurtz Is An Art World Star Since the FBI Came Knocking," *Buffalo News*, April 26, 2008; Matthew Rothschild, "Judge Dismisses Case Against Art Professor," *The Progressive*, April 25, 2008; William Fisher, "Fabricated 'Bioterrorism' Case Collapses," Inter Press Service, May 3, 2008.

27. See Chapter Four, "Growing Up in a Rough Neighborhood."

28. Williams, Galarneau, Rothschild, Fisher.

29. Ibid.

30. Ibid.

31. Lisa Rein, "Police Spied on Activists in Md.," *Washington Post*, July 18, 2008; Lisa Rein and Josh White, "More Groups Than Thought Monitored in Police Spying," *Washington Post*, Jan. 4, 2009.

32. Lisa Rein, "Federal Agency Aided Md. Spying," *Washington Post*, Feb. 17, 2009.

33. "Swine Flu, A False Pandemic," *NewStatesman*, Jan. 13, 2010; "Drug Companies Face European Inquiry Over Swine Flu Vaccine Stockpiles," Jan. 11, 2010; Niko Kiriakou, "Why Pig Flu Didn't Fly: The Full Story," *OpEdNews*, Jan. 30, 2010, http://www.opednews.com.

34. Adrian S. Gibbs, John S. Armstrong, and Jean C. Downie, "From Where Did the 2009 'Swine Origin' Influenza A Virus (H1N1) Emerge?" *Virology Journal*, Nov. 24, 2009, http://www.virologyj.com/content/6/1/207.

35. Kiriakou, "Why Pig Flu Didn't Fly."

36. Nick Rees, "Bipartisan WMD Research Center Established," *BioPrepWatch*, March 17, 2010, http://www.bioprepwatch.com.

Index